FIRE'S EFFECTS ON ECOSYSTEMS

FIRE'S EFFECTS ON ECOSYSTEMS

Leonard F. DeBano
University of Arizona
Tucson, Arizona

Daniel G. Neary
USDA Forest Service
Flagstaff, Arizona

Peter F. Ffolliott
University of Arizona
Tucson, Arizona

John Wiley & Sons, Inc.
New York • Chichester • Weinheim • Brisbane • Singapore • Toronto

This text is printed on acid-free paper. ♾

Library of Congress Cataloging-in-Publication Data

DeBano, Leonard F.
Fire: its effect on soil and other ecosystem resources / Leonard F. DeBano, Daniel G. Neary, Peter F. Ffolliott.
p. cm.
Includes bibliographical references and index.
ISBN 0-471-16356-2 (cloth : alk. paper)
1. Fire ecology. 2. Wildfires. 3. Soil ecology. 4. Fire management. 5. Ecosystem management. I. Neary, Daniel G. II. Ffolliott, Peter F. III. Title.
QH545.F5D435 1998
577.2—dc21 97-35835

Printed in the United States of America

10 9 8 7 6 5 4 3 2 1

Contents

Preface

Fire is a tool that can be used to manage a wide range of natural ecosystems worldwide. As such, it can affect soil, water, vegetation, riparian and wetlands, air, and cultural components of these ecosystems. Fire scientists, managers, and decision-makers need to evaluate fire effects on these components and balance the overall benefits and costs associated with the use of fire in ecosystem management. This book has been written to provide information on fire effects on ecosystem resources that can be used as a basis for planning and implementing fire management activities.

The book emphasizes the effects of fire on soil and water resources to a greater extent than other resources. This emphasis is not to imply that other ecosystem resources are less important, but instead results from a recognition that a comprehensive treatment of fire effects on soil and water has been lacking in the literature. One source of information about fire effects on soils and other resources has been a series of state-of-knowledge papers published by the USDA Forest Service in the late 1970s. These papers have not been updated since their publication, however, and some are no longer in print. This book reviews these early state-of-knowledge papers and also incorporates research that has been reported over the last 20 years pertaining to fire effects on all ecosystem components.

Chapter 1 describes fire regimes in fire-prone ecosystems in the United States and worldwide to illustrate the nature of fire occurrences and consequences in the functioning of these ecosystems. This chapter also presents definitions for common fire terms that are used throughout the remainder of the book. The book is then organized into four parts, with the sequence being information on fire dynamics, responses of the soil resource to fire, responses of other ecosystem resources to fire, and, finally, a discussion of management implications.

- Part I is concerned with fire dynamics and behavior, presenting information on combustion of fuels and heat transfer processes. Heat transfer from burning fuels to the underlying mineral soil is also considered. Knowledge of the combustion process provides background for a discussion on fuels and fire behavior. A conceptual model relating fire severity to a wide range of resource responses is presented. This model serves as the framework for discussing the effect of different fire severities on a wide range of ecosystem resource responses throughout the remainder of the book.
- Part II, a major focus of the book, emphasizes the effects of fire on the soil resource. The reader is introduced to basic features of the soil resource and the relationship of these features to ecosystem functioning. Using this background as a reference, specific information on the physical, chemical, and biological changes associated with soil heating is presented.
- Part III describes the responses of water, vegetation, wetland and riparian systems, air, and cultural resources to fire. This information, coupled with Part II, serves as a framework for more comprehensive analyses of total ecosystem response to fire. Chapters on the effects of fire on wetland and riparian ecosystems and on cultural resources are unique to this book.
- Part IV concludes the book by focusing on fire management concerns, including information on economic considerations and evaluation procedures relevant to fire management decisions. The final chapter discusses the role of fire in ecosystem management in terms of benefits, concerns, and constraints when using fire as a management tool. The book concludes by describing a database of information on fire effects that can be used by students, managers, researchers, and other specialists in fire management to address site-specific questions.

This book has been written as a text for undergraduate and graduate-level courses on forestry, natural resource management, and conservation biology. The book can also serve as a reference for planners, decisionmakers, managers, and technicians who are interested in using fire as a tool in ecosystem management. Management concerns are described throughout the book.

The International System of Units (Système International d'Unites, SI), informally called the metric system (centimeters, cubic meters, grams), is used throughout the book. Exceptions are where the original relationships or data were presented in English units and conversions would be awkward.

The authors wish to acknowledge the assistance of a number of people who have contributed to this book. Technical reviews of individual chapters were offered by Malchus B. Baker, Jr., USDA Forest Service, Flagstaff, Arizona; Kenneth N. Brooks, University of Minnesota, St. Paul, Minnesota; Gerald J. Gottfried, USDA Forest Service, Phoenix, Arizona; Roger Hungerford, USDA Forest Service, Missoula, Montana; Michael Johnson, USDA Natural Resource Conservation Service, Tucson, Arizona; Martin F. Jurgensen, Michigan Technological University, Houghton, Michigan; James O. Klemmedson, University of Arizona, Tucson; Peter Lahm, Arizona Department of Environmental Quality, Phoenix, Arizona; and Deborah Ulinski Potter, USDA Forest Service, Albuquerque, New Mexico. The entire manuscript was reviewed by Malcolm J. Zwolinski, University of Arizona, Tucson. Special appreciation is extended to Ann DeBano, Tucson, Arizona, for her detailed technical and editorial review of individual chapters and the entire book.

Leonard F. DeBano
Tucson, Arizona

Daniel G. Neary
Flagstaff, Arizona

Peter F. Ffolliott
Tucson, Arizona

Chapter

1

Introduction

Fire has long been part of the natural environment of people. Lightning-started fire has been, and remains, a common occurrence in many of the ecosystems occupied by humans. People became increasingly adept from the mid-Pleistocene period onward at using fire to manipulate ecosystems to obtain desired benefits (Pyne et al. 1996). As a consequence, anthropogenic fire overtook lightning as the leading ignition mechanism for biomass combustion worldwide.

The frequent occurrence of wildfire and prescribed burning throughout history has lead to an extensive body of experience and knowledge concerning the effects of fire on natural ecosystems. Some of the more general of these effects are presented in this beginning chapter of the book by describing how fire can impact ecosystems. General relationships between fire occurrence and the processes and functioning of ecosystems affected by burning are illustrated by these selected fire regimes.

With people attaining experience and increased knowledge of fire dynamics, and the effects of fire on natural ecosystems, a vocabulary has also evolved to describe the fire phenomenon and its effects on natural ecosystems. Some of these terms are defined in this chapter to familiarize the reader with such terminology at the outset.

SELECTED FIRE REGIMES

Fire has an irregular incidence. This irregularity contributes to differences in the effect of fire on ecosystem resources (Bond and van Wilgen 1996). Fire varies in how often it occurs (frequency), when it occurs (season), and how fiercely it burns (intensity). Combinations of these three factors are what largely determine a *fire regime*. Selected fire regimes and the associated fire management in the United States and elsewhere are presented next to illustrate the range of variations in global fire ecology.

United States

Fire regimes in the United States are as diverse as the ecosystems represented in the country and the land uses imposed on those ecosystems. Fire regimes in the ponderosa pine *(Pinus ponderosa)* forests of the Southwest, Douglas-fir *(Pseudotsuga menziesii)* forests of the Pacific Northwest, lodgepole pine *(P. contorta)* forests in the Rocky Mountains, loblolly *(P. taeda)* and shortleaf pine *(P. echinata)* forests in the Southeast, chaparral shrubs in southern California, and the tallgrass prairies of the Great Plains are presented to illustrate this diversity.

PONDEROSA PINE FORESTS OF THE SOUTHWEST

Ponderosa pine forests are found in seral and steady-state conditions in the Southwest. Their prevalence in either case depends largely on the occurrence of fire. Fire can lead to a replacement of older ponderosa pine trees with younger seedlings and saplings where these forests occur as a self-perpetuating community (Campbell et al. 1977, Pyne et al. 1996). Regardless of its successional stage, ponderosa pine is adapted to a regime of relatively frequent, low-intensity surface fire.

Surface fire in these forests is fueled by a mixture of grasses and a forb–shrub component in its herbaceous understory, with the mixture of plants depending on the site conditions, climatic features, and past land use. Understory herbaceous vegetation remains dominated by this mixture if the fire frequency is relatively high. A high fire frequency is difficult to sustain, however, as the same fire that eliminates ponderosa pine seedlings and saplings is likely to be the same fire that can also create seedbeds for natural regeneration. The resulting seedlings and saplings, which are vulnerable to fire, must successfully compete with the understory herbaceous plants in the initial 5 to 10 postfire years.

Ponderosa pine trees develop thicker bark, shed their lower branches, and drop their needles to the ground as they mature. As a consequence, the growth of competing herbaceous plants is suppressed, the composition of fuels is altered, and the fire behavior characteristics of these forest communities are changed. As the ponderosa pine trees continue to mature, they cast increasing shade onto the site and cover the mineral soil with their needle depositions. Sites so affected are generally no longer suitable for successful natural tree regeneration. Eventually, the trees die due to fire, disease, or insect pests and, in doing so, create openings necessary for seed germination and seedling establishment, and the cycle repeats itself.

Seedling establishment can be prolific in years when a good seed crop and high precipitation occur simultaneously. Dense and stagnated sapling and pole stands often result if these seedlings remain undisturbed, however. Low-intensity fire can thin these stands if the burning is contained on the ground (Ffolliott and Guertin 1990, Peterson et al. 1994, Pyne et al. 1996). If a fire escalates into the tree crowns, the site is likely to return to the original grass and forb–shrub herbaceous complex. Ponderosa pine forests frequently become uneven-aged in structure as a result of site-specific fire histories, consisting of small and relatively even-aged groupings formed by the surface fire disturbance discussed earlier.

One intent of fire management in ponderosa pine forests can be the substitution of prescribed fire for wildfire when it is possible (Pyne et al. 1996). Opportunities for broadcast burning present themselves because of the tolerance of mature trees to surface fire. While presettlement fire frequencies are estimated to be 5 to 25 years (Swetnam 1990, Swetnam and Baisan 1996), prescribed fire is often scheduled more regularly. Timing of the burning is critical—too early in the fuel accumulation cycle and there might not be sufficient fuels to carry a fire; too late, and the resulting high-intensity fire is likely to cause high tree mortality.

DOUGLAS-FIR FORESTS OF THE PACIFIC NORTHWEST

These Douglas-fir forests, unlike ponderosa pine forests in the Southwest, are found on more mesic sites. Furthermore, Douglas-fir is rarely the sole dominant in the overstory. Douglas-fir trees play multiple successional roles in the Pacific Northwest—they can be seral to other tree species on some sites, or they can be a steady-state dominant on other sites. A thick protective bark, wind-dispersed seeds, and intermediate tolerance to shade enable the species to sustain itself on open or closed sites. Douglas-fir trees, therefore, are able to sustain themselves on sites subjected to a variety of fire frequencies, intensities, and severities.

An arboreal sere in the Pacific Northwest starts with Douglas-fir trees becoming established following a fire, and then being replaced by western hemlock *(Tsuga heterophylla)*, western redcedar *(Thuja plicata)*, red alder *(Alnus rubra)*, or some combination of these species, which (in turn) leads to a steady-state western hemlock forest (Agee 1991, Peterson et al. 1991, Pyne et al. 1996). Douglas-fir and noble fir *(Abies procera)* are often the initial tree species establishing themselves after a fire or other disturbance in the lower subalpine regions of the Cascade Mountains, followed by western hemlock and Pacific silver fir *(A. amabilis)* and, eventually, a steady-state Pacific silver fir forest. These successional cycles are largely determined by the relative shade tolerances of the tree species involved.

Douglas-fir forests in the Pacific Northwest follow these successional paths because of the nature of the fire regimes in the region, the life histories of the tree species, and the wide variations in environmental conditions. Fire management, therefore, is complex. Occurrences of severe wildfire are more likely to correspond to periods of prolonged drought than to patterns of fuel buildups and, as a consequence, can occur at almost any time in the developmental history of the forests (Pyne et al. 1996). The effectiveness of prescribed burning in altering successional patterns can be limited in Douglas-fir forests, largely because its timing must relate to the life histories of more than one tree species, which is not the case with ponderosa pine forests in the Southwest. Prescribed fire is most commonly used for hazard reduction and site preparation following timber harvesting operations.

LODGEPOLE PINE FORESTS IN THE ROCKY MOUNTAINS

Interior lodgepole pine forests are generally associated with the occurrence of high-intensity crown fire. Successful natural regeneration of lodgepole pine trees after a fire is often obtained, since the species possesses cones that split open (dehiscence) when heated by fire (Pyne et al. 1996). Large numbers of seed accumulate on these cones, because the cones do not open to disseminate the seeds without the passage of fire. The seeds then fall onto a postfire seedbed that is conducive to successful germination and seedling establishment.

Successional stages following a high-intensity crown fire start with postfire seedling establishment, leading to spruce–fir *(Picea–Abies)* “climax” forest communities. Crown fire intermediate to the general fire frequency of about 200 to 300 years interrupts this successional progression and, as a consequence, delays the attainment of the spruce–fir forests and returns the system to the starting point (Muir 1993, Pyne et al. 1996). Self-perpetuating lodgepole forests on sites too xeric, too exposed, or otherwise too unsuited to the establishment of spruce and fir are exceptions to this successional cycle. Low-intensity surface fire introduces alternative successional pathways, because tree species compositions and stand structures cannot be set back to earlier seral stages due to the presence of older age classes of trees remaining after the fire. Postfire successional patterns are also a function of species life histories, site conditions, and variations in the frequency of disturbance. Interactions of these factors result in lodgepole pine forests of great diversity.

Fire management in interior lodgepole forests centers on the total suppression of fires in situations where human lives and values are threatened; use of prescribed fire as a tool to satisfy specified management objectives; and, more recently, allowing lightning-ignited fire to burn for the purpose of reducing fuel buildups.

LOBLOLLY AND SHORTLEAF PINE FORESTS IN THE SOUTHEAST

Loblolly and shortleaf pine forests, representing the most extensive forest communities in the Southeast, are characterized by a large variety of successional pathways. Successional sequences on abandoned agricultural fields, referred to as *old-field successions,* are illustrative of these pathways. Herbaceous plants dominate sites in the first few years after abandonment, after which pine seedlings then begin to emerge, gaining dominance for the following 15 to 25 years (Pyne et al. 1966). Annual burning, a common practice in the region, can suppress some of these seedlings. However, the surviving seedlings become resistant to the annual burnings and, as a result, are able to withstand subsequent frequent fires and become dominant on the site. Residual herbaceous plants in the understory decrease in density with the maturing canopy of pine.

A pattern of frequent fires can favor the establishment of longleaf *(P. palustris)* pine on sites where longleaf pine competes with loblolly and shortleaf pine, while infrequent fire on these sites favors the establishment of loblolly and shortleaf pine. Regardless of the species, fire can be used to thin a stand without excessive mortality once the trees have attained diameters of 5 to 10 cm (Wade 1993, Pyne et al. 1996). Shortleaf pine, which can resprout following fire, tends to dominate those sites experiencing fire every 5 years or so. The more shade-tolerant loblolly pine dominates sites with longer fire frequencies. In the absence of fire, an understory of oak and other hardwood species becomes established about 25 years following abandonment of agricultural fields. Occurrence of fire at this successional stage of development can cause high mortality of both pine and the hardwoods.

Fire management in the Southeast is often formulated to maintain the pines and control the invasion of hardwood species. Prescribed fires are ignited in the winter, when intensities should be low, to achieve this end. Scheduling of the burnings varies with the site-specific conditions and management goals (Cain 1993). Prescribed fire is also used for site preparation, slash removal, and fuel reduction once loblolly and shortleaf pines have attained maturity. Maintenance burnings typically occur on a 3 to 6-year cycle.

CHAPARRAL SHRUBS IN SOUTHERN CALIFORNIA

Chaparral (sclerophyllous) shrubs in the mountains of southern California seem to endure with or without fire. Most of these shrub species are vigorous sprouters and some species, like chamise *(Adenostoma fasiculatum),* produce seeds early in the season, which accumulate in large seedbanks. Fire can break the dormancy of seeds in these accumulations, with germination following (Conrad and Oechel 1982, Pyne et al. 1996). Chamise is adapted to the recurrence of fire, because it reproduces by both seed production and sprouting following a fire.

Associated grasses and forbs also produce seeds that lie dormant until fire releases them and prepares a suitable seedbed (Keeley and Keeley 1981). Indigenous forbs grow profusely immediately after fire and, within a year or two of a fire, annual grasses emerge. Shrubs regain their dominance within 5 years, by which time the grasses and forbs have declined in their significance.

This successional pattern accounts for the widespread distribution of sclerophyllous shrubs throughout southern California.

Control of wildfire in chaparral shrub ecosystems is difficult. Conditions of prolonged drought, long and dry summers, large fuel buildups, and hot and dry winds moving westward from inland deserts (the *Santa Ana winds*) combine to create periods of extremely high fire danger. Patterns of recurring wildfire conform to the life cycle of the sclerophyllous shrub species. The coincidence of fuel buildups and fire suggests that prescribed fire might be used to reduce the fuel accumulations and initiate the desired successional development. However, prescribed burning in steep mountainous terrain is difficult to contain. Furthermore, the presence of water-repellent soils, loose surface debris, high watershed and recreational values, and the proximity of housing and other cultural developments limit the widespread use of prescribed burning in many instances.

TALLGRASS PRAIRIES OF THE GREAT PLAINS

Tallgrass prairies of the Great Plains are largely shaped by fire, drought, or combinations of both. People have been (almost certainly) responsible for the occurrence of fire in these ecosystems (Binnema 1996). Fire favors grasses and forbs at the expense of shrubs and trees. Some perennials increase significantly in both numbers and production following a fire, others show lesser increases, and still others are largely unaffected by fire (Hartnett 1991, Pyne et al. 1996). Annual grasses often proliferate briefly after a burn. Tallgrass prairies can succumb to shrubs and trees on sites where fire has been eliminated.

The timing of a fire influences the effects of the fire. Postfire recovery is delayed if a site is burned in a period of drought when soil moisture is low, or if a site is burned in the summer when plants are beginning seed production. Therefore, fires are most commonly prescribed in the spring or fall. Fire is usually prescribed on a 2- to 3-year cycle to sustain managed prairies, although a 5- to 10-year cycle is likely to be adequate to maintain the ecological integrity of the burned sites. With many tallgrass prairies converted to farmsteads and towns in recent years, the occurrence of fire is currently only a concern in small remnants.

Fire management is almost exclusively focused on the maintenance and, when necessary, restoration of these ecological relics (Wright and Bailey 1982). Restoration of tallgrass prairies has rarely been successful without the application of prescribed fire every 1 to 3 years, with the fire being seasonally timed to favor native, cool-season, perennial grasses.

Worldwide

Diversity of fire regimes worldwide is illustrated by those of the bog birch shrubs in the Canadian Rocky Mountains, pine forests in the Sierra Madre of Mexico, the low woodlands of Western Australia, matorral in the Coastal

Lowlands of Southwestern Spain, fynbos vegetation in South Africa, and West African savannas.

BOG BIRCH SHRUBS IN THE CANADIAN ROCKY MOUNTAINS

Naturally occurring fire resulting from lightning, and burning by indigenous people to provide protection from wildfire and maintain open meadows for their horses and village establishment, served to control the spread of bog birch *(Betula glandulosa)* and other shrub species on many of the rangelands in the Canadian Rocky Mountains in the past. However, public agencies responsible for resource management emphasized the suppression of fire in the past to protect timber supplies, recreational pursuits, watersheds, and rural communities (Bork et al. 1996). Ironically, the result of decades of fire suppression left managers searching for a means by which native grasslands can be maintained or restored. One approach to achieving this goal has been through the introduction of prescribed fire.

Prescribed burning in the foothills of the Rocky Mountains in northern Alberta at 3- to 6-year intervals has increased forage production due to the decreased competition from the shrub species and, through removal of the shrub cover, improved livestock access to the forage plants (Bork et al. 1996). Successive prescribed burnings must be timed far enough apart to control the resprouting of the shrub species while minimizing damage to the forage resources. It is also important that burnings be prescribed to maximize their coverage on a rangeland to prevent concentrations of postfire livestock use on sites with improved forage resources. A factor likely to play the most important role in the continuing use of prescribed fire is the willingness of managers to appreciate the importance of fire in maintaining the desired ecosystem conditions.

PINE FORESTS IN THE SIERRA MADRE OF MEXICO

Fire has been, and continues to be, a major factor contributing to the loss of pine forests (mostly *P. reflexa, P. arizonica, P. lumholtzii, P. ayacahuite,* and *P. ponderosa*) in the Sierra Madre Mountains of Mexico, with a high incidence of the fire occurrences attributed to people. It has been estimated that 90 percent of all reported fires are associated with human activities (Secretaria de Agricultura y Recursos Hidráulicos 1985). Fire has been, and still is, used to clear forestlands for agricultural crop production and improve livestock forage resources. Many large burns have also been the result of uncontrolled campfires. Arsonists contribute to the problem by burning forests for revenge or spite (Bojorquez-Tapia 1990). Impacts of these human-caused fires have centered on the destruction of commercial timber resources, alteration of successional cycles, and extensive losses of human and cultural resources. Lightning-ignited burns are relatively few and, in general, affect only a limited area.

Public interest in fire prevention and public involvement in fire suppression have been lacking in the past, largely because people have not recognized the varying roles of forests in providing multiple goods and services on a sustained basis. Furthermore, only limited financial and human resources had been made available by the government for fire presuppression and suppression activities. In attempting to alleviate this situation, more effective forest fire management has evolved in recent years (Bojorquez-Tapia 1990). Recognizing that most of the forest fires in Sierra Madre are related to human activities, these efforts include a better coordination of fire management programs in the governmental agencies; an increasing level of cooperation through partnerships consisting of governmental agencies, the private sector, and the local people; technological developments and training activities; and attempts to place fire in a proper perspective in relation to planned management practices. While early results are promising, sustaining these programs will be crucial to placing the role of fire into a proper perspective relative to the holistic management of forestlands throughout Mexico.

THE LOW WOODLANDS OF WESTERN AUSTRALIA

Wildfires are a regular occurrence in the low woodlands of western Australia, burning extensively through woodlands of small trees (including *Banksia attenuata, B. grandis,* and *B. menziesii*) and a diversity of shrub species. Ignitions of the large accumulations of flammable vegetation in the woodlands, mostly during the very hot, dry, and windy summer months, lead to large and intense burns, altering successional pathways and causing large-scale destruction of human life and property (Lynch 1981, Burrows and McCaw 1990, Hobbs and Atkins 1990). A rapid growth of understory herbaceous plants occurs following fire, although this biomass production then declines to prefire levels within 10 years or more. Postfire fuels also quickly accumulate in the first 4 to 5 years, although their rate of accumulation declines steadily thereafter. Fuel-reduction barriers generally remain effective for only a short time as a consequence of this rapid accumulation of fuels following fire.

Ecosystems of the low woodlands of western Australia, and landscapes elsewhere on the continent, have largely been shaped by the incidence of bushfires. The management and maintenance, when appropriate, of ecosystems throughout Australia, therefore, cannot proceed without considering fire. Bushfire management on public lands is the responsibility of the agencies that administer these lands. At the core of fire management is controlled burning to reduce fuels to the point where fire control is possible (Gill et al. 1987, Pyne et al. 1996). Protection of private lands usually involves a combined effort between urban and rural services, although there can be "breakdowns" of jurisdictions because of the rapidly expanding urban developments.

MATORRAL IN THE COASTAL LOWLANDS OF SOUTHWESTERN SPAIN

Matorral, a term used locally to describe the large scrub cover common in this Mediterranean region, consists of varying compositions of herbaceous plants, shrubs, and small tree species. The matorral vegetation types support shifting agriculture, livestock grazing, and limited wood production, all of which can be threatened by the outbreak of fire nearly every summer, when conditions are ripe for fire ignition. These largely human-caused fires, whether accidental or intended to clear a patch of matorral, range from less than 10 to 1000 ha or more (Garcia Novo 1977). However, the fire usually spreads rapidly, burning only leaves and small branches of the woody plants. Recovery following fire is variable, depending on the plant species involved, intensity and duration of the fire, and site conditions, especially depth to the water table.

Early stages of postfire plant succession are mostly dependent on meteorological conditions, while subsequent stages follow pathways of the individual plant species. Matorral generally recovers to its prefire composition and structure after 5 to 10 years and, therefore, appears well suited to the region. However, the postfire vegetation bears only little resemblance to the original Mediterranean ecosystems from which it was derived, because of the repeated summertime burning cycles experienced.

Similar Mediterranean-type fire regimes are found elsewhere in Spain and other countries in southern Europe (Liacos 1977, Robertson 1977, Susmel 1977, Velez 1977, Ferreira de Almeida and Vilaca y Moura 1992). Fire management throughout this extensive Mediterranean region focuses on containing the devastating wildfires that explode into holocausts from time to time, and attempts to reduce the commonly encountered large fuel buildups by controlled burning. Public enlightenment on the threat of fire occurrence in the summer months, when urban dwellers frequently travel to these fire-prone ecosystems to vacation, is also important.

FYNBOS VEGETATION IN SOUTH AFRICA

Fynbos, an indigenous term for the shrubs of the Cape Province, consists of communities of grass-like plants, short shrubs, and taller shrubs in varying combinations. Fynbos vegetation has burned periodically for 100,000 years or more and, as might be expected, is adapted to survive fires (van Wilgen 1982). More than 4 years must usually elapse since the previous fire to accumulate enough fuel to burn, and the interval between successive fires, caused largely by lightning and human activities, is typically 15 to 20 years.

Much of the fuel buildup remains unburned when fire burns in the cool and moist conditions that exist when most of the deliberately started fires occur. With dry conditions, however, fire burns vigorously in the shrub crowns and consumes much of the accumulated fuel buildup. Intense fire is likely to consume much of the live vegetation and all of the fuels when burning occurs in periods of severe drought. The ability of the fynbos vegetation to recover after severe fire is little understood (Moll et al. 1980). While the vegetation is generally fire-prone, fynbos communities contain nonflammable forest patches that seldom burn.

Fire has an essential ecological role in the life cycles of fynbos plants and, therefore, fire management forms the crux of land management in the Cape Province. Fire management entails the maintenance of fire patterns that result in the persistence of the natural vegetation (van Wilgen et al. 1990, Seydack 1992). Fire management options pursued in meeting this objective include allowing natural fires to burn themselves out, controlling natural and human-caused fire in terms of a set of rules stipulated beforehand, and prescribed burning in blocks or compartments to create a relatively small-scale, equal-sized mosaic of age structures. Other objectives of prescribed burning programs are the reduction of fire hazards and control of undesirable plant species.

WEST AFRICAN SAVANNAS

Lightning is much less frequent in the savannas of West Africa than in the humid equatorial areas. It is not surprising, therefore, that fire due to lightning is negligible in these communities of continuous tall herbaceous plants and scattered shrubs and trees. On the other hand, human-caused fires have been, and still are, the most important factor affecting the composition, structure, and distribution of plant communities (Mounkeila 1984, Swaine 1992). Primitive people indiscriminately set fire to the original dense dry forests to open the landscapes up for hunting purposes. These fires often resulted in large-scale conversions to open savannas. Today, fires are set by farmers to clear lands for agricultural cultivation, by hunters to drive wild animals out of hiding, and by cattlemen to promote the growth of palatable plants. Many of these present-day burnings, along with uncontrolled campfires set by shepherds, farmers, and recreationists, evolve into widespread wildfires causing extensive and frequently unwanted modifications in the ecosystems burned.

Despite the controversy that surrounds its use, controlled fire is widely recognized as a potentially useful tool in the management of natural resources in West Africa. The use of fire in the forests and savanna is widespread, being generally initiated through prescribed burning programs that are designed to meet range and wildlife management objectives (Mounkeila 1984). The state of knowledge on the use of fire as a silvicultural tool is incomplete and, as a consequence, its use in satisfying forest management objectives is more restricted. Managerial aspects of fire are being addressed through the prevention of uncontrolled, human-caused wildfire, development of fire danger rating systems as a prelude to the use of prescribed fire, and applications of the prescribed burning programs mentioned herein.

DEFINITIONS

Some of the more frequently used terms that relate to fire, and the effects of fire on natural ecosystems, are defined here as a prelude to the following chapters in the book (Stoddard and Stoddard 1987, McPherson et al. 1990,

Whelan 1995, Pyne et al. 1996). Other relevant terms are defined in the chapters in which the terms are used.

Fire, as the term is used in this book, refers to the rapid release of the heat energy stored in (living and dead) plants by photosynthesis. One often thinks of flames at the mention of fire. However, there is more to fire than flame. *Fire* is a manifestation of a series of chemical reactions, while *flame* is a gas-phase phenomenon of fire and, therefore, only part of the process. *Flammability* is the relative ease with which a substance ignites and sustains combustion. *Burning* refers to being set on fire. A *burn,* in turn, is the area over which a fire has spread.

Fire behavior is the manner in which a fire reacts to fuel, weather, and topography. Common terms used to describe fire behavior include *smoldering, creeping, running, spotting, torching,* and *crowning.* A *ground fire* is a fire that burns the organic material in the upper soil layer and, at times, the surface litter and low-growing plants. A *surface fire* burns only surface fuels such as litter, other loose debris on the soil surface, and the smaller plants. A *crown fire* is one that advances from top to top of trees or shrubs more-or-less independently of the surface fire.

Fire severity refers to the degree to which a site has been altered or the successional processes disrupted by fire. Fire severity, loosely, is a product of fire intensity and residence time. Fire severity is generally considered to be light, moderate, or high. A *light-severity burn* is one that leaves the soil covered with partially charred organic material. A *moderate-severity burn* results from a burn in which all of the organic material is burned away from the surface of the soil; any remaining fuel is deeply charred. All of the organic material is removed from the soil surface in a *high-severity burn;* organic material below the surface is consumed or charred.

Fire frequency or *fire occurrence* is the number of fires in a specified time and area. *Fire-free interval* or *fire interval* is the time between two successive fires in a designated area; the size of the area should be clearly specified. The *fire cycle,* also called the *fire-return interval,* is the length of time necessary for an area equal to the entire area of interest to burn; again, the size of the area should be specified.

A *natural fire* is a fire of natural origin (lightning, spontaneous combustion, or volcanic activity). A *human-caused fire* is a fire caused directly or indirectly by a person or people. A *wildfire* is a fire that is not meeting management objectives and, therefore, requires a suppression response. *Prescribed burning* (also called *controlled burning* by some fire managers) is the controlled application of fire to fuel buildups, in either their natural or modified state, in specified environmental conditions that allow the fire to be confined to a predetermined area and, at the same time, allow it to produce the fireline intensity and rate of spread required to attain the planned management objectives. A *prescribed fire* is a fire burning with prescription, resulting from planned ignition. A *prescription* is a statement specifying the management objectives to be attained, and the air temperature, humidity, wind direction and wind speed, fuel moisture conditions, and soil moisture conditions in which a fire will be allowed to burn. A *prescribed natural fire* is a fire of natural origin that is allowed to burn as long as it is accomplishing one or more management objectives.

Conversion burning takes place when one vegetative community on a site is replaced by another because of fire. A *stand-replacing fire* is a fire that kills all or most of the overstory trees in a forest and, in doing so, initiates secondary succession or regrowth. More generally, a *vegetation-replacing fire* kills all or most of the living plants (including trees, shrubs, and herbaceous plants) on a site and, as a result, initiates secondary succession or regrowth. A *fire type* is a vegetative type that commonly follows a fire, or otherwise is dependent on the occurrence of fire. A *fire climax* is a plant community maintained by periodic fire.

Fuel-reduction burning removes fuel buildups to reduce the likelihood of ignition or lessen potential damage and the resistance to control of fire when it occurs on a site. *Natural fuels* result from natural processes and, therefore, are not generated by management practices. *Fine fuels* are fast-drying dead fuels, characterized by a high surface area to volume ratio, less than 1 cm in diameter; these fuels (for example, grasses, leaves, and needles) ignite readily and are consumed rapidly.

The accumulation of dead organic plant matter above mineral soil includes three layers of fine fuels—the litter *(L)* layer, consisting of unaltered organic matter, the fermentation *(F)* layer, consisting of partly decomposed matter, and the humus *(H)* layer, consisting of well-decomposed matter. *Duff* is made up of the *F* and *H* layers of organic matter. Therefore, the top of the duff is where leaves, needles, and other castoff vegetation have begun to decompose, while the bottom of the duff is where decomposed organic matter is mixed with mineral soil.

Heavy fuels (snags, logs, large branches, peat) of larger diameter (8 cm or more) ignite and burn more slowly than fine fuels. *Slash* consists of concentrations of downed fuels resulting from either natural events (wind, fire, snow breakage, and so on) or management activities (logging, road construction, and so on). *Slash disposal* is the treatment of slash (by burning or otherwise) to reduce the fire hazard or meet other purposes.

An air *pollutant* is any substance in the atmosphere that is foreign to the natural atmosphere or exceeds its natural concentrations in the atmosphere; the universal connotation of the term is that a pollutant is potentially deleterious. Pollutants can become a problem for people in areas where residences and other human developments form an essentially solid covering of the landscape. The *urban–rural interface* is the line, area, or zone where structures and other human developments meet, or intermingle with, undeveloped wildland areas.

Fire effects, the main focus of this book, are the physical, chemical, and biological impacts of fire on the environment and ecosystem resources. *Fire ecology* is the study of relationships among fire, the environment, and living organisms. *Fire management* refers to all of the management activities required for the protection of burnable values from fire and for the use of fire to meet specified management goals and objectives.

REFERENCES

Agee, J. K. 1991. Fire history of Douglas-fir forests in the Pacific Northwest. In: Ruggiero, L. F., K. B. Aubry, A. B. Carey, and M. H. Huff, technical coordinators. *Wildlife and Vegetation of Unmanaged Douglas-Fir Forests.* USDA Forest Service, General Technical Report PNW-GTR-285, pp. 25–33.

Binnema, T. 1996. Presettlement rangeland management on the northern plains. *Rangelands* 18:217–218.

Bojorquez-Tapia, L. A. 1990. Forest fires in Mexico: causes and strategies. In: Krammes, J. S., technical coordinator. *Effects of Fire Management of Southwestern Natural Resources.* USDA Forest Service, General Technical Report RM-191, pp. 193–196.

Bond, W. J., and B. W. van Wilgen. 1996. *Fire and Plants.* Chapman and Hall, London.

Bork, E., D. Smith, and M. Willoughby. 1996. Prescribed burning of bog birch. *Rangelands* 18:1–7.

Burrows, N. D., and W. L. McCaw. 1990. Fuel characteristics and bushfire control in *Banksia* low woodlands in western Australia. *Journal of Environmental Management* 31:229–336.

Cain, M. D. 1993. A 10–year evaluation of prescribed winter burns in uneven-aged stands of *Pinus taeda* L. and *P. echinata* Mill.: woody understory response. *International Journal of Wildland Fire* 3:13–20.

Campbell, R. E., M. B. Baker, Jr., P. F. Ffolliott, F. R. Larson, and C. C. Avery. 1977. Wildfire effects on a ponderosa pine ecosystem: an Arizona case study. USDA Forest Service, Research Paper RM-191.

Conrad, C. E., and W. C. Oechel, technical coordinators. 1982. *Dynamics and Management of Mediterranean-type Ecosystems.* USDA Forest Service, General Technical Report PSW-58.

Ferreira de Almeida, A. M., and P. V. S. Vilaca y Moura. 1992. The relationship of forest fires to agro-forestry and socio-economic parameters in Portugal. *International Journal of Wildland Fire* 2:37–40.

Ffolliott, P. F., and D. P. Guertin. 1990. Prescribed fire in Arizona ponderosa pine forests: a 24–year case study. In: Krammes, J. S., technical coordinator. *Effects of Fire Management of Southwestern Natural Resources.* USDA Forest Service, General Technical Report RM-191, pp. 250–254.

Garcia Novo, F. 1977. The effects of fire on the vegetation of Donana National Park, Spain. In: Mooney, H. A., and C. E. Conrad, technical coordinators. *Proceedings of the Symposium on the Environmental Consequences of Fire and Fuel Management in Mediterranean Ecosystems.* USDA Forest Service, General Technical Report WO-3, pp. 318–325.

Gill, A. M., K. R. Christian, P. H. R. Moore, and R. I. Forrester. 1987. Bushfire incidence, fire hazard and fuel reduction burning. *Australian Journal of Ecology* 12:299–306.

Hartnett, D. C. 1991. Effects of fire in tallgrass prairie on growth and reproduction of prairie coneflower *(Rabitibda columnifera: Asteraceae)*. *American Journal of Botany* 78:429–435.

Hobbs, R. J., and I. Atkins. 1990. Fire-related dynamics of a *Banksia* woodland in south-western Australia. *Australian Journal of Botany* 38:97–110.

Keeley, J. E., and S. C. Keeley. 1981. Post-fire regeneration of southern California chaparral. *American Journal of Botany* 68:524–530.

Liacos, L. G. 1977. Fire and fuel management in pine forest and evergreen brushland ecosystems in Greece. In: Mooney, H. A., and C. E. Conrad, technical coordinators. *Proceedings of the Symposium on the Environmental Consequences of Fire and Fuel Management in Mediterranean Ecosystems.* USDA Forest Service, General Technical Report WO-3, pp. 289–298.

Lynch, P. 1981. The role of fire in semi-arid vegetations. In: Stanbury, P., editor. *Bushfires: Their Effect on Australian Life and Landscape.* The Macleary Museum, University of Sidney, Sidney, Australia, pp. 85–96.

McPherson, G. R., D. D. Wade, and C. B. Phillips. 1990. *Glossary of Wildland Fire Management Terms Used in the United States.* Society of American Foresters, Washington, DC.

Moll, E. J., B. McKenzie, and D. McLachlan. 1980. A possible explanation of the lack of trees in the fynbos, Cape Province, South Africa. *Biological Conservation* 17:221–228.

Mounkeila, G. 1984. *Fire Ecology and Management in West African Savannas and Steppes.* Master's Thesis, University of Arizona, Tucson, Arizona.

Muir, P. S. 1993. Disturbance effects on structure and tree species composition of *Pinus contorta* forests in western Montana. *Canadian Journal of Forestry* 23:1617–1625.

Peterson, D. L., M. J. Arbaugh, G. H. Pollock, and L. J. Robinson. 1991. Postfire growth of *Pseudotsuga menziesii* and *Pinus contorta* in the northern Rocky Mountains, USA. *International Journal of Wildland Fire* 1:63–71.

Peterson, D. L., S. S. Sackett, L. J. Robinson, and S. M. Haase. 1994. The effects of repeated prescribed burning on *Pinus ponderosa* growth. *International Journal of Wildland Fire* 4:239–247.

Pyne, S. J., P. L. Andrews, and R. D. Laven. 1996. *Introduction to Wildland Fire.* John Wiley & Sons, New York.

Robertson, J. M. S. 1977. Land use planning of the French Mediterranean region. In: Mooney, H. A., and C. E. Conrad, technical coordinators. *Proceedings of the Symposium on the Environmental Consequences of Fire and Fuel Management in Mediterranean Ecosystems.* USDA Forest Service, General Technical Report WO-3, pp. 283–287.

Secretaria de Agricultura y Recursos Hidráulicos. 1985. Estralegas para la prevencion y el combate de incendios forestales 1985–1988. Subsecretaria Forestal, Mexico, D.F.

Seydack, A. H. W. 1992. Fire management options in fynbos mountain catchment areas. *South African Forestry Journal* 161:53–58.

Stoddard, C. H., and G. M. Stoddard. 1987. *Essentials of Forestry Practices.* John Wiley & Sons, New York.

Susmel, L. 1977. Ecology of systems and fire management in the Italian Mediterranean region. In: Mooney, H. A., and C. E. Conrad, technical coordinators. *Proceedings of the Symposium on the Environmental Consequences of Fire and Fuel*

Management in Mediterranean Ecosystems. USDA Forest Service, General Technical Report WO-3, pp. 307–317.

Swaine, M. D. 1992. Characteristics of dry forests in West Africa and the influence of fire. *Journal of Vegetation Science* 3:365–374.

Swetnam, T. W. 1990. Fire history and climate in the southwestern United States. In: Krammes, J. S., technical coordinator. *Effects of Fire Management of Southwestern Natural Resources.* USDA Forest Service, General Technical Report RM-191, pp. 6–17.

Swetnam, T. W., and C. H. Baisan. 1996. Fire histories of montane forests in the Madrean borderlands. In: Ffolliott, P. F., L. F. DeBano, M. B. Baker, Jr., G. J. Gottfried, G. Solis-Garza, C. B. Edminster, D. G. Neary, L. S. Allen, and R. H. Hamre, technical coordinators. *Effects of Fire on Madrean Province Ecosystems: A Symposium Proceedings.* USDA Forest Service, General Technical Report RM-GTR-289, pp. 15–36.

van Wilgen, B. W. 1982. Some effects of post-fire age on the above-ground plant biomass of fynbos (macchia) vegetation in South Africa. *Journal of Ecology* 70:217–225.

van Wilgen, B. W., K. B. Higgins, and D. U. Bellstedt. 1990. The role of vegetation structure and fuel chemistry in excluding fire from forest patches in the fire-prone fynbos shrublands of South Africa. *Journal of Ecology* 78:210–222.

Velez, R. 1977. Environmental difficulties in the fight against forest fires in Spain. In: Mooney, H. A., and C. E. Conrad, technical coordinators. *Proceedings of the Symposium on the Environmental Consequences of Fire and Fuel Management in Mediterranean Ecosystems.* USDA Forest Service, General Technical Report WO-3, pp. 326–333.

Wade, D. D. 1993. Thinning young loblolly stands with fire. *International Journal of Wildland Fire* 3:169–178.

Whelan, R. J. 1995. *The Ecology of Fire.* Cambridge University Press, Cambridge, England.

Wright, H. E., Jr., and A. W. Bailey. 1982. *Fire Ecology.* John Wiley & Sons, New York.

PART

I

Fire Dynamics

The effects of fire on ecosystems result from the release of variable amounts of thermal energy during the combustion of fuels. This energy is transferred almost instantaneously to different components of the ecosystem by radiation, convection, and conduction. The downward movement of heat during the combustion of fuels is complex, involving various processes and pathways. Heat is transferred to the litter and duff surface, or to the mineral soil surface if these organic layers are not present. If litter and duff layers are present, they can be ignited and undergo combustion, which generates additional heat that is transferred downward into the underlying moist or dry mineral soil. The combustion process and heat transfer are described in Chapter 2.

A wide range of aboveground fuels are sources of stored energy that is released during combustion. These fuels are classified into different groups to facilitate describing fire behavior during combustion. Fire behavior characteristics are described for forest, shrubland, and grassland ecosystems. A classification system for fire severities is described, which is used in a conceptual model that relates resource responses to fire behavior. Chapter 3 describes fire behavior and fire severity models.

Chapter

2

Combustion Processes and Heat Transfer

INTRODUCTION

Energy in the form of heat is generated during ignition and subsequent combustion of fuels and, as such, provides the driving force that causes a wide range of responses in ecosystem resources during a fire. Specific responses reflecting the effect of this released energy are discussed in following chapters. However, before discussing these effects, one should become familiar with the different phases of combustion and the ways that energy is transferred into and through the surrounding environment. The combustion of fuels involves a sequence of several well-defined but sometimes overlapping phases. This chapter describes these phases in terms of energy production and the resulting end products. Energy production and transfer are important for determining ecosystem responses.

Soil is an important ecosystem component that is affected by transfer of heat from combusting woody and litter fuels downward into the duff and underlying mineral soil. Therefore, special emphasis is given to heat transfer processes occurring in soils. Processes responsible for heat transfer include radiation, conduction, convection, mass transport, and vaporization and condensation. Heat transferred downward into the soil raises its temperature, begins to thermally decompose organic matter, and then finally ignites and combusts the organic matter. This sequence causes a wide range of changes in physical, chemical, and biological systems in the soil that are dependent on organic matter (see Chapters 4 to 7).

Although heat transfer is complicated by the combustion of different types of fuel, varying fire behavior, and other environmental factors, generalizations about heat production and transfer can be made. This information provides a basis for assessing fire effects on ecosystem resources in later chapters.

COMBUSTION

Combustion is a rapid physical–chemical process, commonly called *fire,* that is in some ways the reverse of photosynthesis. The latter process stores solar energy as organic molecules and compounds in plant material. Combustion releases the solar energy stored in a chemical form in various fuels as heat and a variety of gaseous and particulate by-products. Although flames are the most recognizable and significant phase of the combustion process, other phases can have important impacts on soils and other ecosystem resources.

A wide range of fuels may be available for combustion, varying from a sparse stand of grass to large piles of logs and slash (see Chapter 3). However, a common sequence of physical processes occurs in all these fuels before the energy contained in them can be released and transferred downward, laterally, or upward, where it heats the underlying soil and other ecosystem components.

Components of Fire Ignition

Three components are necessary for a fire to ignite and initiate combustion (Countryman 1975). First, there must be burnable fuel available. Enough heat must then be applied to the fuel to raise its temperature to the ignition point. And finally, there must be enough air to supply oxygen (O_2) needed to keep the combustion process going and maintain the heat supply for ignition of unburned fuel. These three components—fuel, heat, and O_2—form the *fire triangle* (Figure 2.1); all of the components must be present or there will be no fire.

In ecosystems, fuel is usually organic matter that ignites in the presence of heat and O_2 and provides the source of *chemical energy* needed to sustain a fire. The chemical energy of fuels is the solar energy fixed by plants in the synthesis of organic molecules and compounds. Fuel has properties of arrangement, size, density, moisture content, and chemical constituents that allow it to burn. Fuel for a fire can consist of live, recently dead, or decaying tree crowns, branches and stems, shrubs, grasses, forbs, fresh plant litter, decayed litter, duff, roots, organic soils such as peat and muck, geologic strata like coal, wooden structures, and other ignitable materials.

Heat in the fire triangle refers to the original source of fire and the energy

FIGURE 2.1. The fire triangle - the three components that must be present for a fire to ignite are fuel, heat, and O_2.

released by combusting fuels. It is a form of energy called *thermal energy* that results from changes in the molecular activity or structure of a substance. The mere presence of a heat source does not necessarily result in a fire. Before a hot or flaming fire source can ignite the fuel, some of its heat must be imparted to the fuel (Countryman 1976a). In addition, heat must be transferred from the fire to the unburned fuel for a fire to continue to burn and to spread. Heat absorbed while changing the state of water from a liquid to a gas is also an important factor influencing the combustion of fuels. Before a fire can become self-sustaining, sufficient heat must be absorbed by fuels to evaporate much of the water, thus making organic materials flammable. Excess heat-absorbing water or an insufficient heat source can result in a failure of a fire to ignite. Heat can move from one point to another point in fuels by four basic processes—radiation, conduction, convection, and mass transport.

Sufficient O_2 for fire is almost always available in natural ecosystems and fuel is usually plentiful. However, the fuels can be sorted, or packed, in a dense arrangement that restricts the diffusion of O_2. This arrangement might result in a fire not igniting at all, burning very slowly and inefficiently, or being extinguished. Oxygen provides the oxidant in the simplified chemical reaction equation of the combustion reaction (in its most efficient form) that releases the chemical energy of fuel:

$$(C_6H_{10}O_5)_n + O_2 + \text{ignition source heat} >> CO_2 + H_2O + \text{heat}$$

where

$(C_6H_{10}O_5)_n$ = a plant-synthesized cellulose fuel,
O_2 = the reaction oxidant,
CO_2 = carbon dioxide, a combustion by-product,
H_2O = water, a combustion by-product.

Combustion Process

It is important to understand the physical–chemical process of fire to understand how heat is generated by a fire. Fires obviously require an ignition source. Lightning is the primary nonanthropogenic cause of forest, shrubland, and grassland fires. Human-caused fires, both accidental and deliberate, are produced in many ways, including campfires, chimneys, torches, matches, fireworks, cigarettes, vehicle fires, military ordnance, and smoldering slash piles. Fires smoldering in organic soils may reignite surface fires multiple times within a fire season or over many years.

Fires can be viewed as having five phases during their existence. The first phase of a fire, *preignition,* involves fuel heating that results in dehydration and pyrolysis (Ryan and McMahon 1976, Sandberg et al. 1979, Pyne et al. 1996). Once ignited, fire continues as a two-stage thermal process of continuing pyrolysis and combustion, the latter including three phases—*flaming, smoldering,* and *glowing* (Figure 2.2). Pyrolysis continues during flaming, be-

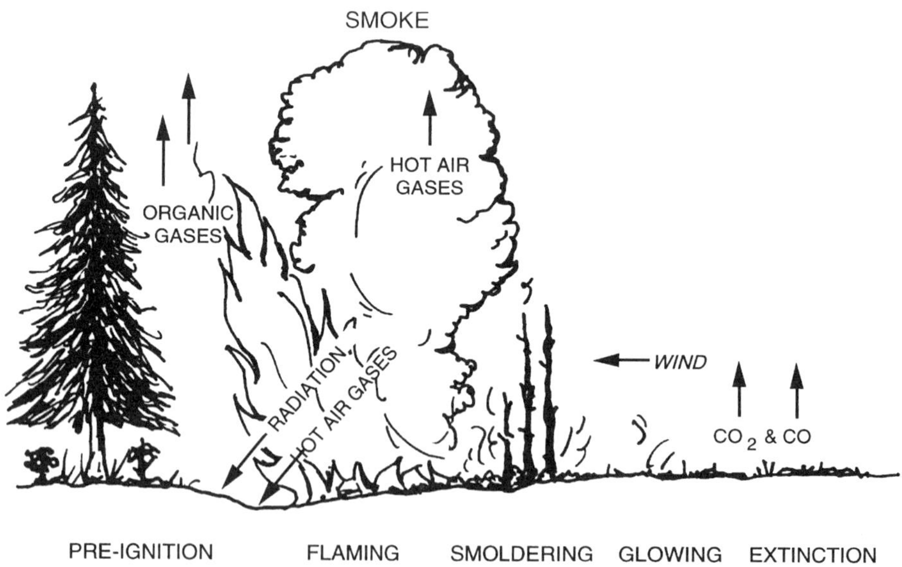

FIGURE 2.2. Phases of a fire's existence. (Adapted from Johansen et al. 1985.)

gins to diminish during smoldering, and virtually ceases in the glowing phase. The last phase, where combustion ceases, is *extinction.*

Example 2.1: Thermal Stages of a Fire: An Explanation

Pyrolysis is an endothermic reaction set off by thermal radiation or convection from an advancing fire front that drives water from the surface of a fuel, elevates fuel temperatures, and then decomposes long-chain organic molecules in plant cells into shorter ones (Shafizadeh 1984). Structurally, plant cell walls consist of primary and secondary walls and the middle lamella. It is the organic compounds in the plant cells and their thermal decomposition pathways that determine the fire emissions and its combustion phase.

The main molecular component of the primary wall and middle lamella is *lignin,* a *polymer* (organic compound formed of repeating structural units like simple sugars) in which the unit structure is a phenolic compound with low oxidation levels. Lignin is the component of woody tissues that provides the hardness and structural rigidity of wood. It makes up 18 to 35 percent of sound wood and 65 percent of rotten wood (Pettersen 1984). Lignin is resistant to both thermal and microbial degradation and provides the large amounts of char needed for the glowing phases of combustion. *Char* is a carbonaceous residue left on the surface of fuels by pyrolysis that is neither intact organic compound nor pure carbon (C). *Cellulose* and *hemicellulose* are long-chain polysaccharides (such as glucose, mannose, galactose, and xylose sugar derivatives) with high oxidation levels that are found in both cell walls, but primarily in the secondary wall. Cellulose constitutes the largest group of carbohydrates in wood (40 to 50 percent) and hemicellulose makes up 25 to 35 percent (Pettersen 1984). When pyrolysis begins, these are the first woody cell compounds to thermally decompose and provide the flammable gases that ignite in flaming combustion.

Fuels such as cellulose are the first to thermally decompose into tars and polymers during the pyrolysis process. *Tar* consists of a complex variety of viscous liquid and semiliquid initial by-products of cellulose pyrolysis (Shafizadeh 1984). The breakdown of cellulose and its immediate by-products also results in the production of char, CO_2, carbon monoxide (CO), water vapor, and highly combustible, high-molecular-weight hydrocarbon vapors and gases through the processes of dehydration, oxidation, depolymerization, and thermal degradation (see below). Some complex organic compounds can be synthesized from smaller free-radical hydrocarbons produced by pyrolysis.

The second thermal stage of a fire is *combustion,* the rapid exothermic oxidation of pyrolysate hydrocarbon vapors released from the surface of the fuel and the slower oxidation of char. Since forest fuels are complex (for example, foliage, twigs, branches, and logs arranged randomly in the same fuel bed), pyrolysis and combustion can proceed at many different rates in these ecosystems. Rates of pyrolysis and combustion are likely to be more uniform in grass fuels because of the more homogeneous nature of the fuels.

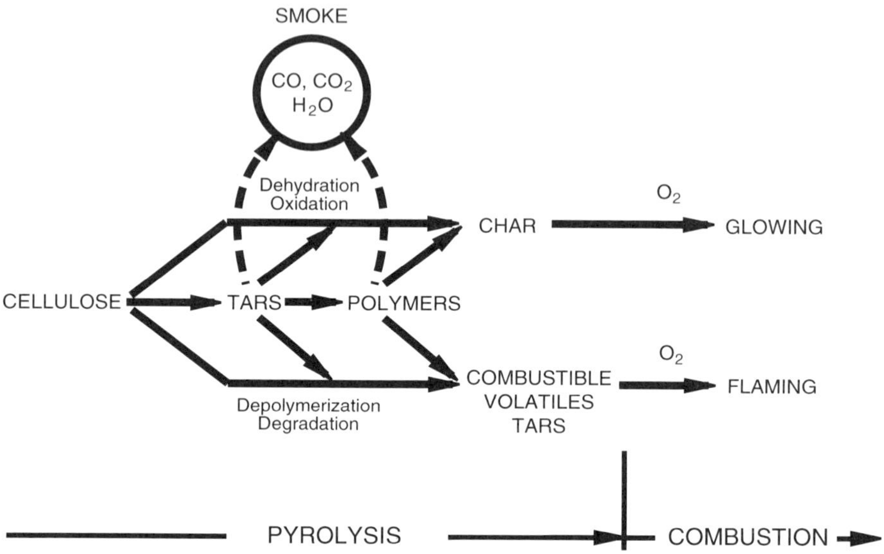

By-products of cellulose obtained in the pyrolysis stage of fire. (Adapted from Philpot 1971. Reprinted with permission from Shafizadeh 1984. Copyright © 1984 American Chemical Society.) ■

Fuels in natural ecosystems do not burst into the flaming combustion phase directly (Countryman 1976b). The fuel must first be converted to gases through *pyrolysis*, a chemical decomposition process brought about by heating (Example 2.1). Pyrolysis and combustion start with *endothermic* (heat-absorbing) reactions and progress to *exothermic* (heat-producing) reactions. In rapidly advancing fires, both pyrolysis and combustion occur simultaneously, but pyrolysis is the initiating stage of chemical decomposition of organic compounds.

PREIGNITION

Dehydration, distillation, and pyrolysis are the predominant processes in the preignition phase. As fuels ahead of the flame are heated by radiation and convection to >100°C, water vapor, noncombustible organics, and volatile extractives are distilled to the fuel surface and driven off into the air (Ryan and McMahon 1976). Radiation and convection can transfer the heat necessary for pyrolysis to the surface of fuels, but transfer of heat to the interior is through conduction.

Therefore, conduction is of major importance in the combustion process, particularly for larger fuels. As the heat flux into woody fuel increases, thermal decomposition of hemicellulose, cellulose, and then lignin produces more combustible organic compounds (pyrolysates). As the hot pyrolysates distill out of woody fuels, they ignite at temperatures of 300 to 600°C in the relatively oxygen-rich air surrounding the fuels. This ignition produces the second flaming phase of combustion.

Distillation of fine fuels (including leaves, needles, and grasses) at temperatures above 100°C produces emissions of water vapor and volatile organic extractives. The latter are principally terpenes and aromatic aldehydes (Hall 1972). As pyrolysis progresses in woody materials, the heating from 200 to 280°C begins to produce water vapor, CO_2, formic and acetic acids, glyoxal, and CO (Browne 1958). Combustible gases and vapors (such as methane, ethane, propane, formaldehyde, formic acid, acetic acid, methanol, CO, and hydrogen (H_2)) are rapidly released from woody fuels at temperatures of 280 to 500°C (Fritschen et al. 1970). Streams of water vapor and CO_2 also transport tars that form smoke (see Example 2.1). At the higher temperatures, low-molecular-weight hydrocarbons evolved from the fuel pyrosynthesize additional tars.

Secondary pyrosynthetic reactions become even more prominent above 500°C, with CO_2 and water reacting with charcoal at the surface of the fuel to produce CO, H_2, formaldehyde, and so on. In the absence of sufficient O_2, condensation of distilled organics into tars and other compounds will occur. Slow heating rates favor charcoal production and reduce the production of flammable gases and tars. Fast heating rates do the opposite, producing flammable gases and tars.

Fuel *ignition,* or the initiation of self-sustaining pyrolysis and flaming combustion, marks the transition point between the mainly endothermic preignition and exothermic flaming phases. At the beginning of a fire, a pilot source (such as lightning, matches, or fusees) provides the heat needed to raise fuels to the 325 to 350°C temperature range needed for ignition. Once a fire is self-sustaining, the heat released by the combustion reaction provides the heat needed both to produce pyrolysis of fresh fuels and to spontaneously ignite the combustible gases released by the pyrolysis process. Ignition of adjacent fuels occurs until the fuels are exhausted, heat release decreases below a level needed to sustain combustion, or O_2 flux to new ignition points declines below amounts needed to sustain the combustion reaction.

FLAMING

Exothermic reactions in this phase cause temperatures to soar from the 300 to 500°C range to 1400°C (Ryan and McMahon 1976). Pyrolysis accelerates and rapid oxidation (flaming) of the combustible gases predominates. Combustible gases and vapors resulting from pyrolysis rise above the fuels, mix with O_2, and burn during the flaming phase. The heat produced from the flaming reaction accelerates the rate of pyrolysis and releases greater quantities of combustible gases. Fire becomes explosive and can literally move with the wind as masses of burning gas in this phase. The volume of combustible gases emitted from the fuels increases markedly. Some pyrolyzed compounds cool and condense below the flame zone, while others are partially burned, producing an enormous spectrum of chemical emissions. Low-molecular-weight hydrocarbon radicals are pyrosynthesized into the high-molecular-weight organic compounds called *polynuclear aromatic hydrocarbons* (PAHs). Some low-molecular-weight volatile organics escape the oxidation process and are transported away in wind. Higher molecular weight compounds, that escape complete oxidation, cool and condense in the wind into tar droplets and soot particles that produce smoke. Flaming is initially vigorous as fuel volatilization increases but then decreases as charcoal formation on the surface of woody fuels reduces the thermal conductivity of the fuel and the flux of heat deep into the burning wood.

Since pyrolysis still occurs, chemicals produced by the flaming phase include many of those produced during preignition. Oxidation of highly combustible organics and other gases in the flame zone produces the largest mass of combustion products such as water, CO_2, sulfur dioxide (SO_2), and nitrogen gas (N_2) and nitrogen oxides (NO_x). (The term NO_x is sometimes used for both nitrous oxide (NO_2) and nitric oxide (NO) combined.) Excess low-molecular-weight organics (such as methane, ethane, acetylene, and propane) that are not immediately oxidized are transported away from the flaming front by air currents. The condensation of soot and tar droplets <1 μm in diameter as smoke is one of the most important processes of the flaming phase. Smoke is the primary air pollutant produced by fires (see Chapter 12). Other products of flaming are PAHs. Extensive amounts of charcoal also form on the surfaces of burning woody fuels.

SMOLDERING

The smoldering phase normally follows flaming combustion. Slow (<3 cm/h in ground fires), flameless combustion becomes the dominant burning process in this phase. It should be pointed out that flaming combustion does not occur at all in fuels such as duff and organic soils. Smoldering is the initial combustion phase in these fuel types. There are two zones that characterize the smoldering phase of combustion—a pyrolysis zone with evolving combustion products and a char zone with no visible combustion product release. The rate of fire spread decreases because the fuels are not able to supply combustible gases in concentrations and at rates needed to support vigorous combustions. Subsequently, heat release declines and temperatures drop, causing larger amounts of gases to condense into smoke. At this point, the heat release is usually insufficient to support a large smoke plume. However, smoke persists near the ground in high concentrations, and atmospheric emissions double or triple over those levels released during the flaming phase (Johansen et al. 1985). By the time the smoldering phase of the combustion process is reached, generation of pyrolysates diminishes substantially and the remaining woody fuels have a black char on the surface. Smoldering generally occurs in fuel beds with finely packed fuels and limited O_2 flow, such as duff, punky wood, and organic soils. An ash layer on the fuel bed and on woody fuels can promote smoldering by separating the reaction zone from the atmospheric O_2.

Particulate product emissions during the smoldering phase are significantly greater than those of the flaming phase. The term *emission factor* (EF) is used to quantify the amount of particulate or gaseous products released by a fire. It is the mass of fire by-products, in kg/Mg of fuel consumed, released into the atmosphere. As shown in Table 2.1, an increasing proportion of smoldering during flaming combustion can result in a doubling or tripling of the particulate EF. When smoldering combustion dominates, the volatile gas and vapor products that are characteristic of pyrolysis in the preignition and the flaming phases cease to evolve. Particulate emissions can then increase tenfold. Charcoal builds up on the surface of woody fuels when smoldering predominates.

TABLE 2.1. Particulate Matter Emission Factors for Fuel Types as a Function of Combustion Phases and Degree of Smoldering

Fuel Type	*Combustion Phase*	*Degree of Smoldering*	*Particulate Emission Factor (kg/Mg)*
Grass	Flaming	None	7.5
Understory plants and litter	Flaming	Light[a]	12.5
	Flaming	Moderate[b]	25.0
	Flaming	Moderate[c]	37.5
Scattered slash	Flaming	None	10.0
	Flaming	Moderate	20.0
Piled-windrowed slash	Flaming	None	12.5
	Flaming	Moderate	25.0
	Flaming	Heavy[d]	37.5
Shrub fuels	Flaming	None	12.5
	Flaming	Moderate	25.0
All fuels	Smoldering	Complete	75.0

[a]Backing or heading fires, light fuels, no duff.

[b]Heading fires in litter, heavy fuels.

[c]Litter fuels, considerable duff combustion.

[d]Large mineral soil incorporation or high fuel moisture.

Source: Johansen et al. (1985).

GLOWING

The glowing phase is the final part of the smoldering process. However, glowing is not smoldering. When a fire reaches the glowing phase, most of the volatile gases have been driven off and O_2 comes into direct contact with the surface of the charred fuel. As a fuel oxidizes, it burns with a characteristic glow and continues this process until the temperature drops so low that combustion can no longer occur or until the fuel is reduced to noncombustible ash. Solid oxidation is the dominating thermal process in the final phase of combustion. The volatile compounds produced by thermal degradation during smoldering are not oxidized to any extent and escape into the surrounding air. Char remaining from the flaming phase, or from the smoldering phase in the case of duff and organic soil fires, is what burns in the glowing phase. The residual fuel burns with a yellow glow. Peak temperatures of the burning fuel are in the 300 to 600°C range. Little or no smoke is produced during glowing combustion.

Products of the glowing phase are mainly CO, CO_2, and ash residue. This phase of the combustion process is probably the most efficient because of its slow rate, better supply of O_2 to combustion sites, and lower volumes of flammable volatiles.

EXTINCTION

A fire finally goes out when all of the available fuel is consumed, or when the heat produced by oxidation in either the smoldering or glowing phase is insufficient for vaporization of the necessary amount of water from moist fuel. Heat absorbed by the fuel moisture, the surrounding air, or inorganic materials (such as rocks and mineral soil) reduces the amount of heat available from combustion, hastening the fire extinction process.

Combustion Efficiency

Moisture increases the amount of heat required to raise the fuel temperature to a point where combustible gases are produced. It absorbs part of the heat available for pyrolysis and it dilutes the gases produced. Therefore, moisture in the fuels increases their ignition time and decreases their burning rate (Countryman 1976b). Basically, dry fuels burn hot, completely, and quickly, while moist fuels either do not burn or do so slowly and at lower temperatures. The effect of fuel moisture on total emissions from combustion ranges from complete (clean) combustion of dry fuels, with lower levels of emissions, to longer durations of smoldering with moist fuels, associated with higher emission levels, but also less fuel burned, leading to lower emission levels. The latter situation can result in a trade-off. Although it is true that emissions per unit of fuel burned are greater at higher fuel moisture contents, total emissions will be lower if larger fuels and the organic accumulations on the soil surface do not burn because they are too wet. Moisture released by fuels absorbs heat in the water vaporization process, reducing flame temperatures. In addition to heat lost in the process of evaporating moisture contained in fuels, the soil mass and surrounding air absorb heat, cooling flame temperatures even further and therefore reducing combustion efficiency.

A major factor affecting the efficiency of the combustion process is the air supply. If air movement into and around flaming combustion is insufficient to allow adequate mixing of O_2 with the flammable gases being released by pyrolysis, oxidation is incomplete and combustion efficiency drops. This phenomenon results in a greater release of emission products such as CO (Figure 2.3). This process can be observed on a larger scale as pulsations in flame fronts produced by temporary O_2 deficits. Flame turbulence also affects combustion efficiency, although the effects of this process on efficiency are only partially known (Ward 1990). As fire intensity increases, turbulence pushes the zone of gaseous fuel and O_2 mixing further from the solid fuels undergo-

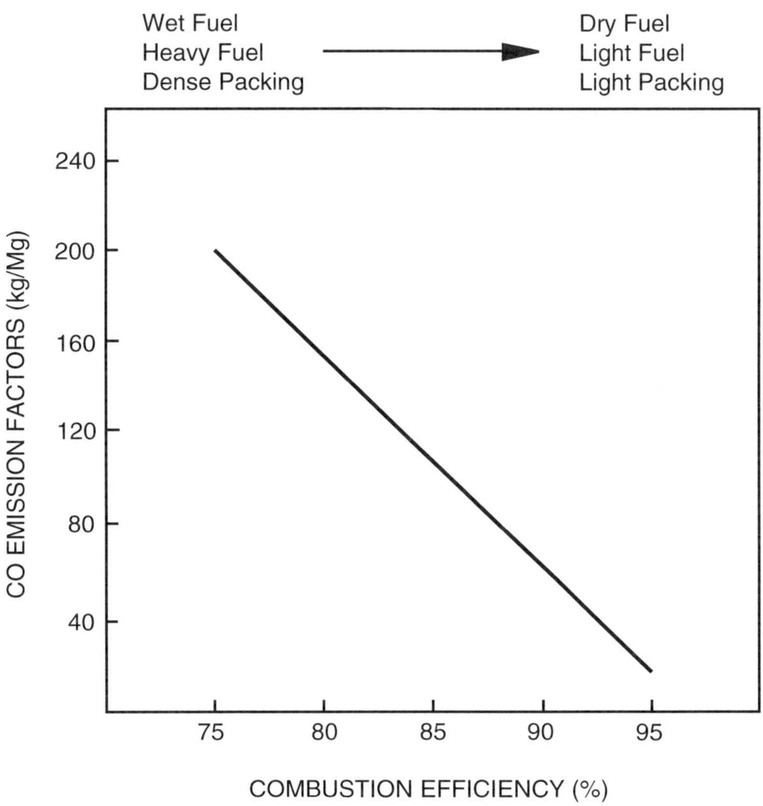

FIGURE 2.3. Relationship between CO emission factors, combustion efficiency, and fuel conditions. (Adapted from Ward 1990. © Springer-Verlag, New York, NY.)

ing pyrolysis, resulting in a decrease in combustion efficiency and an increase in particulate and other gas emissions.

Also related to combustion efficiency are the rate and direction of spread of the flame front. Generally, the faster flames move through a fuel bed, the lower the combustion efficiency. A heading fire can move rapidly with the wind through heavy surface fuels, leaving a large zone of smoldering fuels behind the main combustion zone. Under these conditions, combustion efficiency would be reduced and atmospheric emissions would increase. In contrast, a backing fire usually moves slowly into the wind, resulting in more complete combustion of surface fuels, a smaller zone of smoldering combustion, and a higher combustion efficiency. However, under some conditions, the opposite can be true with a backing fire. Since a backing fire has a longer residence time over any given point, it may cause ignition of the duff or organic soils, thereby producing more smoldering than a heading fire.

Fuel type, arrangement, and loading are other factors that also affect combustion efficiency (see Chapter 3). Anyone who has observed prescribed fires or wildfires is aware that light fuels, such as grasses and fresh litter, ignite more readily and burn faster and more completely than heavier fuels such as logs. The key factor is size of the fuel. The time needed to ignite and consume an individual fuel particle is a function of its surface to volume ratio and mass. Fine fuels require less heat input to pyrolyze and are readily consumed in flaming combustion. Their combustion efficiency is high (>90 percent). On the other hand, heavy, large fuels require large heat inputs and long periods of time to combust. Periods of smoldering combustion are considerable. Consequently, combustion efficiencies are lower and emission products greater.

The arrangement of fuel components is also important in determining combustion efficiency. The tighter fuels are packed together, the higher the *packing ratio* (that is, the proportion of a fuel bed actually occupied by fuel) becomes, and the lower the combustion efficiency because air supply to the fire is restricted by the fuel density. For example, needles from spruce and fir trees lying on the forest floor are usually more slowly and less efficiently burned than ponderosa pine needles because their shorter needles pack together much more tightly than long needles. Densely compacted slash piles also burn less efficiently than loosely piled ones. Lastly, fuel loadings affect combustion efficiency and atmospheric emission products, since heavy fuel loads have long periods of smoldering combustion. With light fuel loads like grasses, all the fuel is quickly consumed during the more efficient flaming phase of combustion.

The term *thermal conductivity* expresses the quantity of heat transferred per unit length per unit time per degree of temperature gradient. (Thermal conductivity is expressed in SI units as W/m/K.) The thermal conductivity of fuels in most natural ecosystems increases as the density of the fuel increases, although it requires more heat per unit volume of fuel to ignite. When the thermal conductivity is high, the available heat is transferred through a deeper layer of fuel in a given time period than when the thermal conductivity is low. There is more material to heat per unit of volume in dense fuels.

As a consequence, fuels with low density tend to ignite more readily and often burn more rapidly than do high-density fuels. Small fuels also ignite more easily and produce heat at a faster rate than large fuels because of their greater ratio of surface area to volume. Their thinner layer of char permits more rapid heat conduction. The chemical composition of fuels also affects their combustion. When small molecules of flammable hydrocarbons are produced, they flame readily and lead to more rapid combustion.

The stage of decomposition is important with large logs (Pyne et al. 1996). Rotten wood has the same specific heat as sound wood below 150°C. However, the large exothermic peak at 300 to 400°C in sound wood is missing in rotten wood. This condition is a result of the lower proportion of cellulose in rotting wood (consumed by microorganisms) relative to lignin. Consequently, it burns mostly by smoldering and glowing combustion.

HEAT TRANSFER

Fuels undergo a series of complex physical and chemical changes during ignition and combustion (see earlier discussion). Heat produced during combustion is transferred to other fuels and to different parts of the ecosystem including the underlying litter, duff, and mineral soil by several processes (radiation, convection, conduction, mass transport, vaporization and condensation). Although heat is transferred in all directions, large amounts of heat are lost into the atmosphere along with smoke, gases, and particulate matter generated by fire. Some of the heat produced is transferred to other unburned fuels to sustain the fire. It has been estimated that only about 10 to 15 percent of the heat energy released during combustion of aboveground fuels is absorbed and transmitted directly downward to the litter and duff, or mineral soil if surface organic layers are absent (DeBano 1974, Raison et al. 1986).

The quantitative prediction of downward heat flow from combusting surface fuels or from smoldering duff to the underlying mineral soil is important when assessing fire effects (Hungerford 1990). The transfer of heat through mineral soil is important because it causes soil heating and produces changes in several soil physical, chemical, and biological properties (see Chapters 5, 6, and 7, respectively).

Heat transfer from burning fuels to duff, through duff, and eventually into and downward through mineral soil involves several heat transfer processes operating separately or simultaneously. Heat can also be transferred along different pathways when moving from fuels to mineral soil.

Processes

It is important to understand the processes by which heat is transferred from place to place when evaluating fire effects in natural ecosystems. Radiation, convection, and conduction are the most common mechanisms by which heat is transferred from one location to another in fuels (Example 2.2). Mass transport by spotting or by materials rolling downslope (for example, conifer cones, small logs, and tumbleweeds) can also be important during intense burning conditions when fire spreads rapidly (Chandler et al. 1983).

When water is present, heat transfer can also occur by vaporization and condensation. Vaporization is important in drying of wet fuels so they will ignite and combust. Vaporization and condensation of both water and organic substances can occur during heat transfer through soils (see Chapter 5).

Example 2.2: An Illustration of Heat Transfer by Conduction, Radiation, and Convection in Woody Fuels (Chandler et al. 1983)

Heat transfer in woody fuels can be illustrated by striking an ordinary wood match. If the match is lit and held vertically with the head on top, the flame will get progressively shorter and will soon go out. In this position, the unburned wood is shielded from most of the flame radiation by the charred residue of the match head and small updrafts produced by the flaming head cool the unburned fuel. The main mechanism for heat transfer in this case is conduction through the match from the glowing char. Because wood is a poor conductor of heat, the flow of heat through the wood in a downward direction is not sufficient to pyrolyze the volatiles to sustain flaming and the fire goes out.

If the match is lit and held horizontally, the flaming behavior is much different. In this case flaming continues, but on the top side of the match stick. Radiation transfer is increased and heat is conducted through the air more easily than when the flame must travel downward. The heat from the glowing char is also conducted into the interior of the match, along with a small amount of convective heat where the base of the flame forms a right angle with the unburned match.

When the match is lit and held vertically with the head downward, all surfaces of the match are exposed to the flame and the whole match stick flames rapidly. In this case, hot gases are distilled from the match and rise along the sides of the unburned match stick, transferring heat by convection. Although both radiation and conduction are still operating, the principal heat transfer process involved is convection. ■

RADIATION

Radiation is heat transfer by electromagnetic waves (Countryman 1976c). Radiation energy is not heat but is comprised of electromagnetic waves that travel at the speed of light (300,000 km/s). Therefore, *radiation* is defined as the transfer of heat from one body to another, not in contact with it, by electromagnetic wave motion. All bodies at temperatures above 0 K produce radiant energy. Radiated energy travels outward in all directions from the emitting substance until it encounters a material capable of absorbing it. The absorbed radiation increases the molecular activity of the absorbing substance, increasing the temperature of the substance and the amount of heat it contains. Heat transfer by radiation does not require an intervening substance between the heat source and the receiving substance. An example of this phenomenon is the sun's radiant energy that is transferred through outer space until it reaches the earth's atmosphere.

Radiation that is generated by fire is not from a point or line source but from a considerable surface area, which can be thought of as a large number of adjacent point sources. Radiation intensity depends on the distance of the source and the angle at which radiation strikes an object. Because radiation travels only in straight lines, the total amount of radiation is spread over a greater area as the distance from the source increases. Since intensity decreases inversely as the square of the distance from a point source of radiation, a log 1 m from a flaming front will receive 100 times the radiant energy as a log 10 m away. Intensity also varies as the angle of radiation varies, with the greatest intensity occurring when the receiving surface is perpendicular to the source. A log laying at a 45° angle to a flame front will intercept less radiant energy than one perpendicular to the flames and will take longer to reach an ignition temperature.

Radiation is an important process affecting fire behavior, particularly fire spread (Countryman 1976d). Fire acts as a high-temperature heat source. Flame temperatures can reach 1400°C and temperatures in the combustion zone can reach 1000 to 1200°C. Glowing material ranges from 400 to 760°C. Therefore, most of the radiant energy from a fire is in the visible wavelengths. The amount of energy released is proportional to the fourth power of the absolute temperature (K) of the radiating body.

Although heat conduction is necessary to keep many fuels burning, conduction is slow in relation to radiated energy. Consequently, heat transfer to unburned fuel ahead of the fire front is accomplished primarily by radiation and convection. Most convection is reduced in the absence of slope, pressure gradient, or fire-induced winds. Therefore, radiation usually becomes the major mode of heat transfer during the advance of a fire in still air. Radiation is the primary heat transfer mechanism during the incipient stages of most forest fires (Chandler et al. 1983).

A large portion of the heat transfer from burning surface fuels to the mineral soil, when organic layers are absent, occurs by radiation and convection. These two heat transfer processes predominate because surface fuels such as grasses, foliage, and small branches are not in extensive contact with the mineral soil.

CONVECTION

Convection is a process whereby heat is transferred from one point to another by the mixing of one portion of a fluid with another fluid (Chandler et al. 1983). Heat transfer by convection is an important mechanism of heat transfer during fire and is responsible for fire's spread rate. Free and forced convection are both common processes during combustion. *Free convection* is when the fluid motion of the gases is dependent on the differences in densities resulting from temperature differences. The formation of a convection column (smoke plume) above a fire or the heating of shrub or tree crowns above a surface fire are examples of heat transfer by free convection. *Forced convection* occurs when external mechanical forces alter the fluid flow from its natural "free" direction and velocity (Chandler et al. 1983). Wind can act as an external mechanical force. Where "strong" convection forces are operating in a fire, more of the heat transfer from the burning fuels occurs by convection. During low-intensity fire in pine litter, the ratio of convective-to-radiant heat is about 3:1, while the ratio is nearly 9:1 in a large, stationary fire in pine slash.

CONDUCTION

Conduction is the transfer of heat by molecular activity from one part of a substance to another part, or between substances in contact, without appreciable movement or displacement of the substance as a whole (Countryman 1976a). Most metals are good conductors, but substances like soil, water, wood, and air conduct heat slowly. The rate at which heat is transferred by conduction depends largely on the ability of a substance to conduct heat and on the temperature gradient. The quantity of heat conducted is also a function of the area through which heat is transferred. Conductive heat losses during fires in natural ecosystems are usually negligible except in heavy fuels like duff, organic soils, or large, dense slash piles with long smoldering times (Chandler et al. 1983).

MASS TRANSPORT

Mass transport of heat occurs during fires by airborne spotting and downslope rolling. *Spotting* involves the physical removal of burning material by thermal updrafts from flaming fuels and subsequent deposition in unignited fuels some meters or kilometers away. This is a predominant mechanism of fire spread in fast-moving uncontrollable fires (Chandler et al. 1983). Spotting distances are normally limited to about 10 km, with most of the spotting occurring within 2 km of the flaming front. The main fire-carrying agents are bark plates, cone bracts, and leaf or needle clusters, although large pieces of woody debris can be carried in fire whorls under extreme burning conditions. This form of heat transfer is important when considering fire effects on air quality. Downslope rolling is of particular importance only on steep slopes. Gravitational forces can move burning conifer cones, logs, and desiccated, ovoid-form plants such as sagebrush (*Artemisia* spp.) downslope into unburned fuels.

VAPORIZATION AND CONDENSATION

Vaporization and condensation are important in fire behavior and serve as a coupled reaction facilitating more rapid transfer of heat through soils. *Vapor-*

ization is the process of adding heat to water until it changes phase from a liquid to a gas. *Condensation* is the opposite process, whereby water changes from a gas to a liquid and releases heat. Water and organic materials are moved through the soil by vaporization and condensation.

A large amount of heat and energy is involved in the change of the physical state of water. The *latent heat of vaporization* of water is 560 cal/g, and this same amount is released during condensation. In addition, 1 cal/g/°C temperature rise is required to raise the temperature of water from ambient temperature to 100°C. Organic compounds found in fuels or released by the pyrolysis process also undergo the same vaporization and condensation process, but their latent heat requirements are not well known because their composition has not specifically been identified.

Heat Transfer Pathways

The pathways by which heat is transferred from combusting fuels to mineral soil, where it increases soil temperatures, are extremely complex. Quantifying the different pathways for heat flow requires the mathematical modeling of fire behavior, duff ignition and combustion, and the transfer of heat downward to and through moist and dry mineral soil (Dimitrakopoulos et al. 1994). The physical and mathematical modeling of these pathways, particularly in duff and mineral soil, is currently under investigation.

The overall pathways of heat flow from combusting fuels to mineral soil can be illustrated by a simplified flowchart (Figure 2.4). The heat produced by combusting fuels is transferred either to the surface of the duff (Path A) or directly to the surface of mineral soil if the organic layers are absent (Path B). In natural ecosystems, heat is usually transferred to an organic layer (Path A) because most vegetative covers produce some litter and duff, although the amounts vary widely. When duff is ignited, it can produce additional heat that subsequently is transferred to the underlying mineral soil (Path C). If duff does not ignite, it does not heat the underlaying mineral soil (Path D). The heat reaching the mineral soil is transferred through either a dry soil (Path E) or a moist soil (Path F). Dry soils are common during wildfire conditions, whereas prescribed fires can be designed to burn over wet or dry soils. The temperature profiles that develop during heat transfer into moist and dry mineral soil are different and consequently have a different effect on the physical, chemical, and biological systems. A discussion of heat transfer pathways, heat transfer processes, soil temperature profiles, and practical implications follows.

HEAT TRANSFER DOWNWARD TO DUFF OR MINERAL SOIL

Heat produced by combustion of fuels can be transferred to the duff (Figure 2.4, Path A) or to the mineral soil surface (Figure 2.4, Path B). The amount

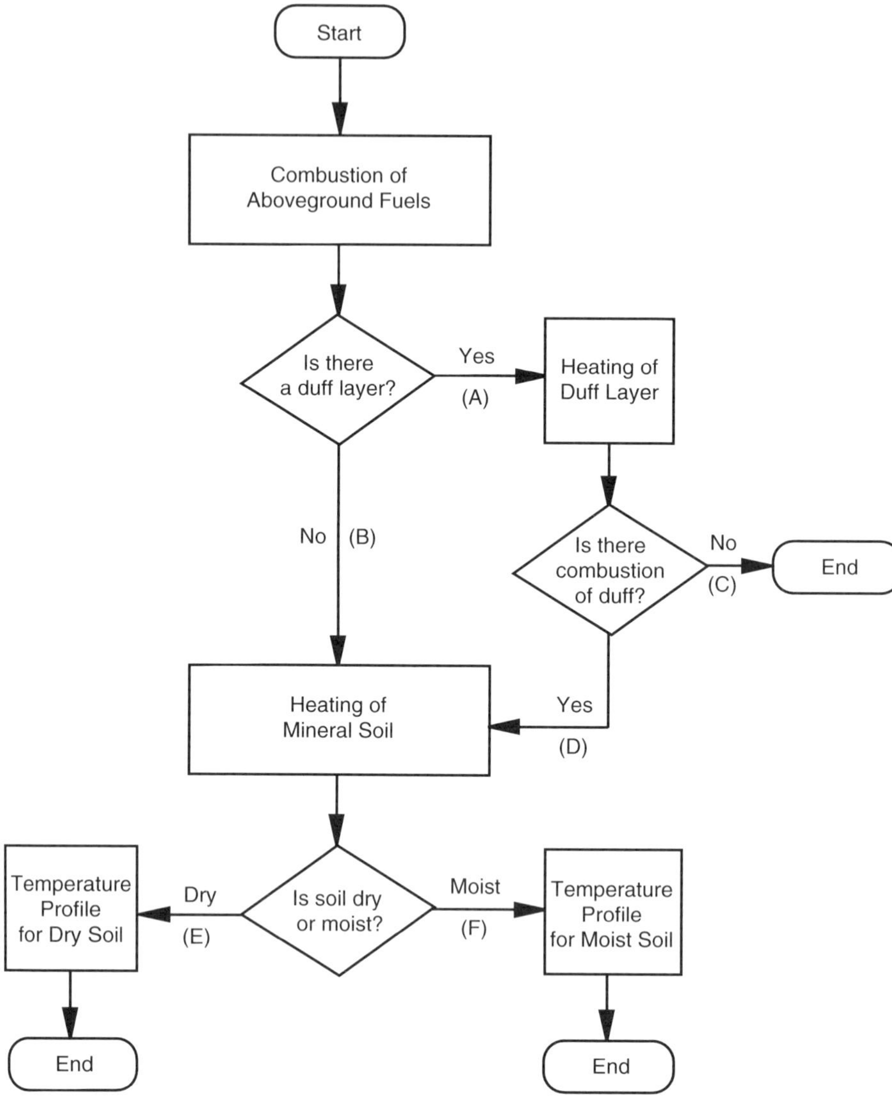

FIGURE 2.4 A conceptual model outlining heat flow pathways from combusting fuels downward through the litter into the underlying mineral soil. (Adapted from Dimitrakopoulos et al. 1994. A simulation model of soil heating during wildfires. In Sala, M., and J. L. Rubion, editors. *Soil Erosion as a Consequence of Forest Fires.* Geoforma Edicones, Logrono, Spain, pp. 199–206. Copyright © 1994 Geoforma Edicones, Longro, Spain, ISBN 84-87779-14-X, DL 7353-1994).

of heat transferred downward during the combustion of aboveground fuels depends on the fuels present and fire behavior (see Chapter 3). Fuel characteristics affecting the total heat generated include loading, size and arrangement, composition, and moisture content. Other environmental conditions such as temperature, wind, and relative humidity also affect combustion and heat production (Brown and Davis 1973, Hungerford 1990). Fire behavior variables (see Chapter 3), such as rate of spread, intensities, flame length, residence time, and rate of heat production, determine the amount of heat that is transferred downward. Attempts to relate duff combustion and soil heating to fire intensity parameters, however, generally have not been successful because of the variability in fire, conditions, and measurement techniques (Hungerford et al. 1991).

Heat is transferred downward from the burning fuels by radiation, convection, and conduction. Radiation and convection are probably the predominant mechanisms operating during the transfer of heat between the combusting fuel and the duff or mineral soil surface, although relatively little is known about the amount of heat transferred by the three processes. Conduction is probably a minor contributor to soil heating because the surface fuels are not generally in direct contact with duff or mineral soil unless large logs or other concentrated fuels (piles) are present.

Emissivity from the combusting fuel surface, temperature of the heat source, distance of the surface from the heat source, and temperature of the surface all determine the radiative *heat flux*. This term is used to describe the rate of heat transfer. Heat flux by radiation can be greater during slow-moving smoldering fires than during those that burn rapidly over the land surface.

HEATING IN DUFF AND ORGANIC SOILS

Heat produced by the combustion of fuels can be transferred to the duff (Path A, Figure 2.4). Duff can act as an insulating layer when it does not combust (Path C), or a heat source when ignited (Path D). Therefore, the amount of heat transferred into the underlying mineral soil depends on whether the duff burns and whether smoldering duff is a source of heat. The ignition and combustion of the duff are complex and attempts to correlate them with heat produced during slash burning have largely been unsuccessful (Albini 1975). Although the duff complicates the heat transfer from the burning aboveground fuels into the underlying mineral soil, some features controlling duff ignition and combustion are known. Important variables needed to describe heat production in duff include depth, amount, density of packing, inorganic content, and moisture content.

If duff does not burn, it provides a barrier to heat flow since the thermal conductivity of organic matter is low (Path D, Figure 2.4). The probability of ignition in organic soils (including duff) depends on both inorganic material and moisture content (Hungerford et al. 1995). Organic soils are not necessarily completely organic matter, but instead they may contain variable amounts of mineral soil as a result of mixing during disturbance. The chances of ignition in organic soils decrease as mineral content increases at any given moisture content. Likewise, the chances of ignition of organic soil decrease as moisture content increases at any given mineral content. In general, duff burns more efficiently when the moisture content is below 30 percent. Varying amounts will burn at moisture contents from 30 to 150 percent, and it is too wet to burn when moisture contents exceed 150 percent (Brown et al. 1985). Moisture affects the thermal conductivity and heat capacity of the duff, which, in turn, affects its ignitability (Hungerford 1990).

Combustion of the duff involves a smoldering reaction, which is initiated by the ignition of a spot or several spots by fire brands, hot ash material, or radiated and conducted heat from the fire front (Pyne et al. 1996). When duff ignites, it can transfer large amounts of heat into the underlying soil by convection, conduction, and radiation and can raise the mineral soil above 350°C for several hours. Therefore, it becomes difficult to quantify this combined heat flow into the underlying mineral soil. When thick layers of duff ignite, glowing combustion can also create an ash layer on the surface of the glowing duff. This ash layer retards heat dissipation upward, thereby causing more heat to penetrate into the soil (Sackett and Haase 1994). As a result, organic layers can transfer 40 to 73 percent of the heat generated during smoldering into the underlying mineral soil (Hungerford and Ryan 1996).

The combustion process in duff (duff burnout) and organic soils has been summarized by Hungerford and others (1995). Ignition is initiated at a single point or several locations on the surface duff (Figure 2.5). Ignition can also occur in cracks or depressions in duff, or be caused by woody material that burns down through the duff (Figure 2.5A). Fire burns both laterally and vertically after ignition (Figure 2.5B). Fire will burn laterally until it encounters incombustible conditions (moist organic matter, rocks, or absence of duff). It burns vertically until it reaches mineral soil or moisture conditions that will not support combustion. During the smoldering of the fire, a hole develops in the burned-out organic layer. Horizontal spread of the fire can leave a thin unburned top crust (Pyne et al. 1996). As the smoldering moves laterally and vertically, it creates a drying zone caused by the heat from the glowing zone, which allows the glowing front to advance until it reaches incombustible conditions (Figure 2.5C).

HEAT TRANSFER THROUGH MINERAL SOIL

Heat produced during the combustion of duff and/or plant canopies and aboveground fuels finally reaches the mineral soil. When duff is absent, soil heating depends largely on heat radiated from the combusting plants and litter and from hot ashes deposited on the soil surface (Pafford et al. 1991). Important ash parameters include initial ash temperature, location of initial contact of ash with the ground surface, and the thickness of the ash layer.

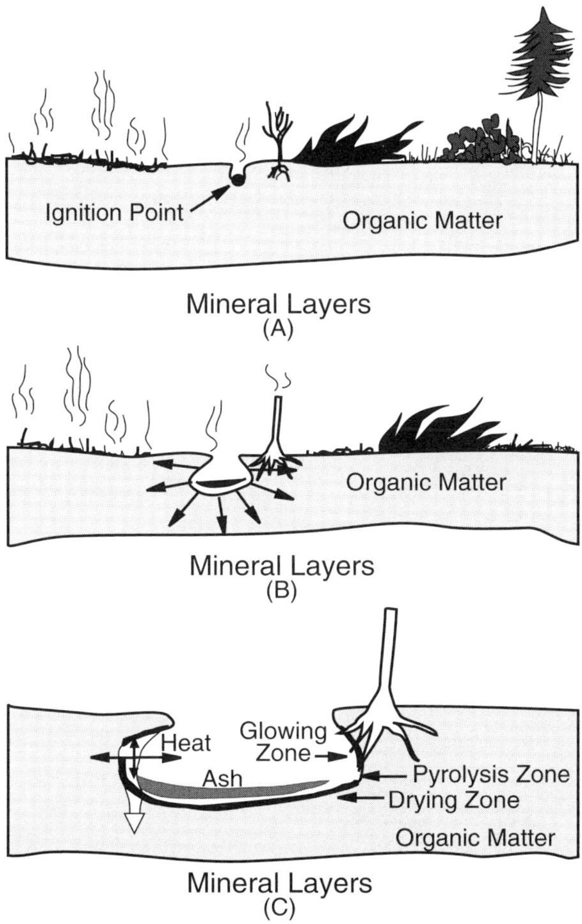

FIGURE 2.5 A schematic diagram of the smoldering process. The initial ignition point is created by the passing fire front. The fire spreads concentrically from the ignition point (A), develops concentric burned areas (B), and finally develops into a large burned-out area (C). (Adapted from Hungerford et al. 1995. Copyright © 1995 Tall Timbers Research Station.)

Categories of factors that affect heat transfer through mineral soil are as follows:

- *Soil moisture content.* As with duff, the amount of water in the soil affects heat transfer.
- *Chemical and physical properties of the soil.* The amount of organic matter, including roots, given an adequate O_2 supply, determines how much combustion occurs below the L layer (duff burnout). Properties

of thermal conductivity, specific heat, and bulk density act together to determine the rate of heat flow through the soil material. Soil parent material, porosity, water content, temperature gradient, hydraulic conductivity, and other factors indirectly influence the rate of heat flow.

The simplest form of heat transfer through mineral soil occurs when soils are dry (Figure 2.4, Path E). Heat transfer through a dry soil is primarily by conduction and the convection of hot gases through the porous matrix of the soil. Under these conditions, the history of the heat input, or the soil surface temperature history, must be known to mathematically model soil temperature profiles (Albini et al. 1996).

In contrast to dry soils, heat transfer in moist soils is complicated by the presence of water (Figure 2.4, Path F). This complexity occurs because water increases both the heat capacity and the heat conductivities of the soil. The increased heat capacity allows soil to absorb large amounts of heat without a subsequent temperature increase until water is vaporized. The heat capacity of water is high compared to other soil constituents (Table 2.2). Moist soils also conduct heat more rapidly because of their increased thermal conductivity (Table 2.2). Therefore, the heating sequence in a moist soil involves heating the soil until water is vaporized and lost (at 95 to 100°C), and then further heating (>100°C) takes place as in a dry soil (Example 2.3).

Example 2.3: Studies on the Effect of Soil Water on Heat Transfer in Mineral Soil

Although free water vaporizes at 100°C, water in soil vaporizes at a lower temperature. For example, during experiments with moist soils, the temperature upon heating initially rose to 95°C and remained there until the water content of the soil at a given depth dropped below 0.02 m^3/m^3 (Campbell et al. 1995). There is also evidence that suggests water may keep the soil at 80°C until the water is lost (Hungerford 1990). Moisture also affects the thermal conductivity of a soil. The thermal conductivity of moist soils (at moisture contents between 0.15 and 0.25 m^3/m^3) was found to be three to five times greater at 90°C than at ambient temperatures (Campbell et al. 1994). This finding suggests that, below about 95°C, the predominant heat transfer mechanism in soils is by conduction through the mineral soil particles and the pore water. However, some heat can be transferred by the movement of water by convection in response to temperature gradients in and between soil pores. There is some evidence to indicate that heat transfer by vapor movement begins in the 84 to 90°C range. Since transfer of heat by vapor movement is a much faster process than conduction, it can be more influential than conduction at temperatures <95°C. The important fact to remember is that the thermal conductivity of the bulk soil medium is responsive to liquid water content when water occupies a significant volume fraction of the medium (Albini et al. 1996). ■

TABLE 2.2 Specific Heat and Thermal Conductivities of Soil Materials

	Specific Heat		***Thermal Conductivity***	
Material	***Oven Dry***	***Fully Saturated***	***Oven Dry***	***Fully Saturated***
	cal/g per °C		***[cal/(cm s °C)] × 10^{-4}***	
Sand	0.20	0.72	7.85	12.85
Clay	0.20	0.82	24.00	16.85
Granite	0.20	—	68.15	—
Charcoal	0.37	0.92	1.80	13.15
Duff	0.42	0.90	1.20	12.85
Water	—	1.00	—	14.45

Source: C. Chandler, P Cheney, P. Thomas, L. Trabaud, and D. Williams. *Fire in Forestry—Volume I: Forest Fire Behavior and Effects.* Copyright © 1983 John Wiley & Sons. Reprinted by permission of John Wiley & Sons, Inc.

The sequence of heat transfer described above is reflected in temperature curves that develop in dry and moist sands at different times and depths below a heat source (Figure 2.6). In the dry sand, heat is transferred into underlying layers so that the pattern of temperature rise is similar at all depths (Figure 2.6A). Conversely, the presence of water in the sand slows the rise in temperature compared to the dry sand. For example, in moist sand, the temperatures at each successive depth downward rise to about 80 to 95°C and remain there until the layer is almost dry (Figure 2.6B). Once the water is evaporated from a sand layer, the temperature of the soil rises similarly to that of a dry sand. As a result, the temperature 25 to 35 mm downward in the moist sand may not rise above 100°C unless there are long residence times (40 to 60 minutes).

Heat transfer and temperature profiles developing in soil would be expected to be similar to those described for a sand medium. The soil temperature regimes developing in dry and moist soils in response to heating have practical implications because temperatures are produced that can cause irreversible changes in soil chemical, physical, and biological properties (see Chapter 4). Temperatures in moist soils do not rise much above 95°C until all the water in a given soil layer is driven off. Therefore, chemical and physical properties of soil are not greatly affected by heating until the soil becomes dry. However, irreversible damage to living organisms near the soil surface is likely to occur, because soil temperatures can easily be elevated above lethal temperatures of 60°C. Lethal temperatures to seeds (see Chapter 9) and microorganisms (see Chapter 5) are lower for moist soils than for those that are dry. Conversely, in dry soil, the temperatures at the surface may increase to

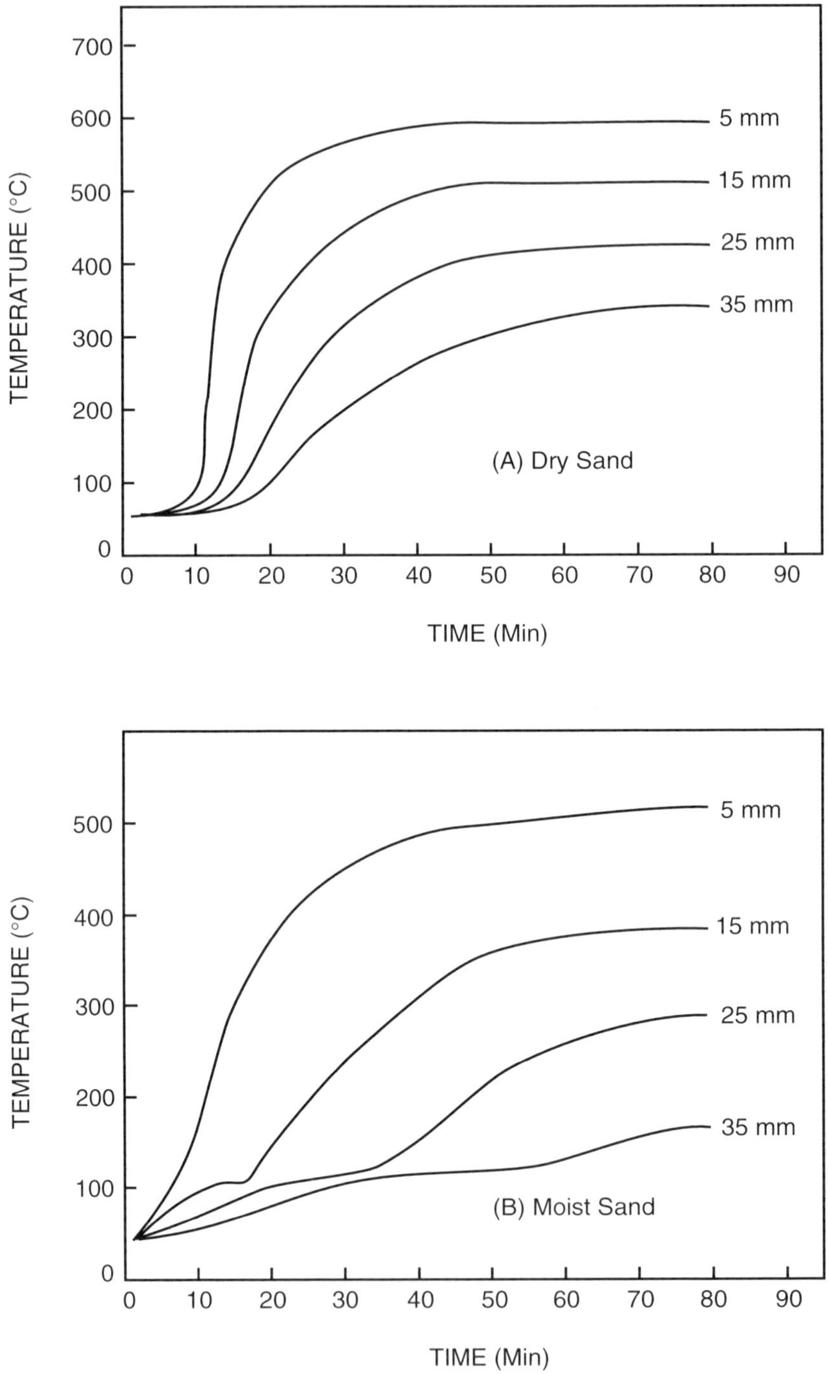

FIGURE 2.6 Temperatures developing at different depths in (A) dry and (B) moist sand as a result of surface heating. (Adapted from Campbell et al. 1995.)

temperatures greater than 100°C almost immediately, and thereby some physical and chemical changes related to organic matter begin to occur (see Chapters 5 and 6).

The depth to which the heat penetrates a moist soil depends on the water content of the soil and on the magnitude and duration of the surface heating during the combustion of aboveground fuels, litter, and duff. For long-duration heating, such as that under a smoldering duff fire (see Chapter 5), substantial heating of the underlying soil to depths of 40 to 50 cm can occur. Where heating is prolonged, temperatures will climb into the lethal range for soil organisms and plant roots. Increased thermal conductivity of moist soil may also produce lethal temperatures at deeper soil depths than if the soil was dry because of the increased thermal conductivity.

Modeling Research

There has been a long-term interest in developing models for quantitatively predicting fire effects (including soil heating) during wildfires and prescribed burns. Considerable detail is known about the combustion process and heat production by burning fuels, as has been described at the beginning of this chapter. Current emphasis on model development is focusing on quantifying heat input to the mineral soil by combustion of aboveground fuels and/or duff (Figure 2.4, Paths B and D) and on describing heat transfer through wet and dry soil once it has reached the mineral soil surface (Figure 2.4, Paths E and F). The end product of this research will be to quantitatively link heat transfer to and through the soil with models that can accommodate the highly variable environmental features that are characteristic of natural ecosystems. A conceptual model of this overall process developed by Dimitrakopoulos et al. (1994) has been simplified in Figure 2.4.

Research on model development is currently being conducted at several levels of emphasis, including development of:

- *General models.* These models predict first-order fire effects that are based on the application of known relationships, equations, and algorithms (Keane et al. 1995).
- *Physical and mathematical models.* These are detailed models describing specific heat transfer processes and pathways in moist and dry soils (Campbell et al. 1995, Albini et al. 1996). The prediction of temperature profiles for dry and moist sands described earlier is part of this ongoing research.
- *Duff burnout models.* These models describe ignition, combustion, and heat transfer capabilities of duff and organic soils. A conceptual model along with quantitative information on the smoldering of duff and organic soils was described earlier.

FIRST-ORDER FIRE EFFECTS MODEL (FOFEM)

A research effort in fire effects modeling is underway to develop a FOFEM (Keane et al. 1995). This model is being developed to predict first-order fire effects, including duff and woody fuel consumption, fire-caused tree mortality, smoke production, and soil heating. First-order effects are those changes produced directly as a result of the fire itself. The current model can predict fuel consumption, tree mortality, and smoke production. It is also anticipated that future models will form the basis for predicting secondary fire effects, including tree regeneration, hydrologic responses, soil erosion, and vegetative succession. Specific versions of the model are being developed for the Pacific West, Interior West, Northeast, and Southeast United States.

HEAT TRANSFER IN SOIL

Models have been developed to describe heat transfer and the simultaneous movement of water in soils. Most of the effort, however, has been concerned with water movement at temperatures experienced during diurnal cycles of heating and cooling, not during the high temperatures and steep temperature gradients produced during a fire. A review of the status-of-knowledge on heat transfer models provides insight on the current status of these modeling efforts in relation to a fire environment (Albini et al. 1996). The review examined the assumptions of different models in terms of the processes of heat transport (including conduction, water vapor flow, and gaseous exchange) and analyzed possible techniques for simplifying and solving the complex differential equations used to describe heat transfer. Some of the models being evaluated include those of deVries (1958, 1963), Lubikov (1975), Aston and Gill (1976), Kaviany (1991), and Campbell et al. (1994, 1995). These models have not been field tested at this time, however. Laboratory verification of assumptions underlying heat flow in soils under temperature regimes representative of fire conditions in natural ecosystems is being explored. The initial results of these studies indicate good agreement between the temperatures predicted by models and those measured in heated soil and sand (Campbell et al. 1995).

BARRIERS AND OPPORTUNITIES

The variability in weather, fuels, living vegetation, organic layers, soils, and topography amplifies the difficulty of quantitatively describing heat transfer resulting from combustion throughout the surrounding ecosystem. For example, duff and litter layers vary widely in the natural environment, from traces of surface organic matter in arid ecosystems to thick layers in more humid forests to several meters thick in peat bogs. Surface organic matter also varies spatially, as when it accumulates under plant canopies but not on the soil surface between plants.

In spite of the variable conditions described, the conceptual framework presented in this chapter to describe complex processes during combustion and heat transfer can be utilized as a framework for understanding and assessing the effect of fire on different components of ecosystems. For example, generalized soil temperature data presented for forests, shrublands, and grasslands (see Chapter 5) can be used in conjunction with fire severity classifications (see Chapter 3) as a basis for making initial estimates of soil responses to fire. Therefore, the information presented in this chapter contributes to a better understanding of the basic processes responsible for changes in the soil resource described later in this book. However, future refinement of physical and mathematical heat transfer models will greatly enhance the ability to predict site-specific changes in field situations and perhaps reduce the need for the more qualitative techniques currently available.

SUMMARY

Fuel, O_2, and heat are necessary for fire ignition. Heat produced by the combustion of the fuels releases a spectrum of compounds, including water vapor, noncombustible gases, volatiles, tars, carbonaceous char, and mineral ash. Effects of fire on natural ecosystems depend on the amount of energy that is released when a fire burns. The main heating processes are radiation, conduction, convection, mass transport, and vaporization–condensation. If the duff or organic soils ignite and burn, additional soil-heating and soil-sterilizing temperatures can occur. At this point in the book, you should be able to:

1. Describe the conditions needed to ignite a fire.
2. Explain the phases of a fire's existence.
3. Understand the meaning and importance of combustion efficiency.
4. Identify and explain the physical processes that result in heat transfer from a fire to fuels, from fuels through duff and organic soils to mineral soil, and through mineral soil.

REFERENCES

Albini, F. A. 1975. *An Attempt (and Failure) to Correlate Duff Removal and Slash Fire Heat.* USDA Forest Service, General Technical Report INT-24.

Albini, F., M. R. Amin, R. D. Hungerford, W. H. Frandsen, and K. C. Ryan. 1996. *Models for Fire-Driven Heat and Moisture Transport in Soils.* USDA Forest Service, General Technical Report INT-GTR-335.

Aston, A. R., and A. M. Gill. 1976. Coupled soil moisture, heat, and water vapour transfers under simulated fire conditions. *Australian Journal of Soil Research* 14:55–66.

Brown, A. A., and K. P. Davis. 1973. *Forest Fire: Control and Use.* McGraw-Hill Book Company, New York.

Brown, J. K., M. A. Marsden, K. C. Ryan, and E. D. Reinhardt. 1985. Predicting duff and woody fuel consumed by prescribed fire in the northern Rocky Mountains. USDA Forest Service, Research Paper INT-337.

Browne, F. L. 1958. *Theories of the Combustion of Wood and Its Control: A Survey of the Literature.* USDA Forest Service, Forest Products Laboratory Report Number 2136, Madison, Wisconsin.

Campbell, G. S., J. D. Jungbauer, Jr., W. R. Bidlake, and R. D. Hungerford. 1994. Predicting the effect of temperature on soil thermal conductivity. *Soil Science* 158:307–313.

Campbell, G. S, J. D. Jungbauer, Jr., K. L. Bristow, and R. D. Hungerford. 1995. Soil temperature and water content beneath a surface fire. *Soil Science* 159:363–374.

Chandler, C., P. Cheney, P. Thomas, L. Trabaud, and D. Williams. 1983. *Fire in Forestry—Volume I: Forest Fire Behavior and Effects.* John Wiley & Sons, New York.

Countryman, C. M. 1975. *The Nature of Heat. Heat—Its Role in Wildland Fire—Part 1.* Unnumbered Publication, USDA Forest Service, Pacific Southwest Forest and Range Experiment Station, Berkeley, California.

Countryman, C. M. 1976a. *Heat Conduction. Heat—Its Role in Wildland Fire—Part 2.* Unnumbered Publication, USDA Forest Service, Pacific Southwest Forest and Range Experiment Station, Berkeley, California.

Countryman, C. M. 1976b. *Heat conduction and wildland fire. Heat—Its Role in Wildland Fire—Part 3.* Unnumbered Publication, USDA Forest Service, Pacific Southwest Forest and Range Experiment Station, Berkeley, California.

Countryman, C. M. 1976c. *Radiation. Heat—Its Role in Wildland Fire—Part 4.* Unnumbered Publication, USDA Forest Service, Pacific Southwest Forest and Range Experiment Station, Berkeley, California.

Countryman, C. M. 1976d. *Radiation and Wildland Fire. Heat—Its Role in Wildland Fire—Part 5.* Unnumbered Publication, USDA Forest Service, Pacific Southwest Forest and Range Experiment Station, Berkeley, California.

DeBano, L. F. 1974. Chaparral soils. In: C. Rosenthal, conference chairman. *Proceedings of a Symposium on Living with the Chaparral.* Sierra Club, San Francisco, California, pp. 19–26.

deVries, D. A. 1958. Simultaneous transfer of heat and moisture in porous media. *Transactions of the American Geophysical Union* 39:909–916.

deVries, D. A. 1963. Thermal properties of soils. In: van Wijk, W. R., editor. *Physics of Plant Environment.* North Holland Publishing Company, Amsterdam, The Netherlands, pp. 210–235.

Dimitrakopoulos, A. P., R. E. Martin, and N. T. Papamichos. 1994. A simulation model of soil heating during wildland fires. In: Sala, M., and J. L. Rubio, editors. *Soil Erosion as a Consequence of Forest Fires.* Geoforma Ediciones, Logrono, Spain, pp. 199–206.

Fritschen, L., H. Bovee, K. Buettner, K. Charlson, R. Monteith, L. Pickford, J. Murphy, and E. Darley. 1970. Slash fire atmospheric pollution. USDA Forest Service, Research Paper PNW-97.

Hall, J. A. 1972. *Forest Fuels, Prescribed Fire, and Air Quality.* Unnumbered Publication, USDA Forest Service, Pacific Northwest Forest and Range Experiment Station, Portland, Oregon.

Hungerford, R. D. 1990. Modeling the downward heat pulse from fire in soils and plant tissue. In: MacIver, D. C., H. Auld, and R. Whitewood, editors. *Proceedings of the 10th Conference on Fire and Forest Meteorology,* Ottawa, Canada, pp. 148–154.

Hungerford, R. D., and K. C. Ryan. 1996. Prescribed fire considerations in southern forested wetlands. In: Flynn, K. M., editor. *Proceedings of the Southern Forest Wetland Ecology and Management Conference.* Consortium for Research on Southern Forested Wetlands, Clemson University, Clemson, South Carolina, pp. 87–92.

Hungerford, R. D., W. H. Frandsen, and K. C. Ryan. 1995. Ignition and burning characteristics of organic soils. In: Cerulean, S. I., and R. T. Engstrom, editors. *Fire in Wetlands: A Management Perspective. Proceedings of the Tall Timbers Ecology Conference No. 19.* Tall Timbers Research Station, Tallahassee, Florida, pp. 78–91.

Hungerford, R. D., M. G. Harrington, W. H. Frandsen, K. C. Ryan, and G. J. Niehoff. 1991. The influence of fire on factors that affect site productivity. In: Harvey, A. C., and L. F. Neuenschwander, compilers. *Proceedings—Management and Productivity of Western-Montane Forest Soils.* USDA Forest Service, General Technical Report INT-280, pp. 32–50.

Keane, R. E., E. D. Reinhardt, and J. K. Brown. 1995. FOFEM: a first order fire effects model. In: Brown, J. K., R. W. Mutch, C. W. Spoon, and R. H. Wakimoto, technical coordinators. *Proceedings of Fire in Wilderness and Park Management, Missoula, Montana.* USDA Forest Service, General Technical Report INT-320, p. 214.

Johansen, R., J. Deeming, M. Long, and D. Ward. 1985. Chapter II. Smoke production characteristics and effects. In: Gardner, F., F. Abeita, L. Bancroft, W. L. McCleese, H. N. Miller, R. Mutch, H. Ryan, J. W. van Wagtendonk, R. Wilson, and F. Zontek, coordinators. *Prescribed Fire Smoke Management Guide.* United States Department of Agriculture, United States Department of Interior, and National Association of State Foresters, NFES Publication No. 1279, pp. 5–10.

Kaviany, M. 1991. *Principles of Heat Transfer in Porous Media.* Springer-Verlag, New York.

Lubikov, A. V. 1975. Systems of differential equations of heat and mass transfer in capillary-porous bodies (review). *International Journal of Heat and Mass Transfer* 18:1–14.

Pafford, D., V. K. Dhir, E. Anderson, and J. Cohen. 1991. Analysis of experimental simulation of ground surface heating during a prescribed burn. *International Journal of Wildland Fire* 1:125–146.

Pettersen, R. C. 1984. The chemical composition of wood. In: Rowell, R., editor. *The Chemistry of Solid Wood.* American Chemical Society, Washington, DC, pp. 57–126.

Philpot, C. W. 1971. The pyrolytic effect of treating cottonwood with plant ash. USDA Forest Service, Research Note INT-139.

Pyne, S. J., P. L. Andrews, and R. D. Laven. 1996. *Introduction to Wildland Fire.* John Wiley & Sons, New York.

Raison, R. J., P. V. Woods, B. F. Jakobsen, and G. A. V. Bary. 1986. Soil temperatures during and following low-intensity prescribed burning in a *Eucalyptus pauciflora* forest. *Australian Journal of Soil Research* 24:33–47.

Ryan, P. W., and C. K. McMahon. 1976. Some chemical and physical characteristics of emissions from forest fires. In: *Proceedings of the 69th Annual Meeting of the Air Pollution Control Association,* Portland, Oregon, Paper Number 76–2.3.

Sackett, S. S., and S. M. Haase. 1994. Soil and cambium temperatures associated with prescribed burning. In: Starr, L., editor. *Natural Resource News Special Edition* (March 1994). Blue Mountains National Research Institute, LaGrande, Oregon, p. 5.

Sandberg, D. V., J. M. Pierovich, D. G. Fox, and E. W. Ross. 1979. *Effects of Fire on Air: A State-of-Knowledge Review.* USDA Forest Service, General Technical Report WO-9.

Shafizadeh, F. 1984. The chemistry of pyrolysis and combustion. In: Rowell, R., editor. *The Chemistry of Solid Wood.* American Chemical Society, Washington, DC, pp. 489–529.

Ward, D. E. 1990. Chapter 18: Factors influencing the emissions of gases and particulate matter from biomass burning. In: Goldammer, J. G., editor. *Fire in the Tropical Biota: Ecosystem Processes and Global Challenges.* Springer-Verlag, Berlin. pp. 418–436.

Chapter

3

Fuels and Fire Behavior

INTRODUCTION

Fire occurring in natural ecosystems begins as a point of ignition, burns outward into circles, and, if the fire escalates, spreads in the direction toward which the wind is blowing and, when burning occurs on uneven terrain, spreads upslope to eventually form itself into broad ellipses. Intensity and rate of spread of the fire are partly controlled by factors such as the vegetation (forests, shrubs, or herbaceous plants) being burned, atmospheric stability and windspeed, and topography. The energy produced by fire is the contributing force for many of the effects that are described in the following chapters of this book. However, the nature of fuels available for burning and the behavior of the resulting fire need to be discussed before moving on to these chapters.

FUELS

Fuels available for burning, along with heat and O_2, represent the components of the *fire triangle,* and all must be present for fire to occur (see Figure 2.1). Fuels available for burning generally consist of all biomass that potentially can be ignited by a heat source. The description, classification, and management of these biomass fuels are considered next.

Description

A number of terms are often used in describing the fuels available for burning in natural ecosystems (Brown and Davis 1973, Martin et al. 1979, Pyne et al. 1996). *Biomass* includes all of the vegetative materials. *Phytobiomass* is the aboveground vegetative material and is often considered to be the *total fuel* available to burning. *Potential fuel* is the material that might burn in intense fire and, therefore, is generally less than the total fuel. The term *available fuel* refers to that amount of fuel available for burning in a particular fire, a value varying widely in magnitude with the environmental conditions on a site.

For simplicity, the terms *fuel* or *fuel available for burning* are used interchangeably throughout this book when referring to the biomass that decomposes as a result of ignition and combustion. The properties, components, and complexes of these fuels are useful in their description.

PROPERTIES

Properties of fuels affect the manner in which a fire burns and, ultimately, the generation of energy and production of air pollutants by the fire. *Physical properties* of interest to a fire manager generally include the quantity (fuel loading), size and shape, compactness, and arrangement.

Fuel loading, or the total dry weight fuel per unit surface area, is a measure of the potential energy that might be released by a fire (Martin et al. 1979, Whelan 1995, Pyne et al. 1996). Natural fuel loading in natural ecosystems ranges from 0.5 to over 400 Mg/ha (Example 3.1). The readily combustible fuels are reduced the most by a prescribed fire, although substantial burning of the finer fuels and the litter layer can also occur. Wildfire usually results in greater reductions in the quantity of fuel available for burning and, in doing so, is likely to produce large quantities of smoke and other atmospheric emissions. A wildfire can also be more erratic, leaving unburned, lightly burned, and severely burned areas in close proximity to each other. Sites with low fuel loadings, which burn at low intensities, generally experience less severe impacts than those with larger fuel loadings, which burn at high intensities.

Example 3.1: Natural Forest Fuel Loading: Two Case Studies (Sackett 1979, Sackett and Haase 1996)

Natural forest fuel loadings cause extreme fire behavior during dry, windy weather experienced during fire seasons in the southwestern United States. Fire severity is also exacerbated from burning heavy fuels, including thick humus layers on the forest floor. Natural loadings of dead fuels in southwestern ponderosa pine forests are heavy, averaging 9.75 Mg/ha. Small fuels less than or equal to 2.54 cm (1 in.) in diameter comprise as much of the total loadings as do large fuels. Although the large fuels are not classified as *heavy fuels,* the fact that most of this larger material is rotten could be important. In dry conditions, rotten material is more easily consumed by burning. It also produces numerous firebrands and provides a more receptive place for firebrands to land and ignite than does sound (not rotten) fuel.

Loadings of dead fuels in southwestern mixed conifer forests, communities of seven coniferous species and quaking aspen *(Populus tremuloides)*, are also relatively heavy, averaging 19.6 Mg/ha of dead fuel. Weight of needle material in the *L* and *F* layers is comparable to that observed in the ponderosa pine forests. The humus is more than twice the weight of that found in ponderosa pine, however. All size classes of fuels are also more than double in weight. ■

Chemical properties of fuels affect their heat content and the types of pollutants emanating from a fire. The complexity of the chemical compounds released into the atmosphere is attributed largely to the diversity of fuels in the ecosystem. Woody fuels are high in cellulose, lignin, and hemicellulose, while foliage of woody plants and the herbaceous fuels contain large amounts of readily combustible extractives.

Oils and resins in the fuel increase the heat release because of their high energy content. Fuel containing concentrations of these chemical compounds would be expected to burn intensely (Whelan 1995). On the other hand, the relatively high concentrations of mineral elements such as those found in leaves and wood (xylem) reduce flammability in some plant species. The range of fire intensities, rates of fire spread, and fire durations encountered in a particular ecosystem result in arrays of combustion conditions, which, in turn, result in a diversity of chemicals being emitted into the atmosphere by burning (see Chapter 11).

COMPONENTS

Fuel components are generally related to the way in which vegetation grows on a site. These components are often delineated with respect to their respective vertical layers (Pyne et al. 1996). This grouping is based largely on differences in the behavior of fire as it burns through each of the layers:

- *Ground fuels* include the *H* layer of duff, roots, and buried or partially buried dead and decaying logs. Fire spread through ground fire is usually slow because of the compactness of ground fuels, with burning by smoldering combustion.
- *Surface fuels* are represented by the physical characteristics of the vegetation itself, that is, the standing trees, herbaceous plants, litter and the *F* layer of duff, and downed woody materials. Most fire in natural ecosystems ignites in, and is carried by, surface fuels.
- *Crown fuels* are tree and shrub crowns. Even when the crowns are not burning, they can affect the behavior of a fire in surface fuels. For example, forest stands with open crowns generally have a faster spreading

surface fire than those with closed (dense) crowns because of the increased airflows in the former.

COMPLEXES

Fuel complexes are associations of fuel components described in terms of vegetative cover or habitat types. A forest fuel complex consists of a combination of the dominant overstory trees and associated understory, which can include herbaceous plants, litter, dead and downed fuels, or some combination of these fuels. A rangeland fuel complex might consist of only grasses or a combination of grasses and shrubs.

It is often difficult to relate fuel complexes directly to the fuels available for burning (Brown and Davis 1973, Martin et al. 1979, Pyne et al. 1996). Fuels available for burning tend to accumulate as mosaics in most natural ecosystems rather than accumulate uniformly over large areas.

Classification

There is infinite variability in fuels in a natural ecosystem. A need to classify fuels for management purposes has evolved as a consequence (Brown and Davis 1973, Martin et al. 1979, Pyne et al. 1996). Fuels can be classified in a number of ways in terms of their condition, vertical position, or composition in responding to this need. These classifications include:

- Whether or not the fuel burns
- Vertical position (not to be confused with fuel component); including:
  Ground fuels—fuels lying on, or within 2 m of, the mineral soil surface
  Aerial fuels—fuels more than 2 m above the mineral soil surface
  Ladder fuels—those fuels continuous between ground fuels and crown fuels, forming a *ladder* by which a fire can spread into tree or shrub crowns
- Type, for example, forest, shrub, herbaceous plants, or litter and duff
- Dominant plant species, such as pine, oak, or chaparral.

Fuel classifications can also be based on the *fuel state,* that is, the moisture condition of the fuel. The amount of fuel available for burning at any given time is largely determined by the fuel's moisture conditions (Martin et al. 1979, Bond and van Wilgen 1996, Pyne et al. 1996). Moisture conditions of live fuels (in temperate regions) generally change seasonally, with spring budding and autumn curing of the living vegetation, while the moisture in dead fuels is mostly related to local weather patterns, such as local precipitation regimes.

One approach to classifying fuels relative to their state is based on the time it takes for the moisture content of the fuel in question to adjust to changes in environmental conditions such as wetting and drying cycles. When environmental changes occur, the resulting changes in fuel moisture move toward a new equilibrium condition. How quickly a fuel gains or loses moisture in response to wetting or drying cycles establishes its response time. *Timelag,* representing the integration of fuel state and weather factors, is a measure of the rate at which a fuel approaches its equilibrium moisture content after experiencing environmental changes. Fuel classification systems based on fuel state include:

- Categories used in the National Fire Danger Rating System of the USDA Forest Service (Martin et al. 1979, Pyne et al. 1996), that is:
  Live fuels—grouped by category as woody or herbaceous fuels.
  Dead fuels—grouped by size class as 1, 10, 100, or 1000 hour timelag classes (see below); size classes often separated into sound or rotten.
- *Timelag constant*—the time for a fuel's moisture content to lose 63 percent (that is, $1 - 1/e$, where e is the base of natural logarithms) of the difference between its moisture content at the beginning of the time period and the new equilibrium moisture content in the conditions to which it is exposed (Fosberg 1970, Pyne et al. 1996). For example, a fuel particle beginning at 20 percent moisture content and exposed to 10 percent equilibrium conditions would drop to $20 - [0.63(20 - 10)] = 20 - 6.3 = 13.7$ percent moisture in one timelag period. If the period is 1 hour, the timelag of the fuel is l hour.
- *Timelag interval,* or response time, is defined as the time required for dead fuel to lose approximately 63 percent of the difference between its initial moisture content and the equilibrium moisture content in an atmosphere of *constant* temperature and humidity (Fosberg 1970, Pyne et al. 1996). Arbitrary time intervals of 0–2, 2–20, 20–200, and 200–2000 hours were appropriate for fire danger ratings in the early 1970s. Class midpoints were subsequently taken as the title for timelag classes (see below).
- *Timelag classes* are used to partition dead fuels in the National Fire Danger Rating System of the USDA Forest Service and in general usage (Table 3.1).

Management

Fuel management is the practice of controlling the flammability of fuels and reducing the resistance of fuels to fire control, through mechanical, chemical, or biological means, or through the application of prescribed fire (Brown and Davis 1973, Martin et al. 1979, Chandler et al. 1983, Pyne et al. 1996). Through management, fuels are modified to modulate fire behavior, ameliorate fire effects, and reduce the costs of fire suppression.

TABLE 3.1. Timelag Class, Timelag Interval, and Fuel Dimensions and Type

Timelag Class	*Timelag Interval (h)*	*Fuel Dimensions (in.)*	
		Roundwood	*Litter and Duff*
1 h	0–2	<0.25	Surface
10 h	2–20	0.25–1	Surface–0.75
100 h	20–200	>1–3	>0.75–4
>100 or 1000 h	>200 or 200–2000	>3	

Source: Martin et al. (1979).

Fuel management can include reducing the loading of available fuel, converting fuels to those with a lower flammability, or isolating or breaking up large continuous bodies of fuels. Fuel management, therefore, is often important in fire protection and is a common objective of prescribed burning (see Chapter 11).

FIRE BEHAVIOR

Fire behavior is the manner in which a fire reacts to its environment, that is, to the fuels available for burning, climate, local weather conditions, and topography. Fire behavior changes in time, space, or both in relation to changes in these environmental components (Table 3.2). The changing status of each of these environmental components in a fire and their interactions with each other determine fire behavior at any point in time (Brown and Davis 1973, Bond and van Wilgen 1996, Pyne et al. 1996). The fuel component rapidly changes in time and space as a fire progresses. Local weather (the most variable environmental component) also changes in time and space. Topography does not vary with time, but it can change greatly in space.

Wind (a weather component) and slope (topography) affect fire similarly because of their effect of tipping the flames toward or away from the fuels ahead of the fire (Pyne et al. 1996). Interactions between the two factors depend on the magnitude and direction of influence of each (Figure 3.1). There is a cumulative effect and the fire moves upslope if the wind is blowing upslope. However, a fast-spreading fire can also move downslope, which often happens when southern California chaparral burns under Santa Ana conditions.

TABLE 3.2 Environmental Components Affecting Fire Behavior

Component	*Effects*
Fuel load	Determines maximum energy available to a fire. Arrangements affect aeration, vertical and horizontal spread of a fire. Size distribution can affect likelihood of initial ignition. Chemistry can increase (for example, resins or oils) or decrease (mineral content) flammability.
Climate	Determines vegetative productivities and, therefore, the rate of fuel accumulation.
Precipitation–humidity	Increased fuel moisture, combined with high relative humidity, decreases likelihood of ignition and rates of combustion and spread of a fire.
Wind	Causes drying of fuel. Increases oxygen available for combustion. Preheats and ignites fuel in advance of a fire front and can produce ignition ahead of the front. Changes in direction can increase the fire front.
Topography	Causes variation in local climate. Permits preheating and ignition for a fire burning uphill. Can provide natural firebreaks. Partially determines distribution of vegetative communities of varying flammabilities.

Source: Whelan (1995).

Types of Fire Behavior

Three types of fire behavior are generally recognized in describing fire behavior in natural ecosystems, with each type likely to produce different impacts on ecosystem resources:

- *Surface fire.* This type of fire consumes litter, low-growing plants, and dead herbaceous plants accumulated on the surface. Surface fire can ignite snags (dead standing trees), can consume shrubs and tree seedlings, and can "torch out" an occasional densely crowned mature tree. It remains a surface fire so long as its rate of spread depends on surface fuels. A ground fire can be initiated where a deep layer of organic soil exists.
- *Ground fire.* Ground fire burns deeply into the organic material on top of, and incorporated into, the soil profile. Ground fire, therefore, is par-

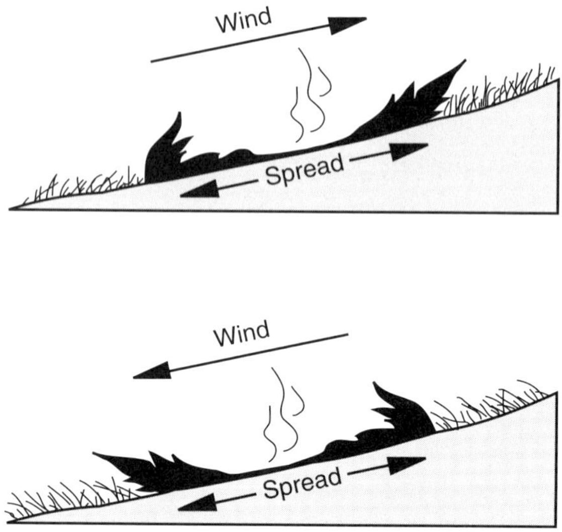

FIGURE 3.1 Influence of wind and slope on a fire. (Adapted from Rothermel and Rinehart 1983.)

ticular to areas in which there are heavy accumulations of organic matter in different stages of decomposition that dry to a point where they can be ignited, such as swamps, bogs, and other wetlands (see Chapter 10). This type of fire generally has a low rate of spread, is difficult to control, and is expensive to extinguish.

- *Crown fire.* Fire of this type spreads rapidly through crowns of trees and shrubs and on the surface. Fire in the crowns races ahead of the fire on the surface in what is called a *running crown fire.* In other cases, fire in the crowns moves only in spurts and is dependent on intense heating from the surface fire; such a crown fire is referred to as a *dependent crown fire.* This type of fire, which generally causes the most severe impact on a natural ecosystem, occurs mostly in coniferous forests and woodlands, and in shrublands comprised of waxy-leaved species.

Fire Intensity

Intensity and rate of spread play significant roles in characterizing fire behavior. *Fire intensity* is a term that is used to describe the rate at which a fire produces thermal energy (Brown and Davis 1973, Chandler et al. 1983, Pyne et al. 1996). While there are several ways of expressing fire intensity (Example 3.2), *fireline intensity* is the most widely used. It is a convenient expression of intensity because of its relation to flame length, which is easily measured.

Fire intensity is influenced by the amount of fuel available for burning, local weather conditions before and at the time of the fire, and the topography of the burning site (Brown and Davis 1973, Whelan 1995, Pyne et al. 1996). The limiting factor in fire intensity is the amount of energy stored in the fuel. As a consequence, the greater the fuel loading, the more intensely a fire is likely to burn. Fuel loads are largely a function of the type and age (development) of the fuel complex represented, fuel properties (including size, arrangement, concentrations of oils and resins) of the fuels, and the time since the last fire.

Example 3.2: Expressions of Fire Intensity (Brown and Davis 1973, Chandler et al. 1983, Pyne et al. 1996)

There are several expressions of fire intensity. *Radiant intensity* is the rate of thermal radiation emission. Radiant intensity is intercepted at (or near) the ground surface, or at some specified distance ahead of the flame front. *Convective intensity* is that part of the total heat output from a fire that lifts gases and entrains air above the flame zone. Convective intensity is not determined directly, but, instead, is the difference between total energy release rate and the sum of conductive and radiant heat losses.

Total fire intensity, the rate of heat output of the fire as a whole, is a function of the rate of area burned, fuel loading, and estimated heat yield. While the expression represents a general characterization of the fire, total fire intensity is not necessarily useful in describing site-specific fire effects. *Reaction intensity,* the total heat release per unit area of fuelbed divided by the burning time, is the time-averaged rate of heat release of the active fire front. Reaction intensity is calculated in the field by estimating the amounts of fuels burned per second and assuming heat yields for the fuels.

A widely used expression of fire intensity is *fireline intensity,* also called *Byram's intensity.* Fireline intensity, the product of the available heat of combustion per unit area of ground surface and the rate of spread of the fire, is calculated by

$$I = 0.007\ HWR$$

where

I = fireline intensity in kW/m
H = heat yield in cal/g
W = fuel loading in Mg/ha
R = rate of spread in m/min

Fireline intensity has been related empirically to flame length, which is easily measured in the field. With knowledge of flame length, the above equation becomes

$$I = 273(h)^{2.17}$$

where h = flame height. This equation is often simplified for field use as

$$I = 300h^2$$

■

The moisture conditions and temperature of fuels at the time of fire occurrence affect fire intensity, because the rate of combustion of moist, cold fuels is slower than that of dry, hot fuels. Fuels accumulated in a vegetative community with a dense canopy tend to retain moisture after each rainfall because of the reduced evaporation of moisture from fuels, while the shading effect reduces fuel temperature.

A fire can generate its own wind. The upward convection of heated gases draws in air from the surrounding burning nucleus, causing the fire front to spread outward into the wind generated by this action. Wind influences fire intensity by providing more O_2 to the burning front, increasing the rate of combustion. Changes in wind direction can change the flank of a fire into the front, also influencing fire behavior.

The influence of slope on a fire is similar to the effect of wind in that flames are brought nearer to the ground surface, and, therefore, more of the fuel accumulated ahead of the front is preheated (Figure 3.1). Also, a fire ignited on a hilltop or ridgeline is likely to burn slowly as it moves downhill, while a fire ignited in a gully downslope will start more rapidly and increase in momentum as it burns uphill. An indirect effect of topography on fire intensity is how it impacts on the occurrence of vegetative communities. Vegetation on a hilltop can be more xerophytic and less dense than the vegetation growing in a gully and, as a consequence, be more flammable.

Rate of Spread

Factors affecting fire intensity also influence the rate of spread of the fire (Brown and Davis 1973, Whelan 1995, Pyne et al. 1996). For example, fire occurring in dry, windy conditions on sites with high fuel loads spreads rapidly because of the rapid rate of combustion. High-intensity fire is also likely to preheat adjacent fuel accumulations rapidly, and the wind generated by the fire results in rapid ignition of the fuels through its influence on flame length and angle.

Vegetative communities inherently vary in their flammability and, as a consequence, in the rate at which fire might spread. Fuel continuity also plays an important role in whether a fire spreads and the resulting "patchiness" of burned vegetation. Fuel continuity is often affected by the time and management practices implemented since the last fire. Topography can create natural firebreaks on a landscape, influencing the spread of a fire.

Fire Behavior Patterns

A fire burns within distinct boundaries, often creating a mosaic of burned and unburned areas. As a consequence, a fire might not affect all of the ecosystem resources within the boundary of the fire in the same way. Furthermore, all fire does not cause the same magnitude of resource impact. While fire effects are largely site-specific, some generalizations can be made about fire behavior patterns that can be expected in forest, shrubland, and grassland ecosystems.

FOREST ECOSYSTEMS

Large differences exist in the behavior of wildfire and prescribed fires in forest ecosystems (Brown and Davis 1973, Connor and Cannon 1991, Pyne et al. 1996). A wildfire frequently evolves into a crown fire over time. More trees burn as the fire becomes more intense, especially where saplings (small trees) grow in dense thickets. Burning at moderate fire intensities can remove some of these smaller trees from the site, providing additional water and nutrients for the remaining larger trees. Wildfire at the upper end of the wildfire intensity spectrum is likely to spread into the tops of the tallest trees in violent and discontinuous surges. Fire that occurs at this severe end of the spectrum responds to its own convective winds, spreading rapidly as sparks from exploding trees ignite other fires many meters away. When fire reaches tree crowns, it often spreads rapidly from treetop to treetop.

Prescribed fire is designed normally to burn only the lighter surface and ground fuels. Intensities of these controlled burns are low to moderate. Low-intensity fire generally burns the ground cover, spreads into the duff and herbaceous plants, and leaves stems and tops of the trees largely unburned. As a prescribed fire increases in intensity, however, it often spreads along the surface in a well-defined flaming front.

SHRUBLAND ECOSYSTEMS

Fire occurrences in southern California chaparral illustrate the nature of fire behavior that can occur in shrubland ecosystems (Mooney and Conrad 1977, Conrad and Oechel 1982, Pyne et al. 1996). Fire commonly concentrates in the canopies of chaparral shrubs and, therefore, is more-or-less continuous depending on the continuity of the prefire shrub cover. Wildfire is frequently difficult to control, because of the large accumulations of highly volatile fuels on steep topography that are subjected to extremely dry conditions. Consequently, wildfire releases large amounts of energy and, in the process, causes large resource responses.

However, it can be difficult to ignite a prescribed fire in chaparral, especially in stands younger than 25 years that have accumulated large amounts of live fuel (Example 3.3). Such a burning usually results in a mosaic burn pattern, with the moister (north-facing) slopes remaining unburned and the drier (south-facing) slopes burned at high intensities. Interspersed among these two extremes can be areas where the shrub canopies are only partially burned and the surface litter and duff unburned.

Example 3.3: Fuels and Fire Behavior in Southern California Chaparral Shrublands (Mooney and Conrad 1977, Conrad and Oechel 1982, Chandler et al. 1983)

Accumulations of dead vegetation are the fuels that carry fire in chaparral shrublands, while live fuels absorb energy until these fuels become dried and then burn. Dead-to-live fuel ratios, therefore, are important in predicting fire behavior. Chaparral rarely burns if the dead fuels make up less than 20 percent of the fuel load. Dead fuels increase slowly with stand age. There is usually less than 20 percent dead material when the stand is 25 years old. When the shrubs are 25 to 50 years old, however, the fuels contain 30 to 45 percent dead material and, in general, are suitable for prescribed burning. Decadent chaparral that has 45 to 100 percent dead materials will burn intensely.

Moisture contents of both the dead and live fuels are also important relative to anticipated fire behavior. Chaparral that is 20 to 30 percent dead should be burned at about 5 to 10 percent dead fuel moisture. The recommended moisture percentage for burning increases as the proportion of dead fuel increases. Shrub stands with 50 percent dead fuel should be burned with dead fuel moisture of 10 to 15 percent. The moisture content of the live fuels should be 60 to 75 percent when burning mature shrubs. Chaparral burns violently when the moisture content of live fuels is below 60 percent. Little burning will occur when the moisture content is above 85 percent without some form of mechanical or chemical fuel treatment. ■

GRASSLAND ECOSYSTEMS

The behavior and effects of fire occurring in grassland ecosystems are more uniform and consistent from site to site than is observed in the case of fire in forest and shrubland ecosystems (Vogl 1979, Pyne et al. 1996). Since grasses and other herbaceous plants possess fine leaves and numerous stems that remain erect, the available fuels in grassland ecosystems ignite easily and burn readily. As a consequence, fire is the primary decomposition agent and the key nutrient recycler in many grasslands.

Fire in grassland ecosystems is carried primarily by the stems and leaves of herbaceous plants, which represent a relatively low fuel loading. As a result of lower fuel loadings, even high-intensity grass fire produces relatively low soil heating and fire impacts. Because the plant canopy carries the fire, smaller differences in impacts on ecosystem resources occur between wildfire and prescribed burning.

FIRE SEVERITY

Fire severity, a term that describes ecosystem responses to fire, can be used to describe the effects of fire on the soil and water system, ecosystem flora and fauna, the atmosphere, and society (Simard 1991). It reflects the amount of energy (heat) that is released by a fire, which affects resource responses. Fire severity is largely dependent on the nature of the fuels available for burning, and the behavior of the fire when these fuels are burned. The relationship of fire severity to soil responses receives emphasis in this chapter, because the most is known about this relationship, and because soil responses (see Chapters 4 to 7) are closely related to hydrologic responses (see Chapter 8) and ecosystem productivity (see Chapters 9 and 10).

Relationship of Fire Severity to Fire Intensity

Fire intensity, as stated before, is concerned mainly with the amount and rate of surface fuel consumption. It is not necessarily related to the total amount of energy produced in the burning process, and, therefore it is not a good measure of the amount of energy transmitted downward into the soil, and the associated changes that occur in physical, chemical, and biological properties of the soil (Example 3.4). For example, it is possible that a high-intensity and fast-moving crown fire will consume little of the surface litter, because only a small amount of the energy released during the combustion of fuels is transferred downward to the litter surface. Little damage of the surface litter takes place in this case. However, if the fire also consumes substantial surface fuels, and the residence time on a site is greater, a "white ash" layer might be the only postfire material left on the soil surface.

Example 3.4: Lack of Agreement Between Fire Intensity and Fire Severity (Hartford and Frandsen 1992)

It is not always possible to estimate the effects of fire on soil, vegetation, and air when these effects are judged by only surface intensity measurements, because other factors overwhelm fire behavior. The general lack of agreement between fire intensity and fire severity can be illustrated by comparing fire intensities and subsurface heating observed in a study where slash fuel accumulations lying on moist and dry duff were prescribe-burned. Data for this study were collected on plots representative of natural fuel loadings in the conifer forests of western Montana and northern Idaho. Needles, twigs, and bark flakes were the major slash fuel components.

Temperatures in the duff and mineral soil were measured under the two conditions of the study. Burning of slash fuels on moist duff produced higher intensities than on dry duff. However, minimal heating occurred in the underlying mineral soil. In contrast, smoldering combustion of slash accumulations on dry duff resulted in a long duration of heating, with nearly complete duff consumption, and lethal temperatures at the surface of the mineral soil. Moisture contents of duff and soil were key factors that determined the relative heating of mineral soil. ■

The relationship of fire intensity to fire severity remains largely undefined because of difficulties encountered in relating resource responses to the burning process. While quantitative relationships have been developed to describe changes in the thermal conductivity of soil, and changes in soil temperature and water content beneath surface fires, these relationships have not been thoroughly extrapolated to field conditions (Campbell et al. 1994, 1995). Nevertheless, it is hoped that relationships such as these can eventually be used to develop models that describe fire-driven heat and moisture transport in soils (Albini et al. 1996). The resource responses discussed in this book relate primarily to different levels of fire severity because models are not available to quantify resource responses in terms of fire intensities, as discussed above.

Some fire managers suggest the term *fire intensity* be used only in a strict thermodynamic sense to describe the amount of energy released, while the term *fire severity* be used to indicate the effects of fire on the ecosystem. Fire severity has been a term used to describe the magnitude of negative fire impacts on natural ecosystems in the past (Simard 1991), but a wider usage of the term to include all fire effects is proposed. Because one can rarely measure the actual energy release of a fire, the term *fire intensity* can have limited practical application when evaluating ecosystem responses to fire.

Classification

Fire severity cannot be expressed as a single quantitative measure that relates to resource impact. Therefore, relative magnitudes of fire impacts, expressed in terms of the percent of soil organisms affected by the fire, proportion of trees and shrubs killed, number of postfire resproutings of the trees or shrubs, or other measures of disturbance, are better placed into broadly defined discrete classes ranging from low to medium to high fire severities. For example, a general fire severity classification based on the postfire appearance of litter and soil has been developed to relate fire severity to the soil resource response (Hungerford 1996). Both the vertical and horizontal dimensions of a fire are used to classify its severity as follows:

- *Low fire severity.* Low soil heating, or light ground char, occurs where litter is scorched, charred, or consumed, but the duff is left largely intact, although it can be charred on the surface. Woody debris accumulations are partially consumed or charred. Mineral soil is not changed. Fire severity in forest ecosystems is low if the litter and duff layers are scorched but not altered over the entire depth. The surface is mostly black in a shrubland or grassland ecosystem, although gray ash can be present for a short time. Soil temperatures at 1 cm are less than 50°C. Lethal temperatures for soil organisms occur down to depths of about 1 cm.
- *Moderate fire severity.* Moderate soil heating, or moderate ground char, occurs where the litter on forest sites is consumed and the duff is deeply charred or consumed, but the underlying mineral soil surface is not visibly altered. Light colored ash is present. Woody debris is mostly consumed, except for logs, which are deeply charred. On shrubland or grassland sites, gray or white ash is present and char can be visible in the upper 1 cm of mineral soil, but the soil is not altered. Soil temperatures at the 1-cm depth can reach 100 to 200°C. Lethal temperatures for soil organisms occur down to depths of 3 to 5 cm.
- *High fire severity.* High soil heating, or deep ground char, occurs, where the duff is completely consumed and the top of the mineral soil is visibly reddish or orange on severely burned sites. Color of the soil below 1 cm is darker or charred from organic material. The char layer can extend to a depth of 10 cm or more. Logs can be consumed or deeply charred, and deep ground char can occur under slash concentrations or burned-out logs. Soil texture in the surface layers is changed and fusion evidenced by clinkers can be observed locally. All shrub stems are consumed and only the charred remains of large stubs may be visible. Soil temperatures at 1 cm are greater than 250°C. Lethal temperatures for soil organisms occur down to depths of 9 to 16 cm.

This system can be extended to classify entire areas by determining the percentage of the total area that is burned at low, moderate, and high fire severities (Wells et al. 1979):

- *Low-severity burn.* Less than 2 percent of the area is severely burned, less than 15 percent moderately burned, and the remainder of the area is burned at a low severity or unburned.
- *Moderate-severity burn.* Less than 10 percent of the area is severely burned, but over 15 percent is burned moderately and the remainder is burned at low severity or unburned.
- *High-severity burn.* More than 10 percent of the area has spots that are burned at high severity, more than 80 percent moderately or severely burned, and the remainder is burned at a low severity.

Classification of fire severity must be developed locally and, as a consequence, is both *resource-* and *site*-specific (Example 3.5).

Example 3.5: Fire Severity Classification System of Madrean Oak in the Southwestern United States and Mexico (Ffolliott and Bennett 1996)

This classification system was developed by monitoring the postfire responses of mostly Emory oak *(Quercus emoryi)* with intermingling Arizona white oak *(Q. arizonica)* for 8 years after a wildfire swept over 485 ha of Madrean oak woodlands in southeastern Arizona. Fire-damaged oak, or lack thereof, and subsequent recovery of the fire-damaged trees were monitored on sites where the fire had burned with low, medium, and high severities. The classification system is as follows:

- *Low fire severity.* At least 50 percent of the trees exhibit no visible damage, with the remainder fire-damaged by scorched crowns, shoot-kill (top kill but sprouting), or root-kill (top kill and no sprouting); over 80 percent of the fire-damaged trees survive.
- *Moderate fire severity.* Twenty to 50 percent of the trees exhibit no visible damage, with the remainder fire-damaged; 40 to 80 percent of the fire-damaged trees survive.
- *High fire severity.* Less than 20 percent of the trees exhibit no visible damage, with the remainder fire-damaged, largely by root-kill; less than 40 percent of the fire-damaged trees survive. ■

Classification systems evolve largely from local experiences in evaluating the effects of fire of differing severities on the resources of concern. Results obtained from such evaluations are the focus of the following chapters.

A Conceptual Model

The previous discussion leads to development of a conceptual model to help planners, managers, and decisionmakers appreciate the full spectrum of resource responses to fire severity. The conceptual model describes fire severity as ranging from low resource responses likely to be experienced with a prescribed fire to high resource responses that could be expected from stand-replacing wildfire in forests (Figure 3.2). Once again, fire severity is not directly related to fire intensity for the reasons discussed earlier.

Prescribed fire conditions are depicted in Figure 3.2 on the left side of the fire severity and resource response curve. These conditions are generally represented by lower temperature–higher humidity burning conditions, where fuel loading is low and fuel moisture can be high. These conditions produce lower fire intensities and, as a consequence, expose ecosystem resources to lower fire severities. Prescribed fire, by its design, usually has minor impacts on these resources.

Fire at the other end of the spectrum (right side of Figure 3.2) more nearly represents conditions that are present during a wildfire, where temperatures, wind speeds, and fuel loadings are high, and humidities and fuel moisture are

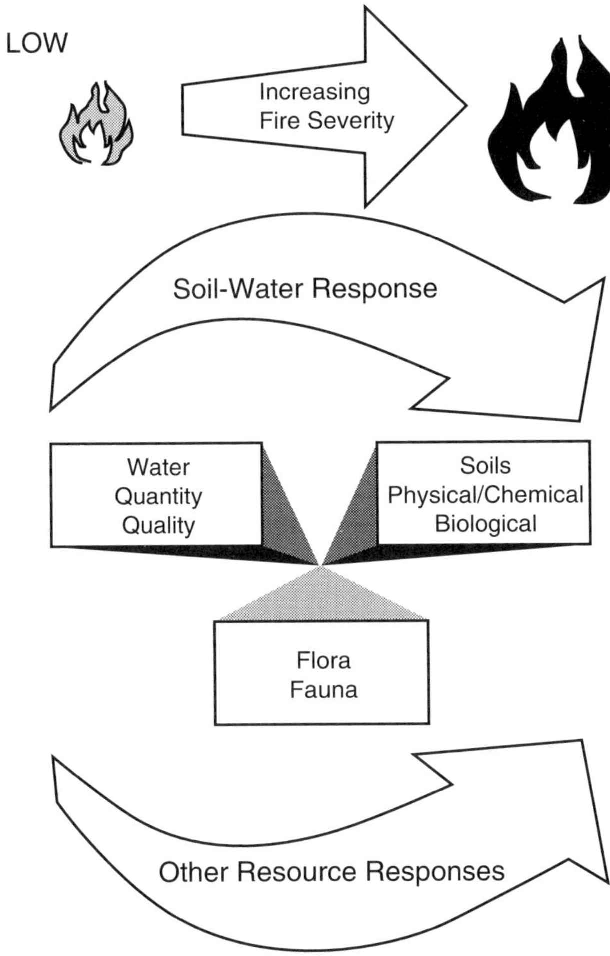

FIGURE 3.2 A conceptual model of the spectrum of ecosystem resource responses to fire severity. (Adapted from DeBano et al. 1995.)

low. In contrast to prescribed burning, wildfire often has a major effect on ecosystem processes, leading to increased sensitivity of the burned site to vegetative loss, eroding forces, and reduced land stability.

Differences in resource impacts along the spectrum between prescribed burning and wildfire depend largely on the extent of the fuel that is burned. For example, only the litter and smaller-diameter surface fuels are ignited and burn when a cool fire is prescribed for southwestern ponderosa pine forests, while there is nearly total canopy consumption in the case of an intense wildfire. Fire in southern California chaparral shrublands is mainly carried by shrub canopies. Therefore, it is more difficult to control the behavior and intensity of chaparral fire so that only minimum impacts to the soil and vegetation occur.

Chapters in the following part of this book are concerned with the effects of fire on soil, water, vegetative, air, and cultural resources of natural ecosystems. However, before moving onto these chapters, it is necessary to describe another type of fire—*prescribed natural fire.* The behavior of prescribed natural fire and the impacts of a prescribed natural fire on ecosystem resources often mimic prescribed burning.

PRESCRIBED NATURAL FIRE

A *prescribed natural fire* is a fire caused by natural ignition, which is allowed to burn so long as it meets prescribed conditions and previously stated objectives. The concept of prescribed natural fire in the United States is largely a product of a report by the Leopold Committee, which was asked by the Department of the Interior to make recommendations on elk habitat management in national parks. The Committee suggested that the National Park Service reconsider its fire suppression policy of that time and consider the ecological benefits of natural fire (Leopold et al. 1963). Areas on which naturally occurring fire might be allowed to continue burning were subsequently established within the Sequoia and Kings Canyon National Parks of California in 1968. Other national parks designated similar natural fire areas in the succeeding years.

National Park Service Fire Policies

Fire management policies of the National Park Service had evolved into support for prescribed natural fire or, when necessary, prescribed fire as a surrogate for natural fire by 1986 (Lorimer 1990, Kilgore and Nichols 1995). These policies stated that occurrence of a natural fire within a particular ecosystem could be recognized as a factor in stimulating, retarding, or eliminating components of the ecosystem. Furthermore, most natural fire is lightning-caused and, as such, should be considered a natural phenomenon, which could be permitted to continue to influence the ecosystem if perpetuation of a truly natural ecosystem was desired.

The 1986 policies also established the use of prescribed natural fire as a preferred means to achieve fire management objectives in the designated prescribed natural fire zones. However, human-ignited prescribed burning could be used on those sites where prescribed natural fire could not meet management objectives (National Park Service 1986). The primary objective of both prescribed burning and prescribed natural fire was simulating the influence of natural fire on the ecosystem to the fullest possible extent.

The Yellowstone fires of 1988 gave impetus to the conclusion of a Policy Review Team, assigned the task of reviewing impacts of these fires, that the objectives of prescribed natural fire policies in national parks and wilderness areas were fundamentally sound (Kilgore and Nichols 1995). This Policy Review Team noted that these policies allowed managers to restore and maintain the natural role of fire when the management objective was perpetuating natural processes and values. It was further recommended that the policies be refined, strengthened, and reaffirmed. As a consequence, the National Park Service made changes in the way that these policies had been implemented.

While the revised policies of 1990 appeared to lessen the earlier National Park Service commitment to allowing natural fire to burn whenever possible, prescribed natural fire was still put forth as a preferred means for achieving management objectives in the areas so designated (National Park Service 1990). These policies further stated that prescribed natural fire could range from creeping surface fire to stand-replacing crown fire, providing that fire behavior was predictable and that the effects were acceptable in terms of the existing prescription and management plan.

Risks

At the time that a natural ignition is being evaluated in relation to the subsequent actions to be taken, regardless of whether the fire is occurring in national parks or elsewhere, there are risks to consider regarding the intensity, rate of spread, direction, and, ultimately, severity of the fire (Bunnell 1995). These risks need close scrutinizing in the decisionmaking process to determine whether the fire should be allowed to burn or not (Figure 3.3). Furthermore, fire managers and resource specialists must articulate and quantify these relative risk levels to the extent possible before decisionmakers are likely to accept the risk of prescribed natural fire decisions. It is also important that decisionmakers appreciate the risks involved in this process.

SUMMARY

The importance of fuels and fire behavior in generating the energy that contributes to how fire affects ecosystem resources should be appreciated after completing this chapter. These fire effects are often described in terms of fire severity, a descriptor that is largely dependent on the nature of the fuels available for burning and the behavior of the fire when these fuels are burned. After reading this chapter, you should be able to:

1. Describe the kinds of fuels available for burning in relation to their properties, components, and complexes.

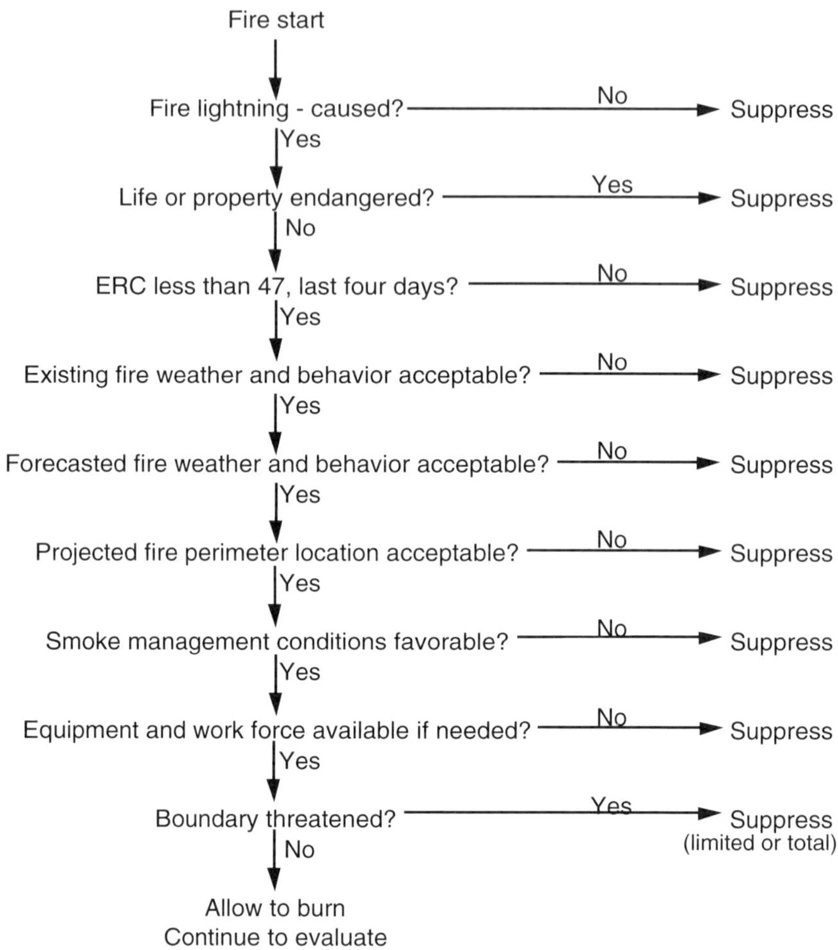

FIGURE 3.3 A decision-making process for determining whether a natural ignition should be allowed to burn as a prescribed natural fire. The energy release component (ERC) is an index from the National Fire Danger Rating System that indicates potential fire intensity. (Adapted from Fischer 1984.)

2. Discuss the ways in which these fuels can be classified for management purposes.
3. Explain the types of fire behavior and the significance of intensity and rate of spread in characterizing the behavior of a fire.
4. Discuss the differences between fire intensity and fire severity in describing how a fire impacts a natural ecosystem.
5. Describe the risks that are taken by a manager when a natural ignition is allowed to burn as a prescribed natural fire.

REFERENCES

Albini, F., M. R. Amin, R. D. Hungerford, W. H. Frandsen, and K. C. Ryan. 1996. *Models for Fire-Driven Heat and Moisture Transport in Soils.* USDA Forest Service, General Technical Report INT-GTR-335.

Bond, W. J., and B. W. van Wilgen. 1996. *Fire and Plants.* Chapman and Hall, London.

Brown, A. A., and K. P. Davis. 1973. *Forest Fire: Control and Use.* McGraw-Hill Book Company, New York.

Bunnell, D. L. 1995. Prescribed natural fire planning considerations: negotiating conflicting goals. In: Brown, J. K., R. W. Mutch, C. W. Soon, and R. H. Wakimoto, technical coordinators. *Proceedings: Symposium on Fire in Wilderness and Park Management.* USDA Forest Service, General Technical Report INT-GTR-320, pp. 55–62.

Campbell, G. S., J. D. Jungbauer, Jr., W. R. Bidlake, and R. D. Hungerford. 1994. Predicting the effect of temperature on soil thermal conductivity. *Soil Science* 158:307–313.

Campbell, G. S., J. D. Jungbauer, Jr., K. L. Bristow, and R. D. Hungerford. 1995. Soil temperature and water content beneath a surface fire. *Soil Science* 159:363–374.

Chandler, C., P. Cheney, P. Thomas, L. Trabaud, and D. Williams. 1983. *Fire in Forestry: Volume II. Forest Fire Management and Organization.* John Wiley & Sons, New York.

Connor, M. A., and K. P. Cannon. 1991. Forest fires as a site formation process in the Rocky Mountains of northwestern Wyoming. *Archaeology in Montana* 32:1–14.

Conrad, C. E., and W. C. Oechel, technical coordinators. 1982. *Proceedings of the Symposium on Dynamics and Management of Mediterranean-type Ecosystems.* USDA Forest Service, General Technical Report PSW-58.

DeBano, L. F., M. B. Baker, Jr., and P. F. Ffolliott. 1995. Effects of prescribed fire on watershed resources. *Hydrology and Water Resources in Arizona and the Southwest* 22–25:39–44.

Ffolliott, P. F., and D. A. Bennett. 1996. Peak fire of 1988: its effect on Madrean oak trees. In: Ffolliott, P. F., L. F. DeBano, M. B. Baker, Jr., G. J. Gottfried, G. Solis-Garza, C. B. Edminster, D. G. Neary, L. S. Allen, and R. H. Hamre, technical coordinators. *Effects of Fire on Madrean Province Ecosystems: A Symposium Proceedings.* USDA Forest Service, General Technical Report RM-GTR-289, pp. 235–237.

Fischer, W. C. 1984. *Wilderness Fire Management Planning Guide.* USDA Forest Service, General Technical Report INT-171.

Fosberg, M. A. 1970. Drying rates of heartwood below fiber saturation. *Forest Science* 16:57–63.

Hartford, R. A., and W. H. Frandsen. 1992. When it's hot, it's hot . . . or maybe it's not! (Surface flaming may not portend extensive soil heating). *International Journal of Wildland Fire* 2:139–144.

Hungerford, R. D. 1996. *Soils. Fire in Ecosystem Management Notes: Unit II-I.* USDA Forest Service, National Advanced Resource Technology Center, Marana, Arizona.

Kilgore, B. M., and T. Nichols. 1995. National Park Service fire policies and program.

In: Brown, J. K., R. W. Mutch, C. W. Soon, and R. H. Wakimoto, technical coordinators. *Proceedings: Symposium on Fire in Wilderness and Park Management.* USDA Forest Service, General Technical Report INT-GTR-320, pp. 24–29.

Leopold, A. S., S. A. Cain, C. M. Cottam, and I. N. Kimball. 1963. Study of wildlife problems in national parks: wildlife management in national parks. *Transactions of the North American Wildlife and Natural Resources Conference* 28:28–45.

Lorimer, C. G. 1990. Behavior and management of forest fires. In: Young, R. A., and R. L. Giese, editors. *Introduction to Forest Science.* John Wiley & Sons, New York, pp. 427–448.

Martin, R. E., H. E. Anderson, W. D. Boyer, J. H. Dieterich, S. N. Hirsch, V. J. Johnson, and W. H. McNab. 1979. *Effects of Fire on Fuels: A State-of-Knowledge Review.* USDA Forest Service, General Technical Report WO-13.

Mooney, H. A., and C. E. Conrad, technical coordinators. 1977. *Proceedings of the Symposium on the Environmental Consequences of Fire and Fuel Management in Mediterranean Ecosystems.* USDA Forest Service, General Technical Report WO-3.

National Park Service. 1986. *Wildland Fire Management Guideline: NPS-18.* National Park Service, Washington, DC.

National Park Service. 1990. *Wildland Fire Management Guideline: NPS-18.* National Park Service, Washington, DC.

Pyne, S. J., P. L. Andrews, and R. D. Laven. 1996. *Introduction to Wildland Fire.* John Wiley & Sons, New York.

Rothermal, R. C., and G. C. Rinehart. 1983. *Field Procedures for Verification and Adjustment of Fire Behavior Predictions.* USDA Forest Service, General Technical Report INT-142.

Sackett, S. S. 1979. Natural fuel loadings in ponderosa pine and mixed conifer forests of the Southwest. USDA Forest Service, Research Paper RM-213.

Sackett, S. S., and S. M. Haase. 1996. Fuel loading in southwestern ecosystems of the United States. In: Ffolliott, P. F., L. F. DeBano, M. B. Baker, Jr., G. J. Gottfried, G. Solis-Garza, C. B. Edminster, D. G. Neary, L. S. Allen, and R. H. Hamre, technical coordinators. *Effects of Fire on Madrean Province Ecosystems: A Symposium Proceedings.* USDA Forest Service, General Technical Report RM-GTR-289, pp. 187–192.

Simard, A. J. 1991. Fire severity, changing scales, and how things hang together. *International Journal of Wildland Fire* 1:23–34.

Vogl, R. J. 1979. Some basic principles of grassland fire management. *Environmental Management* 3:51–57.

Wells, C. G., R. E. Campbell, L. F. DeBano, C. E. Lewis, R. L. Fredriksen, E. C. Franklin, R. C. Froelich, and P. H. Dunn. 1979. *Effects of Fire on Soil. A State-of-Knowledge Review.* USDA Forest Service, General Technical Report WO-7.

Whelan, R. J. 1995. *The Ecology of Fire.* Cambridge University Press, Cambridge, England.

PART

2

Soil Responses

Soil is an integral component of all terrestrial ecosystems and, as such, plays a central role in ecosystem functioning and productivity. Changes in soil properties during fire and soil heating are reflected in a wide range of interrelated resource responses throughout the surrounding ecosystem (water, vegetation, air, and cultural resources). Soils are unique because they can contain large amounts of organic matter, which constantly interacts with mineral particles to provide nutrients necessary for sustainable plant growth.

Heat radiated downward in the soil is the driving mechanism for increasing soil temperatures in the soil. These temperature increases initiate a wide range of responses in the soil physical, chemical, and biological systems. Soil heating and temperature regimes that develop during fires are presented for forest, shrubland, and grassland ecosystems in Chapter 5.

Because soil is an important component of ecosystems that are affected by fire, a review of soils and their properties is presented first (see Chapter 4) to prepare the reader for a detailed discussion of specific fire effects on different soil properties. The properties of soils affected by fire and soil heating are separated into physical, chemical, and biological systems, and detailed responses of these three systems to fire are presented in Chapters 5, 6, and 7, respectively.

Chapter

4

Soil Resource

INTRODUCTION

Soil is an unconsolidated, thin, variable layer of mineral and organic matter on the earth's crust that has formed as a result of several physical, chemical, and biological processes operating simultaneously over long periods of time (Hendricks 1985, Singer and Munns 1996). Soil is formed where there is continual interaction between the soil system and the biotic (faunal and floral), climatic (atmospheric and hydrologic), and topographic components of the environment. Soil interrelates with other ecosystem resources in several ways. It supplies air and water, nutrients, and mechanical support for the sustenance of plants. Soil also absorbs water during infiltration and, by doing so, provides storage for water as well as acting as a conduit that delivers water slowly from upstream slopes to channels, where it surfaces as streamflow. When the infiltration capacity of the soil for rainfall is exceeded, organic and inorganic soil materials are eroded and become a major source of sediment, nutrients, and other pollutants in streams. There is also an active and ongoing exchange of gases between the soil and the surrounding atmosphere. Soil also provides a repository for many cultural artifacts, which can remain in the soil for thousands of years without undergoing appreciable change.

Fire-related changes in soils produce a wide array of changes in the surrounding ecosystem because of the interdependency between the soil and other ecosystem components. Immediate, short-term and long-term responses to fire occur (Figure 4.1). Immediate effects of fire arise as a result of the heat

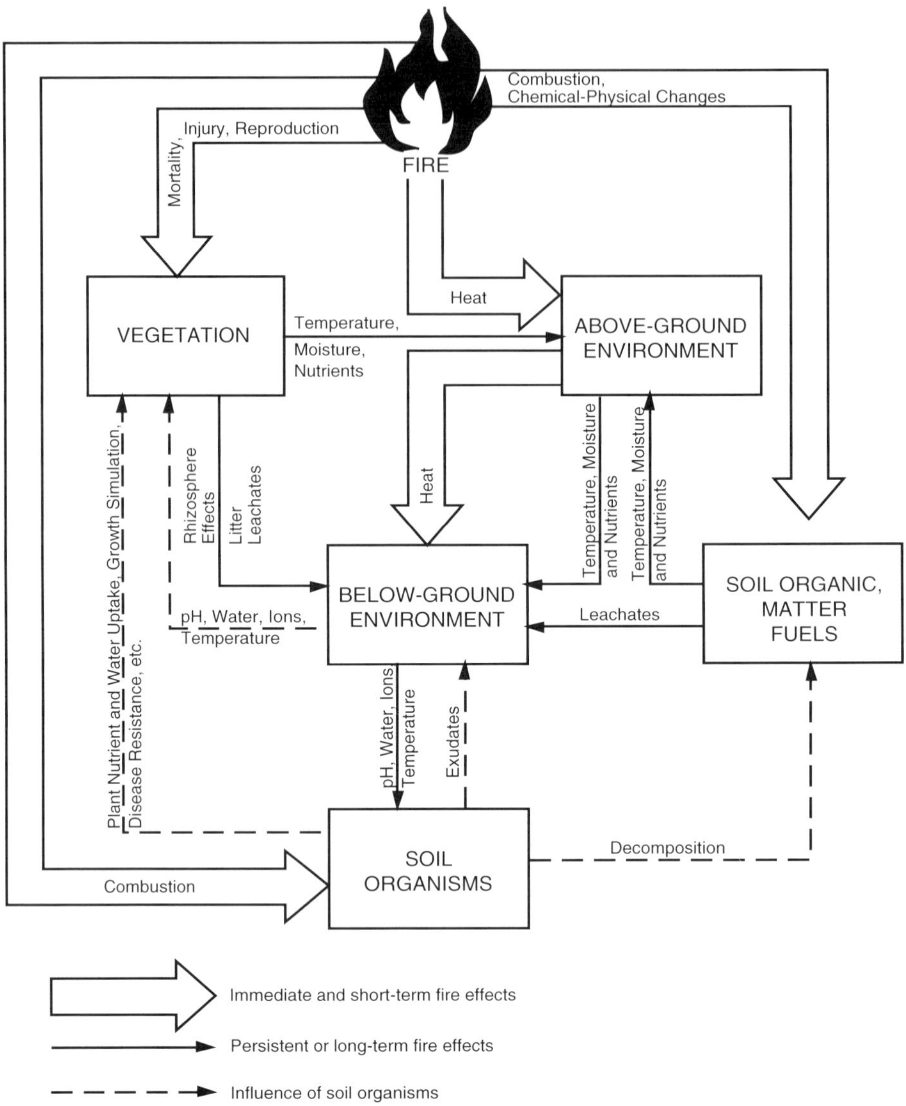

FIGURE 4.1 Effects of fire on immediate short-term and long-term relationships between soil and other ecosystem components. (Adapted from Borchers and Perry 1990. In: *Natural and Prescribed Fire in Pacific Northwest Forests,* edited by J. D. Walstad, S. R. Radosevich, and D. V. Sandberg. Copyright © 1990 Oregon State University Press. Reproduced by permission.)

released during a fire and directly impact the soil and surrounding ecosystem. The response of biological components (soil microorganisms and vegetation) of the ecosystem to fire is rapid. Another immediate effect is the release of gases and other air pollutants by the combustion of soil organic matter during a fire. The long-term fire effects are usually more subtle, but they can persist for years following the fire or be permanent, as occurs when cultural resources are damaged. Other long-term fire effects arise from the relationship between nutrient cycling and long-term productivity.

NATURE OF THE SOIL RESOURCE

Soil has four major constituents—minerals, organic matter, air, and water. Associated with these constituents is a wide range of living organisms. The mineral component includes unaltered minerals, which are part of the parent rock, and secondary minerals formed by the recombination of substances produced during weathering. Secondary minerals are mainly *clays,* which are minerals less than 0.002 mm in diameter that have highly reactive, negatively charged surfaces that adsorb positive ions (cations).

Organic matter is the decaying plant and animal material that is broken down and decomposed by the actions of animals and microorganisms living in the soil. An important component of organic matter is *humus,* the colloidal soil organic matter (particles smaller than 0.001 to 0.005 mm in diameter) that decomposes slowly. Humus provides negative adsorption sites, similar to clay minerals, and also acts as an organic glue that helps hold mineral soil particles together to form aggregates and thus contributes to soil structure. Inorganic and organic particles are arranged so that pore space is present in soil for the movement of air and water. Soil water and its dissolved ions (cations and anions) are called the *soil solution.*

SOIL PROFILE

The variable amounts of minerals, organic matter, air, and water are manifest in a wide range of physical, chemical, and biological properties of a soil that can be altered during a fire. These properties are not randomly distributed but occur in an orderly arrangement in the soil profile. Understanding this arrangement is necessary before discussing fire effects on different soil properties.

If a trench is dug in a soil body, the vertical face of this exposed soil could exhibit a sequence of nearly horizontal layers called *soil horizons.* These horizons are formed over long time periods as the result of weathering, decomposition, synthesis of new substances, and translocation of materials vertically in the soil by water moving through the soil.

The arrangement of soil horizons is called a *soil profile* (Figure 4.2). Most

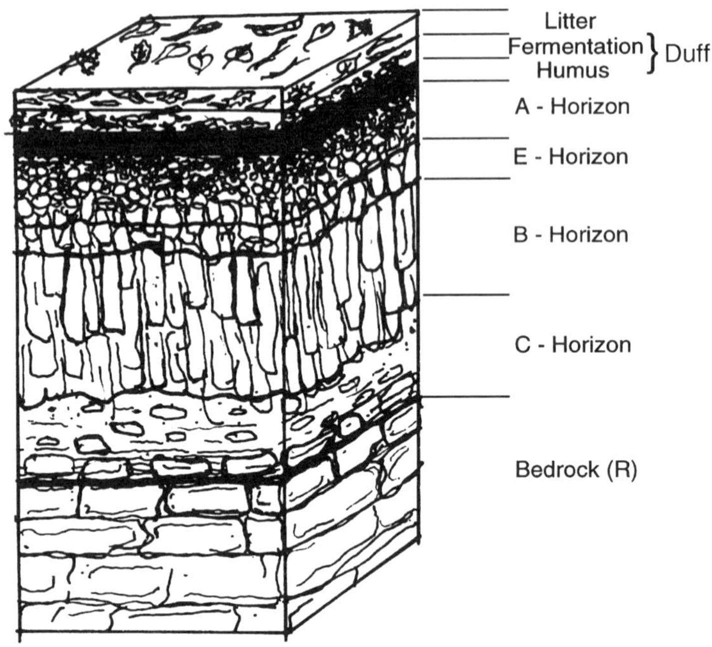

FIGURE 4.2 A soil profile showing organic matter accumulation (litter, fermentation, humus, and duff layers) and mineral soil horizons (*A, E, B, C,* and *R* horizons).

of the horizons are present in older, well-developed soil. However, in younger, less-well-developed soil, there are fewer horizons; or there may be no vertical differentiation other than the accumulation of organic matter on the mineral soil surface or stratification of mineral horizons that reflect the original parent material (depositional strata in alluvial or aeolian soils). Organic and mineral horizons have been classified in a number of ways (Example 4.1), but surface organic matter in this book focuses on three horizons referred to as the litter (*L*), fermentation (*F*), and humus (*H*) layers. Fire managers commonly refer to the combination of the *F* and *H* layers as the duff (*D*) layer (see Chapter 1).

Example 4.1: Additional Classification of Horizons in a Soil Profile

Soil horizons are classified by several systems. For example, the organic horizons are referred to as the *O* horizon according to current soil taxonomic terminology. The *L* layer is referred to as the O_i or O_1 horizon. The *F* layer is designated as the O_e horizon or part of the O_2 horizon. The *H* layer is denser than the *L* and *F* layers and is designated as the O_a or O_2 horizon. The *A* horizon can be made up of several transition layers between the *A* horizon and the overlying *O* horizon and the underlying *B* horizon.

A mineral *E* horizon is formed immediately below the *A* horizon in some soils. The main feature of the *E* horizon is the loss of silicate clay, iron (Fe), or aluminum (Al), leaving a concentration of sand and silt particles that are resistant to further weathering. The *B* horizon is characterized by the obliteration of the original rock structure and the accumulation of silicate clay, Fe, Al, carbonate, gypsum, or silicon. The *C* and *R* horizons have not been affected appreciably by soil formation, although initial weathering of primary minerals can occur. The *C* horizon consists of unconsolidated parent material, while the *R* horizon is hard consolidated bedrock. ■

The *A* horizon consists mainly of minerals and is found immediately below the *L, F,* and *H* layers. The upper part of the *A* horizon is characterized by an accumulation of humified organic matter intimately mixed with mineral particles. The *B* horizon is below the *A* or (if present) *E* horizon. The *C* and *R* horizons are located below the *B* horizon. Roots and associated microorganisms can be found throughout the *A, E, B,* and *C* horizons. Shallow roots are frequently found in, and immediately below, the *H* layer and, as a consequence, can be exposed to high temperatures during combustion of *L* and *D* layers.

The *L, F,* and *H* layers are important when discussing fire effects, because they are directly subjected to heat produced by burning of surface fuels and contain a large proportion of the organic matter found in soil profiles. Changes occurring in organic matter as a result of heating and combustion are described in detail later (see Chapter 6). Most of the fire effects produced during surface fires occur in the upper organic horizons, or in the top part of the *A* horizon. Heating of the *B* horizon or below occurs only when roots burn and create localized subsurface heating.

SOIL PROPERTIES

Physical soil properties are those characteristics, processes, or reactions of a soil that are caused by physical forces that can be described by, or expressed in, physical terms or equations (Soil Science Society of America 1997). Therefore, physical properties reflect the influence of the mineral component as it interacts with the other two components (chemical and biological systems). Important physical properties of soil are texture, clay content, bulk density, porosity (amount and size), structure, infiltration rate, soil temperature (ambient and fire generated), and water repellency. *Soil texture* is based on the relative proportions of *sand* (0.05 to 2 mm in diameter), *silt* (0.002 to 0.05 mm in diameter), and *clay* (<0.002 mm in diameter) particles in the soil. The proportions are used to specify several textural classes. Clays are small-diameter silicate minerals having complex molecular structures, which contribute to both the physical and chemical properties of a soil. *Bulk density* is the mass of dry soil per unit bulk volume (expressed in g/cm^3). *Porosity* is the volume of pores in a soil sample (nonsolid volume) divided by the bulk volume of the sample. *Infiltration rate* is the rate of downward entry of water into the soil through the soil surface (see Chapter 8). *Soil structure* is the arrangement of primary soil particles into aggregates having distinctive patterns (columnar, prismatic, blocky). *Water repellency* is a property of the soil that limits the rate at which water infiltrates into a soil.

Soil chemical properties are those characteristics, processes, or reactions arising from the chemical constitution, chemical properties, and chemical reactions occurring in soils. The most important soil chemical properties affected by fire are organic matter, cation exchange capacity, pH, buffer power, and nutrient-related processes (including species, availability, losses, and cycling rates). The role of organic matter was described earlier. *Cation exchange* is the interchange between a cation in solution and another cation on the surface of any negatively charged material, such as a clay or organic colloid (humus). *Cation exchange capacity* is the sum of the exchangeable cations found on organic and inorganic soil colloids. It arises from the quantities of charged particles, mainly negatively charged clay particles and colloidal organic matter, contained in the soil. The relative contribution of clay particles and organic matter to the cation exchange capacity of the soil depends largely on the proportion of the two components and the total quantities of each present (Tate 1987). The cation exchange capacity will also depend on the type of clay and organic matter present. Some clay materials such as montmorillonites have larger exchange capacities than do kaolinitic clays. Likewise, the degree of humification of the organic matter will affect cation exchange capacity with the well-decomposed colloidal organic material being the most important. Soil *pH* is a measure of the hydrogen ion activity in the soil. The degree of acidity is determined at specified moisture contents. Neutral soils have a pH of 7, acidic soils are less than 7, and basic soils are greater than 7. *Buffer power* is the ability of ions associated with the solid phase to buffer changes in ion concentration in the soil solution.

Soil biological properties are considered to reflect a wide range of living organisms that inhabit the soil and contribute directly to the productivity and sustainability of terrestrial ecosystems worldwide (Borchers and Perry 1990). These living soil organisms consist of an assemblage of animals and plants ranging in size from microscopic bacteria and fungi to small vertebrates that remain below ground during all or part of their life cycle. Insects are also important biological organisms. Other biological components include plant roots and seeds. The effect of fire on the biological system is discussed in Chapter 7, with the focus on the living organisms that remain predominantly underground or near the soil surface.

Classification of some soil properties is arbitrary, because they are borderline between two of the three general groups of properties. For example, *water repellency* is sometimes listed as a physiochemical property instead of strictly a physical property because, while it affects infiltration (a physical property), it is also a chemical property since it is made up of organic compounds. Water repellency is discussed as a physical property in this book.

CHANGES IN SOIL PROPERTIES DURING HEATING

Effects of fire on soil physical, chemical, and biological properties are complex. However, general relationships describing the response of soil properties to soil heating can be established. These relationships show the following:

- Biological properties of soil are most sensitive to soil heating, with fatal temperatures for most living organisms occurring below 100°C.
- Complete dehydration of soil occurs when the temperature reaches 220°C, although it does not significantly affect the physical or chemical properties of the soil (Raison 1979, Giovannini and Lucchesi 1993).
- Heating between 220 and 460°C combusts soil organic matter and, in doing so, affects soil properties dependent on organic matter. The destruction of organic matter can be beneficial, because it provides large amounts of readily available plant nutrients. However, it has the disadvantage of destroying soil structure, with the extent of damage depending largely on the rate at which organic matter replenishment occurs following fire. It could cause long-term damage in arid environments where organic matter is replenished slowly.
- Heating above 460°C drives off hydroxyl (OH) groups from clays and thus disrupts carbonate structure (Giovannini et al. 1988). These irreversible changes produce a soil that is less porous, less plastic, and less elastic, and one that is highly erodible.

The soil responses described above can be used to develop a list of soil physical, chemical, and biological properties and their threshold temperatures (Table 4.1). *Threshold temperatures* are defined as those temperatures where nutrients are volatilized or a soil property is irreversibly changed (DeBano 1991). This list contains soil properties that are changed if temperatures exceed 460°C (relatively insensitive), are between 100 and 400°C (moderately sensitive), or are less than about 100°C (sensitive).

The arrangement and properties of the horizons in a soil profile (Figure 4.2) are significant when evaluating fire effects. Organic matter and associated biological activities are concentrated on or near the soil surface. The soil surface is also in closest proximity to burning fuels and, as a consequence, receives the greatest heat flux during the combustion of aboveground and surface fuels. As a result, the most sensitive soil properties are directly subjected to the greatest flux of heat radiated downward during a fire and consequently reach the highest temperatures.

ASSESSING FIRE EFFECTS

An important outcome arising from the knowledge of fire effects on soil properties is the ability to assess soil changes resulting from different severi-

TABLE 4.1 Threshold Temperatures (°C) for Insensitive, Moderately Sensitive, and Sensitive Soil Properties

Soil Property	*Threshold Temperature*	*Source*
Relatively Insensitive		
Manganese	1962	Raison et al. 1985
Calcium	1484	Raison et al. 1985
Magnesium	1107	DeBano 1991
Clay alteration	460–980	DeBano 1990
Phosphorus and potassium	774	Raison et al. 1985
Moderately Sensitive		
Sulfur	375	Tiedemann 1987
Soil structure	300	DeBano 1990
Soil wettability	250	DeBano and Krammes 1966
Nitrogen	200	White et al. 1973
Organic matter	100	Hosking 1938
Sensitive		
Protein coagulation	60	Precht et al. 1973
Bacteria—wet	100	Dunn and DeBano 1977
Bacteria—dry	120	Dunn and DeBano 1977
Nitrosomonas bacteria—wet	80	Dunn and DeBano 1977
Nitrosomonas bacteria—dry	90	Dunn and DeBano 1977
Fungi—wet	60	Dunn et al. 1985
Fungi—dry	80	Dunn et al. 1985
VAM	94	Klopatek et al. 1988
Plant roots	48–54	Hare 1961
Seeds—wet	70	Martin et al. 1975
Seeds—dry	90	Martin et al. 1975
Small mammals	49–63	Lyon et al. 1978

ties of burning (see Chapter 3). Assessing fire effects requires being able to quantify the effect of fire and associated soil heating on soil properties. This assessment involves three steps:

- It is necessary to estimate the level of energy that is radiated downward during combustion of fuels (see Chapter 2). Although only a small amount of the total energy released during combustion is radiated downward, this energy provides the driving force responsible for changes in soil properties. The amount of change in soil properties is largely dependent on the amount of energy radiated onto the soil surface and transferred downward into the underlying duff and mineral soil. This radiated heat increases the temperature and initiates changes in organic matter and other soil properties. Appearance of vegetation, litter, duff, and upper soil horizons can be used to estimate the amounts of heat radiated downward. A technique for characterizing severity (described in Chapter 3) can be used for classifying fire severity as low, moderate, or high. The assumption made is that during low severity-fires only a limited amount of heat is radiated downward during combustion. The amount of heat radiated downward increases as severity increases from low to moderate to high.

- Fire severity is then related to soil heating by relating soil temperatures to different severities of burning. Soil heating (measured by temperature) is not easily predicted, however, because the relationship between fire intensity and soil heating has not been modeled adequately, particularly under field conditions (see Chapter 2). Nevertheless, some generalizations concerning the relationship between fire severities and soil temperatures have been developed to bridge this knowledge gap and provide guidelines for quantifying soil heating in terms of temperature regimes for forest, shrubland, and grassland ecosystems (see Chapter 5).
- Finally, soil temperature information is used to predict changes that can be expected in specific soil properties. Specific changes in physical, chemical, and biological properties occurring at different temperatures are discussed in Chapters 5, 6, and 7, respectively, and summarized in Table 4.1.

SUMMARY

A basic understanding of a soil profile and its physical, chemical, and biological properties is important when describing and evaluating fire effects on the soil resource. Before moving on to a more detailed discussion of these effects in the following chapter, you should be able to:

1. Explain the significance of different soil horizons in terms of their response to fire.
2. Identify soil properties that are relatively insensitive, moderately sensitive, and sensitive to temperatures generated in soil during fire.
3. Outline a procedure for assessing the effect of fire on different soil properties.
4. Explain how the soil resource is related to other ecosystem components such as water, vegetation, and air.

REFERENCES

Borchers, J. G., and D. A. Perry. 1990. Effects of prescribed fire on soil organisms. In: Walstad, J. D., S. R. Radosevich, and D. V. Sandberg, editors. *Natural and Prescribed Fire in Pacific Northwest Forests.* Oregon State University Press, Corvallis, Oregon, pp. 143–157.

DeBano, L. F. 1990. Effects of fire on the soil resource in Arizona chaparral. In: Krammes, J. S., technical coordinator. *Effects of Fire Management of Southwestern Natural Resources.* USDA Forest Service, General Technical Report RM-191, pp. 65–77.

DeBano, L. F. 1991. The effect of fire on soil properties. In: Harvey, A. C., and L. F. Neuenschwander, compilers. *Proceedings—Management and Productivity of Western-Montane Forest Soils.* USDA Forest Service, General Technical Report INT-280, pp. 151–156.

DeBano, L. F., and J. S. Krammes. 1966. Water repellent soils and their relation to wildfire temperatures. *International Association of Hydrological Sciences* 2:14–19.

Dunn, P. J., and L. F. DeBano. 1977. Fire's effect on biological and chemical properties of chaparral soils. In: Mooney, H. A., and C. E. Conrad, technical coordinators. *Proceedings of the Symposium on the Environmental Consequences of Fire and Fuel Management in Mediterranean Ecosystems.* USDA Forest Service, General Technical Report WO-3, pp. 75–84.

Dunn, P. H., S. C. Barro, and M. Poth. 1985. Soil moisture affects survival of microorganisms in heated chaparral fire. *Soil Biology and Biochemistry* 17:143–148.

Giovannini, G., and S. Lucchesi. 1993. Effects of fire on soil physio-chemical characters and erosion dynamics. In: Trabaud, L., and P. Prodon, editors. *Fire in Mediterranean Ecosystems.* Commission of the European Countries, Ecosystem Research Report 5. Brussels, Belgium, pp. 403–412.

Giovannini, G., S. Lucchesi, and M. Giachetti. 1988. Effect of heating on some physical and chemical parameters related to soil aggregation and erodibility. *Soil Science* 146:255–262.

Hare, R. C. 1961. Heat effects on living plants. USDA Forest Service, Occasional Paper 183. Southern Forest Experiment Station, New Orleans, Louisiana.

Hendricks, D. L. 1985. *Arizona Soils.* College of Agriculture, University of Arizona, Tucson, Arizona.

Hosking, J. S. 1938. The ignition at low temperatures of the organic matter in soils. *Journal of Agricultural Science* 28:393–400.

Klopatek, C. C., L. F. DeBano, and J. M. Klopatek. 1988. Effects of simulated fire on vesicular-arbuscular mycorrhiazae in pinyon-juniper woodland soil. *Plant and Soil* 109:245–249.

Lyon, J. L., H. S. Crawford, E. Czuhai, R. L. Fredriksen, R. F. Harlow, L. J. Metz, and H. A. Pearson. 1978. *Effects of Fire on Fauna: A State-of-Knowledge Review.* USDA Forest Service, General Technical Report WO-6.

Martin, R. E., R. L. Miller, and C. T. Cushwa. 1975. Germination response of legume seeds subjected to moist and dry heat. *Ecology* 56:1441–1445.

Precht, J., J. Chrisphersen, H. Hensel, and W. Larcher. 1973. *Temperature and Life.* Springer-Verlag, New York.

Raison, R. J. 1979. Modification of the soil environment by vegetation fires, with particular reference to nitrogen transformations: a review. *Plant and Soil* 51:73–108.

Raison, R. J., P. K. Khanna, and P. V. Woods. 1985. Mechanisms of element transfer to the atmosphere during vegetation fires. *Canadian Journal of Forest Research* 15:132–140.

Singer, M. J., and D. N. Munns. 1996. *Soils: An Introduction.* Prentice Hall, Upper Saddle River, New Jersey.

Soil Science Society of America. 1997. *Glossary of Soil Science Terms.* Soil Science Society of America, Madison, Wisconsin.

Tate, R. L. III. 1987. *Soil Organic Matter: Biological and Ecological Effects.* John Wiley & Sons, New York.

Tiedemann, A. R. 1987. Combustion losses of sulfur and forest foliage and litter. *Forest Science* 33:216–223.

White, E. M., W. W. Thompson, and F. R. Gartner. 1973. Heat effects on nutrient release from soils under ponderosa pine. *Journal of Range Management* 26:22–24.

Chapter

5

Physical Soil System

INTRODUCTION

Fire affects soil physical properties mainly by destroying organic matter, which is essential for maintaining soil structure. Destruction of organic matter by fire destroys the soil structure and reduces bulk density and porosity, which decreases infiltration and increases runoff and erosion (see Chapter 8). The magnitude of the change in physical properties depends largely on the severity of a fire, the proportions of overstory and understory vegetation destroyed, forest floor consumed, heating of the soil, proportion of area burned, and length of fire intervals (Wells et al. 1979).

SOIL TEMPERATURES

Temperature regimes in the soil (reflections of the magnitude and duration of soil heating) are the end product of the combustion and heat transfer processes (see Chapter 2). The duff and soil temperatures are the ultimate gauges of the changes that can be expected in postfire soil properties. Unfortunately, soil temperatures are difficult to quantify for different severities of fire, because the heating of soils is site-specific and depends on soil conditions at the time of burning, particularly the prefire soil water content. In spite of the large variability from site to site, information on maximum ground surface temperatures during fire in different ecosystems worldwide is available (Figure 5.1). Furthermore, specific data sets for soil temperatures over time are

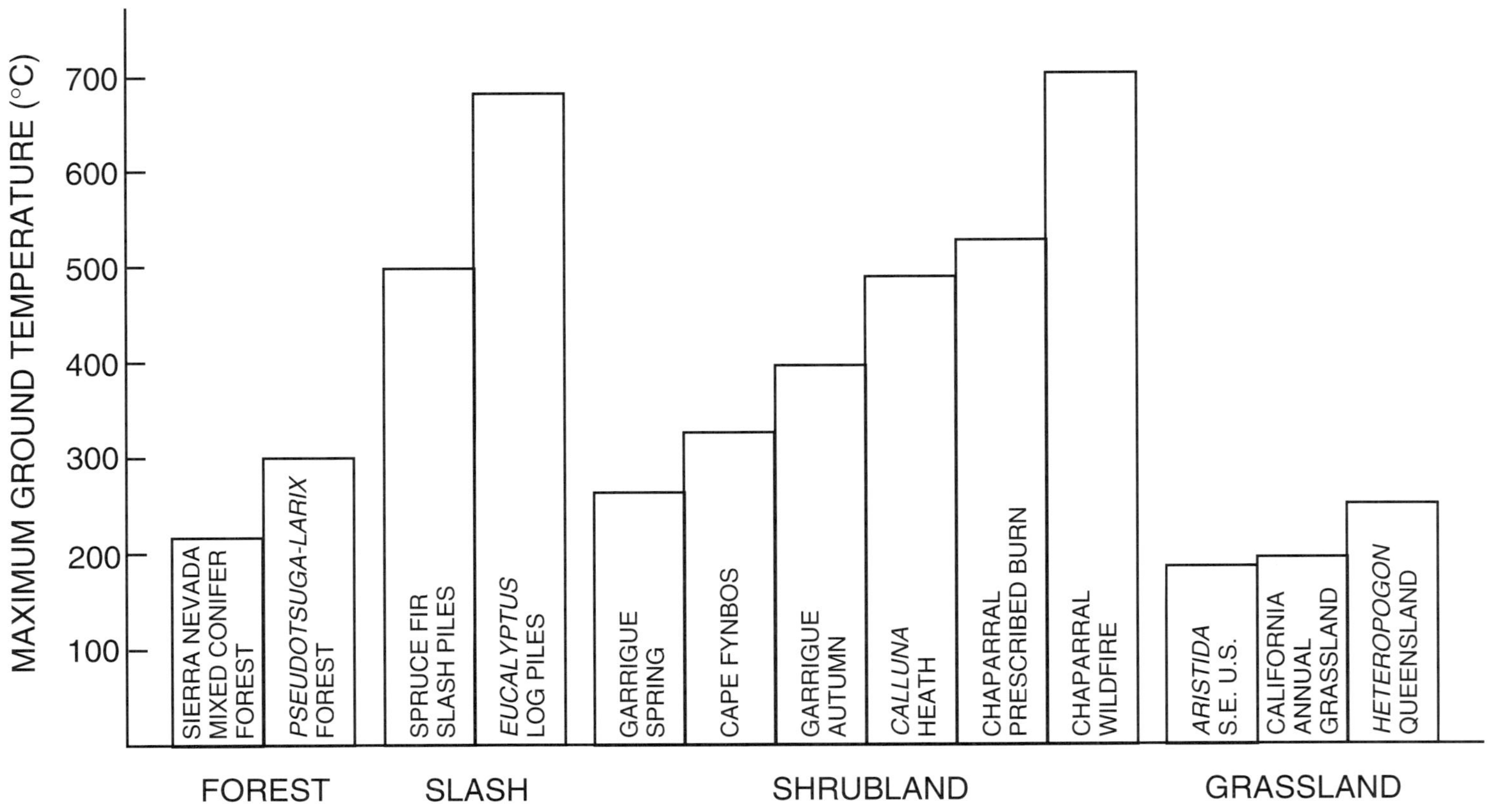

FIGURE 5.1 Maximum ground surface temperatures during fires in different ecosystems worldwide. (Adapted from Rundel 1983.)

also available for different severities of burning in forest, shrubland, and grassland ecosystems.

Forest Ecosystems

Soil temperatures resulting from fires in forest environments vary widely depending on whether the fire is a cool-burning surface fire or a stand-replacing wildfire. Important factors affecting fire intensities are microrelief, wind speed and direction, soil and fuel moisture content, and the kind and spatial distribution of fuels, particularly duff, litter, and slash accumulations (Raison 1979).

Low-severity fires burning only surface fuels do not heat the soil. For example, a cool-burning prescribed fire in a mixed conifer forest in California heated the soil only slightly above ambient temperatures at 5-cm depth, and these temperatures were sustained for less than an hour (Figure 5.2A). Surface temperatures in the fire reached only 100°C, and at 5 cm in the soil the temperatures were only 50°C. Conversely, when large accumulations of organic matter in the *D* layer of a ponderosa pine forest were ignited, the duration of heating increased and produced soil temperatures that are lethal to plant roots deep in the soil (Figure 5.2B). The ponderosa pine forest studied had a large accumulation of dry litter, which smoldered for days. The resulting temperature at the interface of the *A* and *D* horizons reached a maximum temperature of 270°C, and temperatures greater than 100°C were present at 10 cm in the soil for about 24 hours.

More typically, soil temperatures do not rise substantially in forests where repeated cool-burning fires are used to regularly reduce fuel buildup. Soil temperatures produced while burning longleaf pine forests in the southeastern United States seldom reached 50°C for more than 15 minutes at shallow depths of 3 to 6 mm (Heyward 1938). Individual maxima were below 120°C. The highest temperature that occurred 2.5 cm below the soil surface was about 65°C.

A management tool used by forest managers is slash burning. Slash is piled, windrowed, or scattered on the land surface before burning. The slash is then burned to produce a mineral soil seed bed for regeneration or for other silvicultural purposes (see Chapter 14). When slash is burned in large piles, intense soil heating can occur on localized areas where the slash is piled. Similar results can be obtained when burning large piles of logs (Example 5.1). Although major soil damage occurs when concentrated piles of logs are burned, the area damaged by the fire is limited to small areas where the logs were piled. Fuels concentrated in windrows can also cause severe soil heating, particularly when large amounts of material are burned. Scattering slash and then burning it mitigates damage to soil to some extent, although localized hot spots may occur under larger branches.

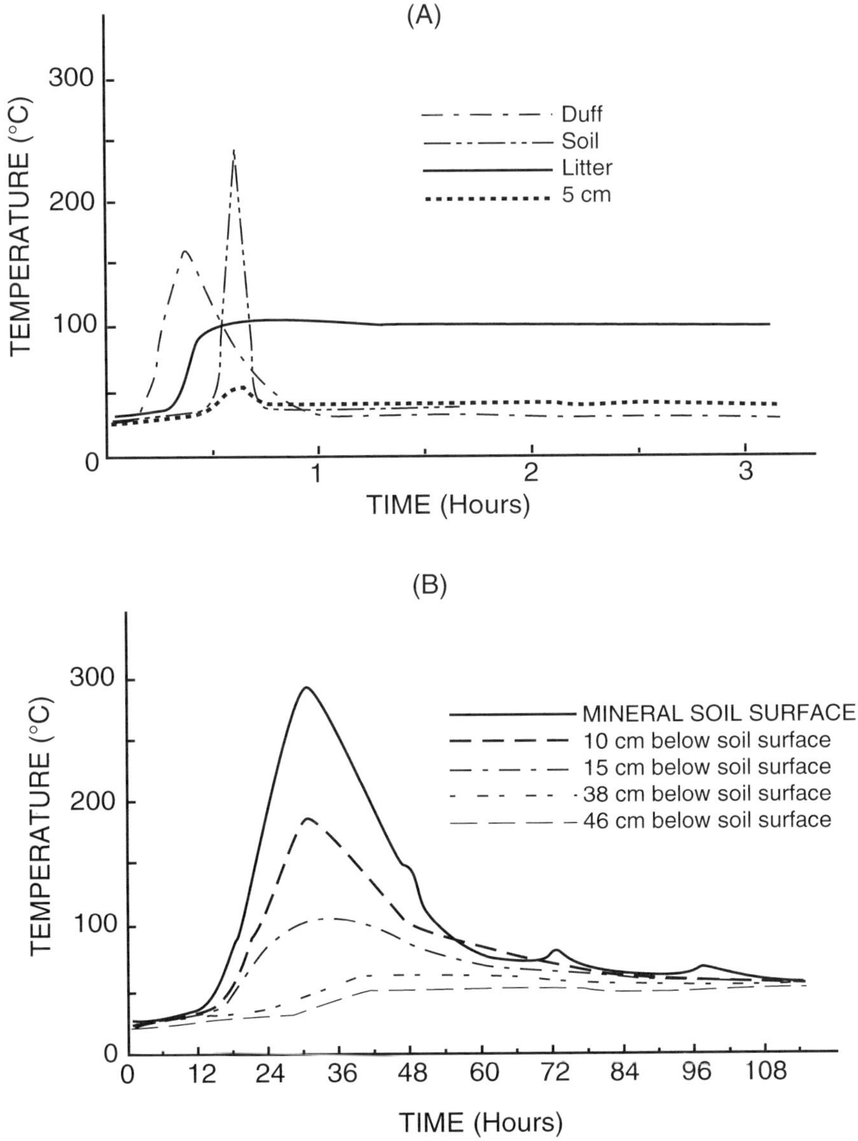

FIGURE 5.2 Surface and soil temperatures recorded under (A) a cool-burning prescribed fire in mixed conifer forest (adapted from Agee 1973, University of California, Water Resources Center, Contribution 143) and (B) a severe duff fire in a ponderosa pine forest (adapted from Sackett and Haase 1992) in California.

Example 5.1 Burning Concentrated Piles of Logs Produces High Soil Temperatures for Long Periods of Time

Burning a large pile of *Eucalyptus* logs in Australia illustrates that high temperatures can develop under large piles of fuel (see figure below). A log pile 4.3 m by 2.5 m by 2 m high, made up of logs 13 to 18 cm in diameter, was burned. The prefire moisture content of the logs averaged about 15 percent. Maximum temperature at the soil surface was about 660°C. Temperatures of 100°C were maintained at the 16.5-cm soil depth for 21 hours. Clearly, this type of fire would sterilize the soil to a depth of at least 10 to 15 cm and have long-lasting effects on site productivity.

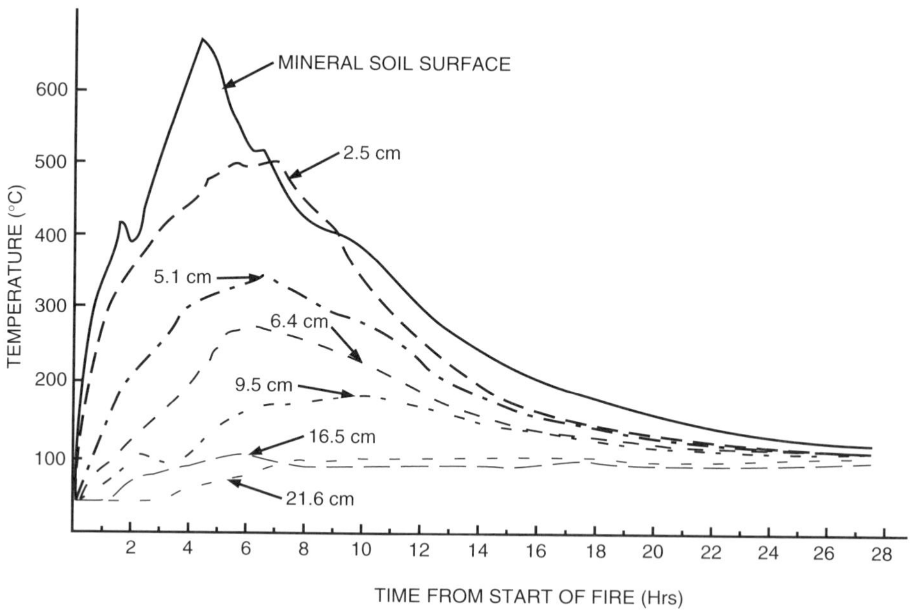

Soil temperatures under a pile of burning *Eucalyptus* logs. (From W. B. Roberts. 1965. Soil temperatures under a pile of burning logs. *Australian Forest Research* 1(3):21–25.) ■

Shrubland Ecosystems

Soil temperatures measured during fires in southern California chaparral are presented to illustrate the nature of soil heating in shrubland ecosystems. Soil temperatures have been measured on several study sites burned at different severities (DeBano et al. 1979). Low-severity burns were characterized by a charred leaf litter produced when the poorly aerated litter layer was not totally incinerated by the heat radiated downward from a burning brush canopy. Some grayish ash was present immediately after the fire but soon became inconspicuous. Maximum soil temperatures during a burn producing "black ash" conditions were about 225°C at the soil surface and 125°C at 2.5- cm depth (Figure 5.3A).

When a moderately severe burn occurred, a "bare-soil" seedbed was produced. The bare-soil condition occurred when fire was hot enough to consume the leaf litter and fine woody material on the ground surface. Some charred material remained, but it was sparse in coverage of the site. Immediately after the fire, ash from incinerated litter and branches was inconspicuous and soon disappeared. Maximum temperature at the mineral surface was almost 430°C, and at 2.5 cm it was about 200°C (Figure 5.3B).

Areas that burned at a high severity were characterized by a "white ash" seedbed. This seedbed was identified by a fluffy ash layer, where large branches or main stems of trees or shrubs had burned. White ash layers were most extensive where dense stands of heavy brush had been burned to stubs. Temperatures at the surface were just over 700°C, and nearly reached 250°C at 2.5 cm (Figure 5.3C).

Grassland Ecosystems

Fuel in grasslands consists mainly of finely divided leaves and stems of grass plants, which can burn robustly. Therefore, a grass fire is characterized as a narrow zone of flames advancing across a relatively homogeneously dispersed fuel. The moisture content of grass fuels changes rapidly in response to changes in relative humidity and other weather conditions (Raison 1979). When grass plants are dry, they burn rapidly and completely as a fire front passes. Intensities can be high, as was reported for a grass fire in Australia where the intensity was 20,000 kW/m (Noble 1991). Although this fire was intense, residence time was short because the fire moved at a measured 6.4 m/s. Therefore, the soil surface was exposed only briefly to high temperatures.

Soil temperatures produced during grass fires are variable and largely dependent on fuel distribution, moisture content, and prevailing weather conditions. While soil surface temperatures can be high, the underlying soil is usually not heated much above ambient temperature, particularly under low-intensity fires. Soil temperatures at the soil surface can range from about 95 to 720°C, although these temperatures generally rise rapidly and then return to prefire levels within several minutes after the passage of the fire front (Raison 1979). Increased temperatures in the upper 3 to 4 cm in the soil are usually less than 50 to 80°C and persist for only a few minutes. Soil heating is likely to be more severe and longer in duration under denser stands of grass having larger accumulations of litter (Daubenmire 1975). The effect of fire on the physiology and phenology of grass plants at different seasons of burning is probably more important than the changes in soil properties.

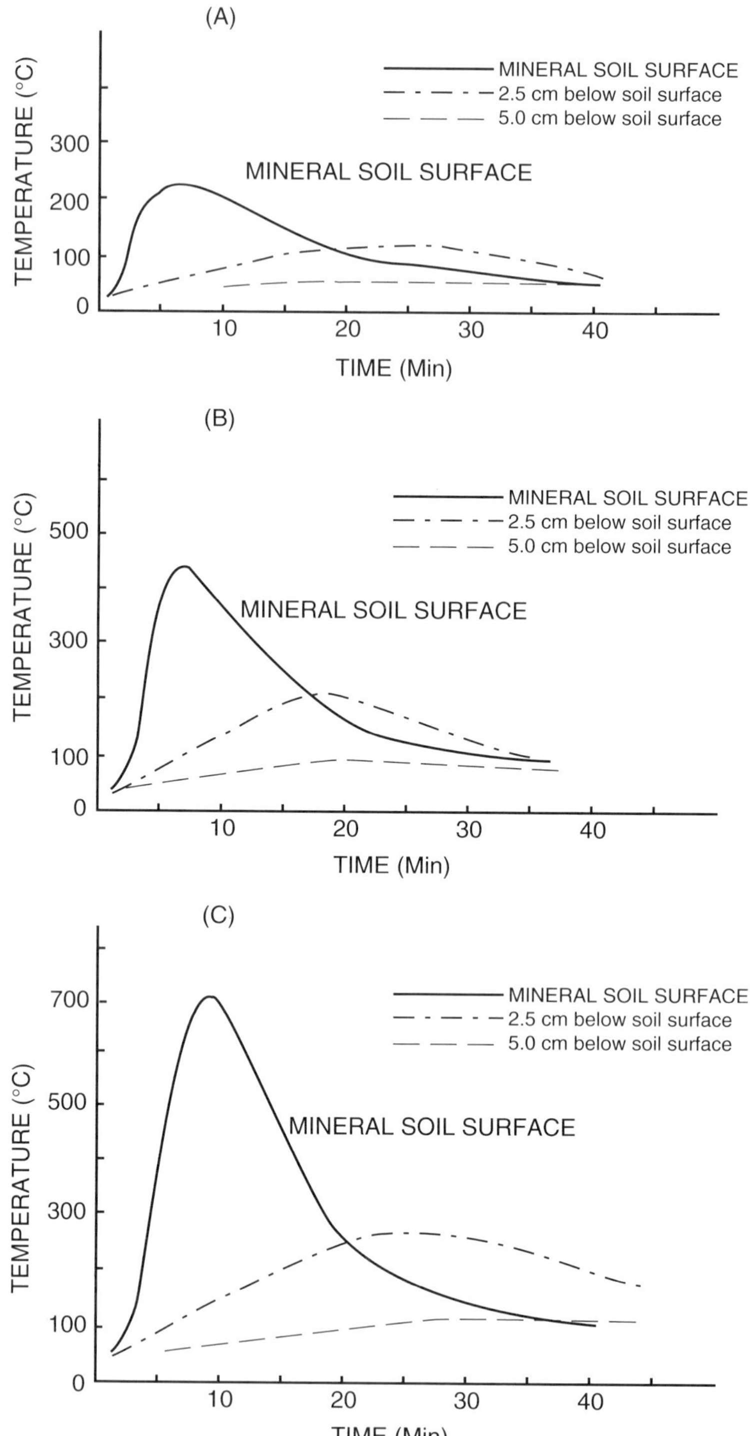

FIGURE 5.3 Surface and soil temperatures in California chaparral during (A) a low-severity fire, (B) a moderately severe fire, and (C) a high-severity fire approaching wildfire conditions. (Adapted from DeBano et al. 1979.)

SOIL PHYSICAL PROPERTIES

Soil physical properties affected by fire include the soil structure, total porosity and pore size distribution, and water retention. Water repellency is another physical property that is affected by fire on hydrophobic soils, where its formation is fire-dependent.

Soil Structure

The structure of soils results from the aggregation of mineral soil particles by organic matter. At the upper part of the mineral soil, at the duff–upper *A* horizon interface, organic matter has a dominant role in soil structure. Further downward in the soil profile (in the *B* horizon), soil structure is more dependent on clay minerals and the composition of cations in the soil solution.

The aggregation of individual mineral particles in the soil enhances porosity and structure. As a result, bulk densities of well-aggregated soils are generally lower than soils having less organic matter. Fire can affect both the clay mineral and organic components of soil, but in different ways.

Clay minerals in soils are usually not significantly altered during a fire for two reasons:

- Clay minerals can tolerate relatively high temperatures before becoming irreversibly damaged (see Table 4.1). Heating must be at a temperature of 460°C or above to produce a loss of OH groups from the clays (Giovannini et al. 1988). Irreversible changes in water of hydration occur at higher temperatures of 980°C or greater (DeBano et al. 1977).
- Clay content in the first few centimeters of surface soils tends to be less than 5 percent. These low clay contents occur at the surfaces because clay minerals are formed deeper in the soil profile during soil formation (see Chapter 4).

When clays are present, larger percentages are found in the *B* horizon, which is far removed from the mineral soil surface. Therefore, only soil profiles that have truncated *B* horizons (where the *A* horizon has been removed by some disturbance such as erosion) or immature soils that have clays deposited on the surface would be affected, and any damage would occur only during severe fires (for example, burning of piled logs).

The effect of fire on the organic component of soil structure is different than for clay particles because:

- The destruction of organic matter begins at low temperatures, starting at 200°C, and is complete at 500°C (see Chapter 6).

- Organic matter is concentrated in the upper part of the soil profile (*D* layer and upper *A* horizon), where it is exposed more directly to heat radiated downward during combustion of surface fuels. Because of the location and sensitivity of organic matter to heating, some changes in structure of surface soil horizons can occur during low-severity fires. The degree of alteration increases directly with increasing fire severity.

Durations of the effects of fire on soil structure range from 1 year to many decades, depending on the severity of a fire, rate of recovery as influenced by natural conditions, postfire land use, and remedial measures applied (Wells et al. 1979). Recovery from low-severity surface fires in the relatively humid eastern United States has been reported to be rapid (Bower 1966). In contrast, soil recovery from intense fires tends to be slower in higher elevations of the northwestern United States or in the more arid areas of the southwestern United States (Anderson 1976).

Porosity

Pore space in soils controls rates of water (soil solution) and air movement through the soil. Well-aggregated soils contain a balance of macropores (>0.6 mm in diameter) and micropores (<0.6 mm in diameter) (Singer and Munns 1996). This balance allows a soil both to transmit water and air rapidly through macropores and to retain water by capillarity in micropores (see later discussion). Macropores in the surface soil horizons are especially important pathways for infiltration of water into the soil.

Fire and associated soil heating can destroy soil structure, affecting both total porosity and pore size distribution in the surface horizons of a soil. The result is a decrease in total porosity and pore size. Loss of macropores in the surface soil reduces infiltration rates and produces overland flow (see Chapter 8). Alteration of organic matter can also lead to a water-repellent soil condition, which further decreases infiltration rates.

Water Retention

Water is retained in soil pores by capillarity. The smaller the pore size (that is, the greater the micropore space), the greater the water-holding capacity. Small pore sizes in clay-sized inorganic materials have large water-holding capacities per unit volume and, as a consequence, permit retention of significant amounts of water for plant growth. Organic matter also glues mineral soil particles together to produce aggregates that enhance soil structure and improve water retention. Therefore, a loss of soil organic matter and structure during a fire has an adverse effect on water retention by soil.

Water Repellency

Water repellency is a physical soil property that can be affected by fire (DeBano 1981). Most dry soils have a high absorptive capacity for water. Usually, when water is applied to a dry coarse-textured soil, it will readily penetrate and infiltrate into the soil. Conversely, water droplets will not be absorbed in water-repellent soils but, instead, will "ball up" and remain on the soil surface, even in coarse textured soils. This phenomenon occurs because the attraction between individual water molecules is greater than it is between water and the hydrophobic surface on soil particles. Water repellency occurs because of the following:

- Irreversible drying of organic matter can induce water repellency (Jemison 1942, Gilmour 1968).
- Organic substances are leached from plant litter material and coat min-

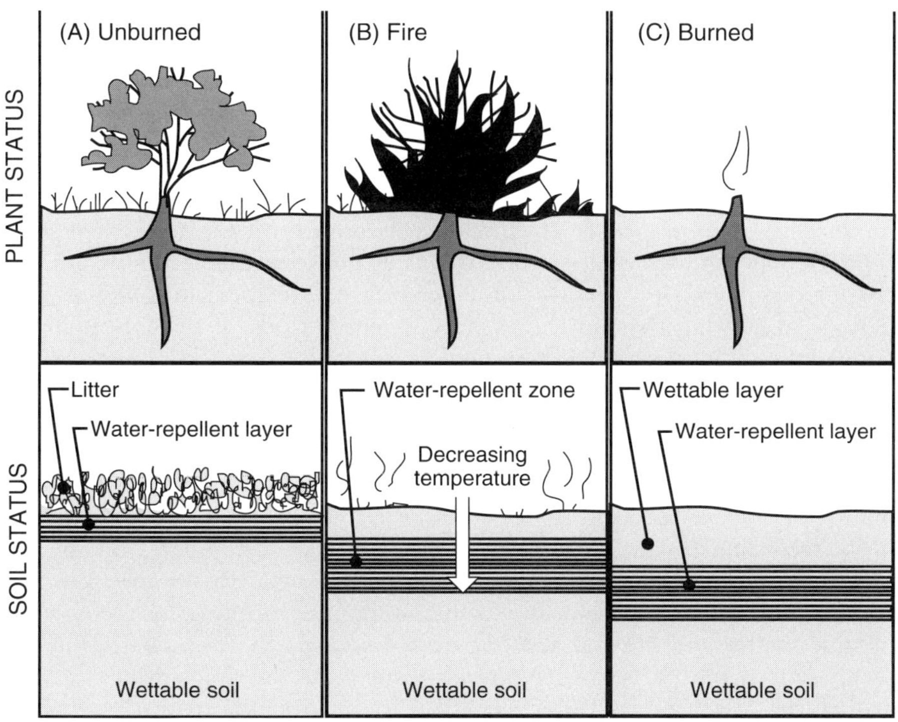

FIGURE 5.4 Soil water repellency is altered by fire. (A) Hydrophobic substances accumulate in the litter and duff layers and in the mineral soil immediately beneath them before a fire; (B) fire then burns the vegetation, litter and duff layers, causing hydrophobic substances to move downward along temperature gradients; and (C) a water-repellent layer is present below and parallel to the soil surface on the burned area. (Adapted from DeBano 1981.)

eral particles, thus inducing water repellency in underlying coarse-textured soils and sand (Roberts and Carbon 1972, DeBano 1981).
- Hydrophobic by-products of microbial mycelium can induce water repellency in coarse-textured soils and sands (Bond and Harris 1964, Savage et al. 1969b).
- Water repellency can occur as a result of the intermixing of hydrophobic organic matter and mineral soil particles (Das and Das 1972, McGhie and Posner 1980, DeBano 1981).
- Heating organic matter on the soil surface or mixtures of hydrophobic soil and mineral soil particles can impart intense water repellency in some soils as described later (DeBano 1966, DeBano et al. 1976).

WATER REPELLENCY AND FIRE

After fire, water repellency can be found as a discrete layer of variable thickness on the soil surface or a few centimeters below the mineral soil surface, and parallel to the soil surface. If found in mineral soil, water repellency is usually covered by a layer of severely burned soil or an ash layer. This layer has been described as the "tin roof" effect (DeBano 1969).

A hypothesis describing the formation of a water-repellent soil layer in soils has been developed by DeBano (1981). This hypothesis evolved from the results of several laboratory and field studies. An initial study showed that water repellency could be intensified by heating a soil–organic matter mixture in a muffle furnace at different temperatures for different lengths of time (DeBano and Krammes 1966). It was demonstrated that a more efficient coating of mineral soil particles occurred at lower temperatures and at short periods of heating than in the case of longer periods of heating at higher temperatures that destroyed the organic substances responsible for the water repellency.

Laboratory tests of changes in water repellency resulting from different times and temperatures of heating were combined with measured temperatures during prescribed fires and wildfires to develop the hypothesis on formation of a water-repellent layer beneath the soil surface during a fire (Figure 5.4). According to this hypothesis, organic matter accumulates on the soil surface during intervals between fires. During these intervals, upper soil horizons become water repellent due to the intermixing of partially decomposed organic matter and mineral soil (Figure 5.4A) and due to leachate from decomposing plant parts, which are deposited on the soil surface. Fungal growth can also produce water repellency in upper soil horizons.

Heating during a fire intensifies this initial water repellency. When plant canopies and surface fuels are combusted, litter and upper soil horizons are exposed to intense heating, particularly during a severe burn (Figure 5.3C). High levels of energy can be released if the canopies are combusted during wildfire conditions. Although less than 10 percent of the combustion energy is normally radiated downward, this energy is sufficient to ignite the surface litter and transfer heat downward into the underlying soil (DeBano 1974). Temperatures in the canopy of burning chaparral brush can reach 1090°C (Countryman 1964). These temperatures can reach about 850°C at the soil–litter interface (Dunn and DeBano 1977). Temperatures at 5 cm in the mineral soil probably do not exceed 150°C because dry soil is a good insulator (DeBano et al. 1979).

The combination of combustion and heat transfer leads to large temperature gradients developing in the upper part of the mineral soil. Heat produced by combustion of the litter layer on the soil surface vaporizes organic substances, which are then moved downward in the soil along the steep temperature gradients until they reach cooler underlying soil layers, where they condense (Figure 5.4C). Incipient water repellency at different soil depths can also be intensified in place by heating, because organic particles are heated to the extent that they coat and are chemically bonded to nearby mineral particles. Movement of hydrophobic substances downward in the soil occurs mainly during the fire (Savage 1974). After fire has passed, the continued heat movement downward through the soil can revolatilize some of the hydrophobic substances, resulting in thickening the water-repellent soil layer. Relationships between water repellency and temperature include the following:

- Little change in water repellency occurs when soils are heated below 176°C (DeBano 1981).
- Intense water repellency is formed when soils are heated between 176 and 204°C (DeBano 1981).
- Water repellency is destroyed when soils are heated above 288°C (Savage 1974, DeBano et al. 1976).

A water-repellent layer of varying thickness remains after a fire has passed through an area. The depth of the resulting water-repellent layer allows water to infiltrate only a limited depth into the soil. Depth and thickness of the organic layer depend on several factors including:

- *Severity of fire.* The more severe the fire, the deeper the layer is found unless the fire is so hot as to completely destroy all organic matter on the surface or in the soil (DeBano et al. 1976).
- *Type and amount of organic matter present.* There must be sufficient organic matter to provide hydrophobic substances. Most vegetation types and mycelium materials will induce water repellency (DeBano 1981).
- *Temperature gradients developing in the upper mineral soil.* Steep temperature gradients in dry soil enhance translocation (DeBano et al. 1976, DeBano 1981).
- *Texture of the soil.* Sandy and coarse-textured soils are more susceptible

to water repellency (Bond and Harris 1964, Gilmour 1968, DeBano 1981), although clays have been reported to become water repellent (McGhie and Posner 1980).
- *Water content of the soil.* Soil water interferes with heat transfer and condensation of hydrophobic substances (DeBano et al. 1976, Robichaud 1996).

Organic substances also contribute to water repellency in both fire and nonfire environments. The chemical composition of these organic substances is not clear, but several substances have been identified (Example 5.2).

Example 5.2: Substances Causing Water Repellency in Soil

These substances include:

- Long-chain aliphatic hydrocarbons (Savage et al. 1972, Dinel et al. 1991)
- Microbial generated humic acids (Savage et al. 1969a, Adhikari and Chakrabarti 1976, Piccolo and Mbagwu 1989)
- Organic radicals adsorbed on exchange sites of mineral soil particles (Chakravarti and Mukherjee 1965)
- Polyvalent metals linked to organic matter (Giovannini et al. 1983)
- A complex ester between phenolic acids and polysaccharide-like substances (Giovannini and Lucchesi 1984)
- Organic matter with certain physical and structural attributes (Hudson et al. 1994), particularly factors such as the degree of dispersion and roughness of the mineral particles (Grelewicz and Plichta 1985) ■

WATER REPELLENCY AND AGGREGATE STABILITY

One should not be left with the impression that water repellency is always an undesirable feature in soil. To the contrary, water repellency that is not severe can be beneficial and contribute to aggregate stability in soil. This benefit is particularly the case for soil in nonfire environments (Example 5.3).

Example 5.3: Role of Organic Matter and Hydrophobicity in Aggregate Stability and Soil Structure

Soil structure has long been recognized as an important characteristic that enhances productivity and water relations in both agricultural and wildland soils. Improved infiltration into aggregated soil can reduce soil erodibility. The interaction of organic matter with mineral soil particles has an effect on cation adsorption capacities of a soil. The capacity of soils to adsorb cations is a function of the total exchange sites on both organic matter and soil minerals.

Organic matter binds mineral soil particles together and enhances soil structure. Living and dead organic matter undergoes microbial decomposition, which produces organic substances that can aggregate individual mineral soil particles. Physical processes such as deformation and compression by freezing and thawing, and wetting and drying, can also cause aggregation. The abundance of cations in saline and alkali soils can provide this aggregating effect, leading to a strong prism-like structure. Hydrophobic substances in soil can be beneficial to soil aggregation because they function similarly to other types of cementing agents (Giovannini et al. 1983). Hydrophobic compounds contained in organic matter also enhance the stability of aggregates by increasing their resistance to disintegration (slaking) when wetted (Coughlan et al. 1973, Giovannini and Lucchesi 1983). Although individual aggregates exhibit some water repellency, the creation of macropore spaces provides pathways for rapid infiltration and water conduction throughout the soil profile. At the other end of the spectrum, if individual soil particles (particularly sand) are coated with hydrophobic substances, the entire soil system can become highly water repellent. This severe water repellency limits infiltration and can cause excessive runoff and erosion. ■

Water repellency produced by low- to moderate-severity fires is usually of shorter duration than that produced by high-severity fires. For example, water repellency produced by a low-severity burn in late spring in the forests of southwestern Oregon began to allow infiltration at a nearly normal rate after the rains began in fall (McNabb et al. 1989). Dyrness (1976) found that wettability of soil on areas that burned at low severity recovered more rapidly than for those soils in severely burned areas. Wettability of soils on sites burned at both low and high severity approached that of an unburned soil by the sixth year after fire. Conversely, 3 years after passage of a fire on a experimental plot in Sardinia, subsurface layers (in which translocated hydrophobic matter had accumulated) showed these hydrophobic substances to be unaltered, but they were more strongly cemented because the translocated hydrophobic organic matter complexed with polyvalent ions (Giovannini et al. 1987).

Effect on Infiltration and Runoff

Two important erosion processes usually occur following fire—dry ravel and rill formation. Dry ravel occurs during and immediately following the fire on steep slopes where gravity moves loose materials released during the fire downslope. Rill formation occurs when rainfall exceeds infiltration rates and surface runoff occurs. Dry ravel is well illustrated in the southwestern United States where loose soil and rock material is held on steep slopes by shrubs during the interval between fires. High-severity wildfires burn the shrubs, releasing stored material that moves downslope by gravity. Dry ravel accumulates in channels at the base of steep slopes and remains there until increased streamflow moves it downstream (Wells 1987).

A striking feature on freshly burned watersheds during the first postfire rainstorms is the formation of an extensive network of rills (Wells 1981). Formation of these rills appears to be related directly to water repellency (Figure 5.5). Observations of rills after the first rainstorms on recently burned watersheds show that downcutting of the rill ends at the bottom of the water repellent layer (Wells 1987).

It is hypothesized that water infiltrates into the wettable surface soil during storms and percolates until it encounters a water-repellent layer (Wells 1981). This process occurs uniformly over the landscape so that when the wetting front reaches the water-repellent layer, it can drain neither downward nor laterally. As rainfall continues, water fills all available pore space until the wettable soil layer becomes saturated. Because pores cannot drain, pressures build up, particularly immediately above a water-repellent layer. This increased pressure reduces intergranular stress among soil particles and, as a result, decreases shear strength in the soil mass. This increased pore pressure

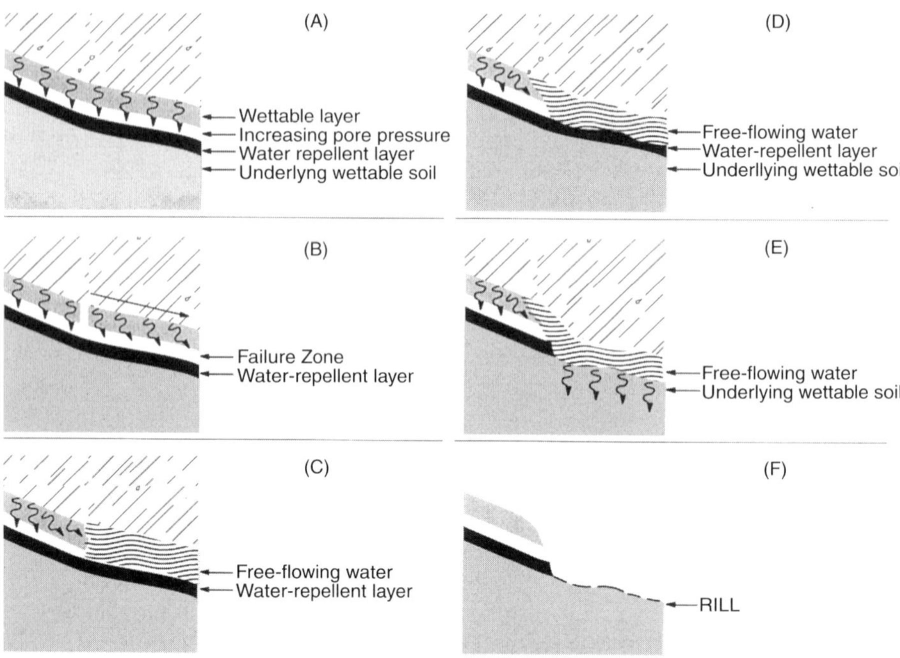

FIGURE 5.5 Sequence of rill formation on a burned slope with a water-repellent layer includes (A) saturation of wettable surface area, (B) development of a failure zone in wettable surface layer, (C) free flowing water over the water-repellent layer, (D) erosion of the water-repellent layer, (E) removal of the water repellent layer and infiltration into underlying wettable soil, and (F) resultant rill. (From W. G. Wells II. The effects of fire on the generation of debris flows in southern California. *Reviews in Engineering Geology* 7:105–114. Modified with permission of the publisher, the Geological Society of America, Boulder, Colorado USA. Copyright © 1987 Geological Society of America.)

produces potential zones of failure at the boundary between the wettable and water-repellent layers, where pore pressures are greatest (Figure 5.5A). Pore pressure continues to increase and shear strength decreases until it is exceeded by the shear stress of gravity acting on the soil mass. When this happens, a failure occurs and a portion of the wettable soil begins to slide downslope (Figure 5.5B). If the soil is coarse textured, initial failure causes a reorientation of the soil particles in the failure zone and causes them to momentarily lose contact with each other. The loss of intergranular contact further reduces shear strength and extends the failure zone downslope. When most of the soil grains lose contact, a condition develops in which the shearing soil is almost fluid. This fluid condition produces a miniature debris flow in the upper wettable soil layer, which propagates downslope to the bottom of the slope or until it empties into a channel.

Water in the wettable soil layer adjacent to the debris flow is no longer confined and can flow out into the rill formed by the debris flow (Figure 5.5C). Water in this rill still cannot infiltrate into the water-repellent soil and, therefore, flows down the debris flow track as free water in an open channel (Wells 1981). As the water flows down the track, turbulent flow develops, which erodes and entrains particles in the water-repellent layer (Figure 5.3D). Eventually, the flow cuts completely through the water-repellent layer and begins infiltrating into the underlying wettable soil. Flow then diminishes, turbulence is reduced, and downcutting ceases (Figure 5.5E). The final result is a rill that stabilizes immediately below the lower edge of the water-repellent layer (Figure 5.5F). Individual rills develop into a network that can extend the length of a small watershed.

SUMMARY

Effects of fire on the soil physical system are complex and result from changes in organic matter during soil heating. Soil temperatures occurring in the soils of forest, shrubland, and grassland ecosystems vary widely during fire. After reading this chapter, you should be able to:

1. Describe and quantify soil temperatures that can occur in soils during forest, shrubland, and grassland fires.
2. Explain the importance of soil organic matter to soil physical properties.
3. Explain how water-repellent soils are formed.
4. Outline and discuss the effect of fire on infiltration of water into soils.

REFERENCES

Adhikari, M., and G. Chakrabarti. 1976. Contribution of natural and microbial humic acids to water repellency in soil. *Journal of Indian Society Soil Science* 24:217–219.

Agee, J. K. 1973. Prescribed fire effects on physical and hydrologic properties of mixed-conifer forest floor and soil. Report 143, University of California Resources Center, Davis, California.

Anderson, H. W. 1976. Fire effects on water supply, floods and sedimentation. In: *Proceedings, Tall Timbers Fire Ecology Conference No. 15*. Tall Timbers Research Station, Tallahassee, Florida, pp. 249–260.

Bond, R. D., and J. R. Harris. 1964. The influence of the microflora on physical properties of soils: I. Effects associated with filamentous algae and fungi. *Australian Journal of Soil Research* 2:111–122.

Bower, D. R. 1966. Surface soil recovers quickly after burn. USDA Forest Service, Research Note SO-46.

Chakravarti, S. K., and S. K. Mukherjee. 1965. Organophilic characteristics and water proofness of clay–organic complexes. *Journal of Indian Chemistry Society* 42:289–296.

Coughlan, K. J., W. E. Fox, and J. D. Hughes. 1973. A study of the mechanisms of aggregation in an Krasnozem soil. *Australian Journal of Soil Research* 11:65–73.

Countryman, C. M. 1964. Mass fires and fire behavior. USDA Forest Service, Research Paper PSW-19.

Das, D. K., and B. Das. 1972. Characterization of water-repellency in Indian soils. *Indian Journal of Agricultural Science* 42:1099–1102.

Daubenmire, R. 1975. Ecology of fire in grasslands. *Advances in Ecological Research* 9:209–266.

DeBano, L. F. 1966. Formation of non-wettable soils . . . involves a heat transfer mechanism. USDA Forest Service, Research Note PSW-132.

DeBano, L. F. 1969. Observations on water-repellent soils in western United States. In: DeBano, L. F., and J. Letey, editors. *Proceedings of a Conference on Water Repellent Soils*, Riverside, California, pp. 17–29.

DeBano, L. F. 1974. Chaparral soils. In: Rosenthal, C., conference chairman. *Proceedings of a Symposium on Living with the Chaparral*. Sierra Club, San Francisco, California, pp. 19–26.

DeBano, L. F. 1981. *Water Repellent Soils: A State-of-the Art*. USDA Forest Service, General Technical Report PSW-46.

DeBano, L. F., and J. S. Krammes. 1966. Water repellent soils and their relation to wildfire temperatures. *International Association of Hydrological Sciences* 2:14–19.

DeBano, L. F., S. M. Savage, and D. A. Hamilton. 1976. The transfer of heat and hydrophobic substances during burning. *Soil Science Society of America Journal* 40:779–782.

DeBano, L. F., P. H. Dunn, and C. E. Conrad. 1977. Fire's effect on physical and chemical properties of chaparral soils. In: Mooney, H. A., and C. E. Conrad, technical coordinators. *Proceedings of the Symposium on the Environmental Consequences of Fire and Fuel Management in Mediterranean Ecosystems*. USDA Forest Service, General Technical Report WO-3, pp. 65–74.

DeBano, L. F., R. M. Rice, and C. E. Conrad. 1979. Soil heating in chaparral fires: effects on soil properties, plant nutrients, erosion and runoff. USDA Forest Service, Research Paper PSW-145.

Dinel, H., M. Levesque, and G. R. Mehuys. 1991. Effects of long-chain aliphatic compounds on the aggregate stability of a lacustrine silty clay. *Soil Science* 151:228–239.

Dunn, P. H., and L. F. DeBano. 1977. Fire's effect on biological and chemical properties of chaparral soils. In: Mooney, H. A., and C. E. Conrad, technical coordinators. *Proceedings of a Symposium on the Environmental Consequences of Fire and Fuel Management in Mediterranean Ecosystems.* USDA Forest Service, General Technical Report WO-3, pp. 75–84.

Dyrness, C. T. 1976. Effect of wildfire on soil wettability in the High Cascades of Oregon. USDA Forest Service, Research Paper PNW-202.

Gilmour, D. A. 1968. Water repellence of soils related to surface dryness. *Australian Forestry* 32:145–148.

Giovannini, G., and S. Lucchesi. 1983. Effect of fire on hydrophobic and cementing substances of soil aggregates. *Soil Science* 136:231–236.

Giovannini, G., and S. Lucchesi. 1984. Differential thermal analysis and infrared investigations on soil hydrophobic substances. *Soil Science* 137:457–463.

Giovannini, G., S. Lucchesi, and S. Cervelli. 1983. Water-repellent substances and aggregate stability in hydrophobic soil. *Soil Science* 135:110–113.

Giovannini, G., S. Lucchesi, and M. Giachetti. 1987. The natural evolution of a burned soil: a three-year investigation. *Soil Science* 143:220–226.

Giovannini, G., S. Lucchesi, and M. Giachetti. 1988. Effect of heating on some physical and chemical parameters related to soil aggregation and erodibility. *Soil Science* 146:255–262.

Grelewicz, A., and W. Plichta. 1985. The effect of the physical state of the surface of organic material on its wettability. *Forest Ecology and Management* 11:245–256.

Heyward, F. 1938. Soil temperatures during forest fires in the longleaf pine region. *Journal of Forestry* 36:478–491.

Hudson, R. A., S. J. Traina, and W. S. Shane. 1994. Organic matter comparison of wettable and nonwettable soils from bentgrass sand greens. *Soil Science Society of America Journal* 58:361–367.

Jemison, V. C. 1942. The slow reversible drying of sandy surface soils beneath citrus trees in central Florida. *Soil Science Society of America Proceedings* 7:36–41.

McGhie, D. A., and A. M. Posner. 1980. Water repellence of a heavy-textured western Australian surface soil. *Australian Journal of Soil Research* 18:309–323.

McNabb, D. H., F. Gaweda, and H. A. Froehlich. 1989. Infiltration, water repellency, and soil moisture content after broadcast burning a forest site in southwest Oregon. *Journal of Soil and Water Conservation* 44:87–90.

Noble, J. C. 1991. Behaviour of a very fast grassland wildfire on the riverine plain of southeastern Australia. *International Journal of Wildland Fire* 1:189–196.

Piccolo, A., and J. S. C. Mbagwu. 1989. Effects of humic substances and surfactants on the stability of soil aggregates. *Soil Science* 147:47–54.

Raison, R. J. 1979. Modification of the soil environment by vegetation fires, with particular reference to nitrogen transformations: a review. *Plant and Soil* 51:73–108.

Roberts, F. J., and B. A. Carbon. 1972. Water repellence in sandy soils of south-western Australia. *Australian Journal of Soil Research* 10:35–42.

Roberts, W. B. 1965. Soil temperatures under a pile of burning logs. *Australian Forest Research* 1(3):21–25.

Robichaud, P. R. 1996. *Spatially-Varied Erosion Potential from Harvested Hillslopes After Prescribed Fire in the Interior Northwest.* Ph.D. Thesis, University of Idaho, Moscow, Idaho.

Rundel, P. W. 1983. Impact of fire on nutrient cycles in Mediterranean-type ecosystems with reference to chaparral. In: Kruger, F. J., D. T. Mitchell, and J. U. M. Jarvis, editors. *Mediterranean-type Ecosystems: The Role of Nutrients.* Springer-Verlag, New York, pp. 192–207.

Sackett, S. S., and S. M. Haase. 1992. *Measuring Soil and Tree Temperatures During Prescribed Fires with Thermocouple Probes.* USDA Forest Service, General Technical Report PSW-131.

Savage, S. M. 1974. Mechanism of fire-induced water repellency in soil. *Soil Science Society of America Proceedings* 38:652–657.

Savage, S. M., J. P. Martin, and J. Letey. 1969a. Contribution of humic acid and a polysaccharide to water repellency in sand and soil. *Soil Science Society of America Proceedings* 33:149–151.

Savage, S. M., J. P. Martin, and J. Letey. 1969b. Contribution of some soil fungi to natural and heat-induced water repellency in sand. *Soil Science Society of America Proceedings* 33:405–409.

Savage, S. M., J. Letey, J. Osborn, and C. Heaton. 1972. Water repellency in soils induced by forest fires. *Soil Science Society of America Proceedings* 26:674–678.

Singer, M. J., and D. N. Munns. 1996. *Soils: An Introduction.* Prentice Hall, Upper Saddle River, New Jersey.

Wells, C. G., R. E. Campbell, L. F. DeBano, C. E. Lewis, R. L. Fredricksen, E. C. Franklin, R. C. Froelich, and P. H. Dunn. 1979. *Effects of Fire on Soil: A State-of-Knowledge Review.* USDA Forest Service, General Technical Report WO-7.

Wells, W. G. II. 1981. *Erosion and Sediment Transport in Pacific Rim Steeplands.* IASH Publication No. 132 (Christchurch, 1981), pp. 305–342.

Wells, W. G. II. 1987. The effects of fire on the generation of debris flows in southern California. *Reviews in Engineering Geology* 7:105–114.

Chapter

6

Chemical Soil System

INTRODUCTION

The chemical soil system contains both organic and inorganic components that interact with each other continually. Soil chemical properties involved in this interaction include organic matter, cation exchange capacity, exchangeable cations, buffer capacity, and pH. Processes regulating nutrient storage and cycling, nutrient availability, and losses by volatilization and leaching are also key features of the chemical system. Fire affects both the soil properties and the processes they regulate. This chapter discusses these changes.

Burning affects the chemical properties of soils by ashing the organic matter contained in the aboveground vegetation and litter and, under higher fire severities, the duff and upper mineral soil horizons. Several things can happen to the ash material produced by a fire. Precipitation can dissolve the ash and carry the chemical elements into the soil, where they remain as dissolved ions in the soil solution (Wells et al. 1979). Variable amounts of the ash can also be lost from the site aerially during the fire or be carried off-site by wind following the fire (Clayton 1976). Surface runoff, and sediments in the runoff, can carry the elements in the ash and soil downstream (DeBano and Conrad 1976). Soluble ions moving into and through the soil can be lost by leaching (Grier 1975).

ORGANIC MATTER

Organic matter is found both aboveground and below-ground. Major components of aboveground organic matter are standing dead and live plants, and dead organic debris (including leaves, stems, twigs, and logs) accumulating on the soil surface. Organic matter in forest soils can be found as six components (Harvey 1982):

- The *L* layer composed of recognizable plant material
- The *F* layer containing partially decomposed but recognizable plant litter
- The *H* layer consisting of extensively decayed and disintegrated organic materials, some of which may be mixed with the mineral soil at the soil–humus interface
- Decayed wood remaining after the decaying of woody material on the soil surface or buried in the forest floor
- Charcoal, or extensively charred wood, that is mixed into the forest floor and mineral soil
- Upper mineral soil horizon (the *A* horizon), which contains a mixture of mineral soil, organic matter, roots, and microbial biomass

All of these components may not be present in shrubland and grassland soils, however. For example, the woody component is lacking in most grassland soils except for those that are partially occupied by trees or are found in streambank (riparian) areas.

The amount of organic material produced by plant canopies varies widely among natural ecosystems throughout the world, depending on the type of vegetation occupying a site and the prevailing temperature and moisture regimes. Only a few grams of aboveground plant material can be produced annually in dry arid deserts, in contrast to that found in productive coniferous forest ecosystems, which can deposit up to 6 Mg/ha of material annually (Tate 1987). Likewise, the relative amounts of organic matter found in the aboveground and below-ground components of ecosystems, vary in different parts of the world (see Figure 6.1). Large portions of the C pool are located below the soil surface in all of these ecosystems, and particularly in savannas and tundra ecosystems.

Organic matter deposited on, and incorporated into, the soil surface is decomposed by soil organisms. Rates of decomposition and accumulation are regulated by soil temperature and moisture conditions on a site. Organic matter decomposition is slow in colder climates, and, consequently, large amounts of organic matter can accumulate in peat bogs and tundra. Roots and microbial biomass are also important sources of soil organic matter. Organic matter produced by roots can be concentrated in the upper meter of soil in grasslands, or extend to several meters in areas covered by forest and shrub vegetation. The root environment, called the *rhizosphere,* also provides a favorable environment for soil microorganisms and, therefore, is an important site of microbial activity.

Surface organic matter provides a protective layer for the soil surface, and organic matter in duff and mineral soil improves soil structure and water movement into the soil. Additional organic matter found deeper in the soil is made up of living and dead roots, or soluble organic matter that has been

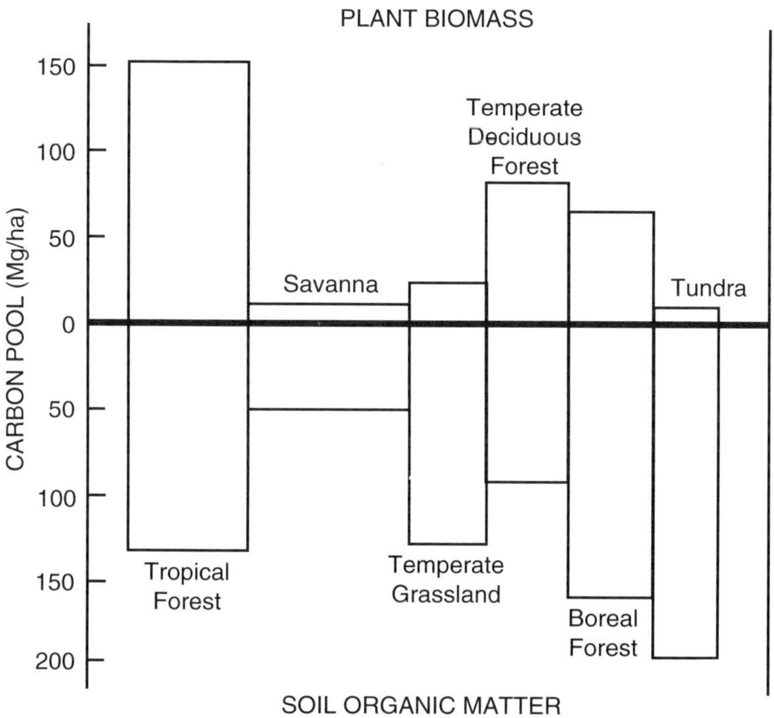

FIGURE 6.1 Distribution of C and soil organic matter (including litter) in major ecosystem types of the world. (Adapted from J. M. Anderson, 1991. The effects of climate change on decomposition processes in grassland and coniferous forest. *Ecological Applications* 1:326–347. Copyright © 1991 Ecological Society of America.)

leached or dislocated from the surface horizons. This organic matter is less exposed to the direct effects of fire.

Ecological Significance

Organic matter in soil systems plays two major roles (Tate 1987):

- It serves as a primary medium for sustaining numerous chemical and biological transformations. Biogeochemical cycling of nutrients and the synthesis of living tissue are examples of this role.
- Its presence enhances other physical and chemical properties. Examples of the latter role are soil properties that are improved by the presence of organic matter, such as soil structure and cation exchange capacity.

Organic matter not only affects the physical and chemical properties of the soil but also is necessary for sustaining the biological activity of soils (see

Chapter 7). The livelihood of soil microorganisms is dependent on a reliable, readily available source of soil organic matter. Organic matter provides both a suitable habitat and an energy source for soil microorganisms. Both of these functions are critical for maintaining the nutritional quality and moisture-holding capacity of forest soils (Harvey et al. 1987). Organic matter serves as a powerful aggregating agent and, as such, plays an important role in creating and maintaining a well-aggregated soil. Soil aggregation improves soil structure, creating macro pore space and improving aeration. Aggregated soils also have higher infiltration rates than nonaggregated soils containing less organic matter.

Nutrient-Holding Capacity

Macronutrients and micronutrients are two major categories of nutrients required for plant growth. *Macronutrients* are those nutrients that are needed in the largest concentrations, such as phosphorus (P), nitrogen (N), sulfur (S), iron (Fe), calcium (Ca), potassium (K), and magnesium (Mg). *Micronutrients*, such as zinc (Zn), manganese (Mn), cobalt (Co), molybdenum (Mo), and nickel (Ni), are only needed in trace amounts for plant growth, although they are indispensable. Before these micro- and macronutrients can be incorporated into cellular biomass, they must be released as inorganic ions into the soil solution (Tate 1987). Solubility of these ions is controlled by the following:

- The physical and chemical properties of the soil
- Interactions with organic matter
- Microbial activity

Microorganisms in nonfire environments enhance cycling of both macro- and micronutrients by modifying the pH, synthesizing chelating agents, secreting biological products that form ionic associations with cations, and decomposing organic matter. Nutrients stored in organic matter are released slowly through decomposition, providing a readily available source of nutrients that keeps leaching losses at low levels. However, fire accelerates the decomposition rates immensely (St. John and Rundel 1976). Nutrient changes that normally occur over a period of many years by microbial decomposition in nonfire environments can occur within a few seconds during a fire.

Organic matter plays an important role in storing and cycling nutrients. Organic matter is the primary source of most available P and S, and virtually all of the available N. The storage of large amounts of the ecosystem N in organic matter assures the continued high productivity of soils, which depends on large supplies of available N. Organic matter in the *H* layer and upper part of the *A* horizon also provides chemically active cation exchange sites that retain the important cations, including ammonium (NH_4), K, Ca, and Mg. These cations are a readily available source of cations that are absorbed by living plants and soil microorganisms. Soil organic matter in the upper part of the *A* horizon can provide over 50 percent of the cation exchange capacity of some forest soils. Organic matter in the *A* and *H* horizons includes active chelating agents that retain many essential micronutrients.

Responses to Fire

The distribution of organic matter above and below the soil surface is important when evaluating fire effects. Typical soil profiles have most of the organic matter concentrated near the surface (see Figure 4.1). Because a large proportion of organic matter is either on the soil surface or in the upper soil layers, relatively large amounts of organic matter in most terrestrial ecosystems are exposed directly to the heat radiated downward during a fire (see Chapter 4). Different chemical components found in organic matter are lost when soil temperatures increase during heating (Hosking 1938):

- Losses of organic matter can occur at temperatures below 100°C.
- Volatile constituents are lost at temperatures up to 200°C.
- Destructive distillation results in loss of about 85 percent of the soil organic matter at temperatures between 200 and 300°C.
- Above 300°C, the greater part of the residual organic matter consists of carbonaceous material, which is finally lost upon ignition.
- Heating the soil to 450°C for 2 hours, or to 500°C for ½ hour, removes about 99 percent of the organic matter.

Some of the specific chemical changes occurring in organic matter during soil heating have been identified (Example 6.1). More importantly, however, loss of organic matter is also closely linked to volatilization of N.

Example 6.1 Alteration of Organic Matter During Heating Involves Specific Chemical Changes

Primary losses from organic matter at lower temperatures are functional groups, while decomposition of the nuclei occurs at higher temperatures (Schnitzer and Hoffman 1964). Both phenolic OH and carboxyl (COOH) groups were lost between 250 and 400°C, although phenolic groups are the more stable of the two.

A study of thermal changes in the *H* layer under an evergreen oak forest in Spain showed that oxygen-containing functional groups found in humic and fulvic acids were altered (Almendros et al. 1990). Humic acids were transformed mostly into alkali-insoluble substances, which quantitatively contributed to soil humus. Simultaneously, fulvic acids were transformed into acid-insoluble polymers. The incompletely burned part of plant biomass contained both alkali-soluble lignin materials and brown products formed by dehydration of carbohydrates. Production of lignin compounds increased resistance to chemical and biological degradation in the resulting organic matter. However, the increased stability of organic matter was compensated for by a massive incorporation of ligno-cellulosic material that enhanced recovery of soil microorganisms following fire. ■

NUTRIENT CYCLING AND LOSSES

Importance of Nutrient Cycling

Sustained productivity of natural ecosystems depends on a regular and consistent cycling of nutrients essential for plant growth. Nutrient cycling in nonfire environments involves a number of complex pathways (see Figure 6.2). Nutrients enter the soil by precipitation, dry fall, N fixation, and geochemical weathering of rocks. Nutrients in soil organic matter are transformed into forms available to plants by decomposition and mineralization (see Chapter 7). Biologically regulated decomposition predominates in nonfire environments, and the rate of decomposition varies widely depending on moisture, temperature, and type of organic matter. Nutrient losses from unburned ecosystems are usually low, although small losses can occur by volatilization, erosion, leaching, and denitrification.

Fire alters these nutrient cycling processes dramatically, with the magnitude of these changes depending mostly on fire severity. Fire volatilizes nutrients both in vegetation and from surface organic soil horizons. Heat transferred into the soil affects biological processes such as decomposition and mineralization (see Chapter 7). Increased runoff and erosion carry nutrients off the site after fire.

Many nutrients are essential for plant growth, although only a few are important when assessing fire effects. Important nutrients affected by fire are N, P, S, and some cations. Nitrogen is likely the most limiting nutrient in natural systems (Maars et al. 1983), followed by P. Sulfur can become deficient in some natural ecosystems. Cations are important because of their effect on soil pH following fire and their immobilization of P. Little is known about the role of micronutrients on productivity of natural ecosystems in relation to fire effects, except that over one-half of the selenium in burned laboratory samples can be recovered in the ash residue (King et al. 1977).

Nutrient Loss Mechanisms

The following fates await nutrients contained in plant biomass and soil organic matter during and following a fire:

- *Direct gaseous volatilization into the atmosphere during fire.* Nitrogen can be transformed into N_2 along with other nitrogenous gases (DeBell and Ralston 1970).
- *Particulate loss in smoke.* Phosphorus and cations are frequently lost into the atmosphere as particulates during combustion (Clayton 1976, Raison et al. 1985a).
- *Deposition in ash on the soil surface.* These highly available nutrients can subsequently be leached into the soil or lost as wind erosion (Christensen 1973, Kauffman et al. 1993).
- *Losses by surface runoff and erosion* (DeBano and Conrad 1976, Raison et al. 1993).
- *Leaching into and through the soil profile* (Grier 1975).
- *Nutrients remaining on site in incompletely burned vegetation and detritus* (Boerner 1982).

Nutrients respond differently to heating during combustion of organic matter. Nutrients such as N are readily volatilized (low temperature thresholds) and lost during fire, while other nutrients (including Ca, Mg, and Na) remain as oxides in the ash layer following fire. Threshold temperatures for plant nutrients and physical, chemical, and biological properties are discussed in Chapter 4 (Table 4.1).

Nitrogen Cycling and Losses

Nitrogen contained in unburned ecosystem components (water, vegetation, and air) is transformed among the different components solely by biological processes, specifically N fixation, mineralization, and denitrification, and, therefore, is referred to as being regulated by *biochemical cycling* (McGill and Cole 1981). Because of the close relationship between C and N, C:N ratios are related to decomposition rates of organic matter and, therefore, control the rate at which N and other nutrients are released and cycled (Turner 1977).

NITROGEN RESPONSE TO BURNING

There are differing opinions on the effect of burning and soil heating on soil N. Total and available soil N concentrations have been reported to increase, decrease, or remain the same (Wells et al. 1979, Chandler et al. 1983, Carballas et al. 1993). These conflicting sources are not reviewed individually in this book, because the majority of the conclusions depend on the method of calculating N concentrations (as percent of the sample weight) rather than representing actual changes (kg/ha) in total N in response to fire. This enigma

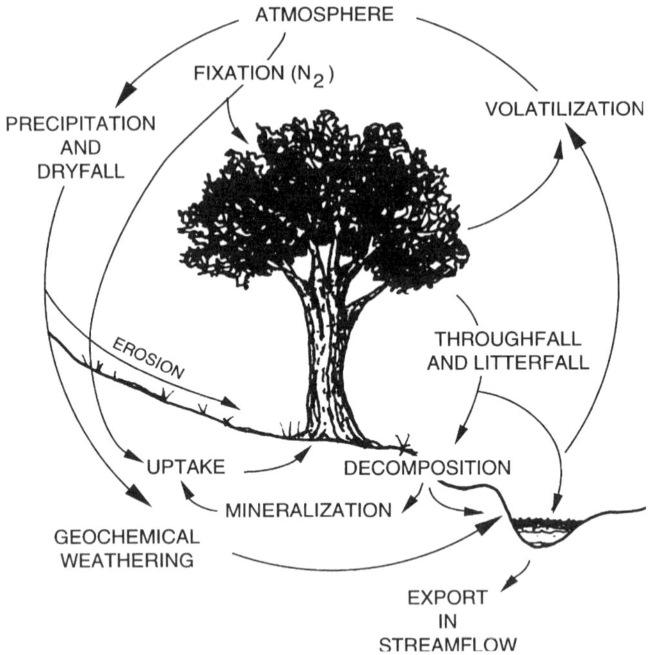

FIGURE 6.2 Nutrient cycling in natural environments. (Adapted from Brown 1980.)

concerning N changes during combustion usually arises because the dry weight of ash is used as a basis to calculate percent N after fire (Example 6.2).

Example 6.2: Concentrations of N Following Fire Can Increase, Decrease, or Remain the Same Depending on the Method of Calculation

A common way of expressing N is as a percent of the ash weight remaining after fire, rather than expressing the amount of N as a percent of the weight of the organic matter that was initially combusted (Knight 1966). Basing percent N on ash weights, or combusted organic matter, usually results in increased N concentrations following fire, because not all the N is lost, and the remaining N makes up a greater percentage of the burned ash sample (see figure below, line A) than does the N in the original unburned organic matter (see figure below, line B). The disparity between percent N in a burned sample (based on ash weight) and the original N percent starts at 100°C and increases to about 500°C, after which it decreases and reaches a percent N at 700°C that is similar to the percent N in the unburned sample (line A). Conversely, when the percent N is based on the initial weight of the sample, percent N starts decreasing at 200°C and continues to decrease upon further heating (line B).

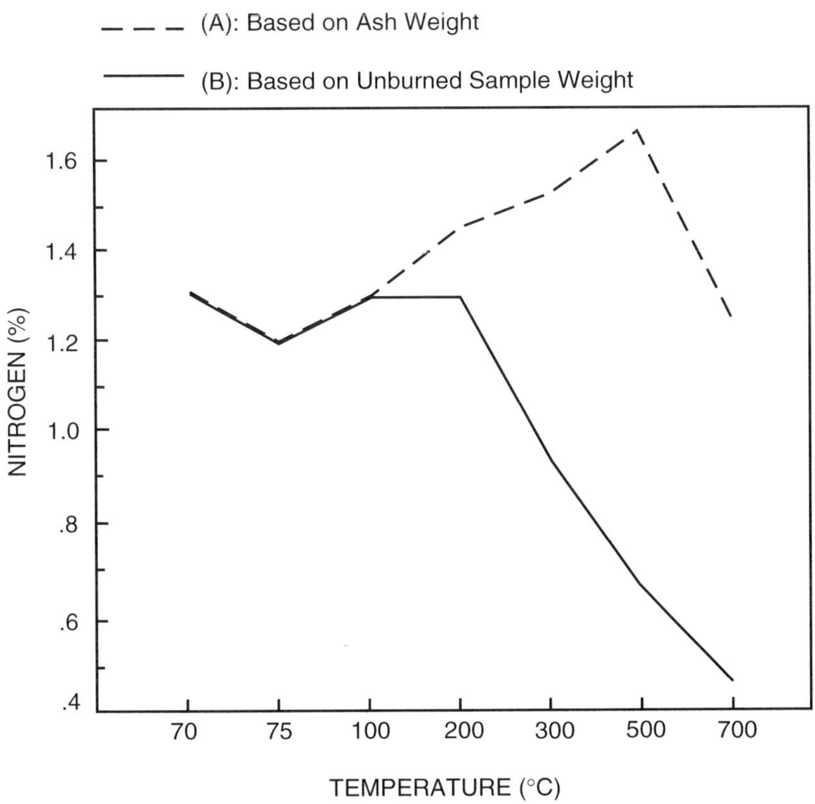

Nitrogen can increase, decrease, or stay the same depending on whether (A) the percent N is based on the weight of ash after burning or (B) percent N is based on the initial weight of the organic matter before burning. (Adapted from H. Knight, 1966. Loss of nitrogen from the forest floor by burning. *Forestry Chronicle* 42:149–152.) ■

Nitrogen compounds remaining after fire, particularly ammonium nitrogen (NH_4-N), are also increasingly available (Christensen 1973, DeBano et al. 1979, Carballas et al. 1993). This increased N availability enhances postfire plant growth and gives the impression that more N is present after fire. However, studies have shown that this increased fertility decreases after a year or two.

NITROGEN LOSS BY VOLATILIZATION

Losses of N at different temperatures are well understood. Laboratory heating studies on total N in forest litter showed that little N is lost below 200°C

(Knight 1966, White et al. 1973), although above 100°C some secondary amide groups in proteins can be converted to ammonia and retained as NH_4 on the soil cation exchange sites provided by clays or humus materials (Russell et al. 1974). Relationships between N loss from plants and litter in relation to burning temperatures determined by laboratory studies are (White et al. 1973):

- Complete loss (100 percent) of N at temperatures above 500°C
- Nitrogen losses between 75 and 100 percent at temperatures of 400 to 500°C
- Nitrogen losses between 50 and 75 percent at temperatures of 300 to 400°C
- Nitrogen losses between 25 and 50 percent at temperatures of 200 to 300°C
- No N loss at temperatures below 200°C

The amount of total N volatilized during combustion is directly proportional to the amount of organic matter combusted, because almost all soil N is contained in organic matter (Raison et al. 1985a). Up to 99 percent of this volatilized N reverts to N_2 (DeBell and Ralston 1970). This relationship might not necessarily hold at lower temperatures (Grier 1975), because organic matter can also decompose without volatilizing N. In this case, N remains on the site in uncombusted fuels or as highly available NH_4-N in the soil (DeBano et al. 1979, Covington and Sackett 1986, Kutiel and Naveh 1987, DeBano 1991).

Potential N losses are not fully realized when fire burns in natural ecosystems, because irregular burning patterns occur on the landscape and overall combustion is less efficient (see Chapter 2). For example, only 10 percent of the total N from plant, litter, and upper soil layers was lost during a prescribed burn in southern California chaparral (DeBano and Conrad 1978). A higher percentage loss occurred in aboveground fuels and litter on the soil surface. Other studies in dry and moist soils showed that about two-thirds of the total N is lost during burns over dry soils, contrasted to only 25 percent when the litter and soil are moist (DeBano et al. 1979). Nevertheless, N losses during fire can be important because N is most likely to be limiting in many natural ecosystems (Maars et al. 1983). Therefore, losses of N during a fire could adversely affect long-term productivity of many ecosystems, particularly if N replenishment mechanisms are not provided during postfire management.

NITROGEN LOSSES AND ECOSYSTEM PRODUCTIVITY

Although concentrations of N increase in soil following fire (see Example 6.2), total N is lost from the ecosystem. This loss of N is proportional to the amount of litter, duff, and biomass combusted. The effect of N losses during fire on ecosystem productivity depends on the proportion of total N lost for a given ecosystem. Even small losses of N by volatilization can impact long-term productivity in ecosystems that are N-limited (Example 6.3).

Example 6.3: Effect of N Loss During Fire on Site Productivity in Neotropical Forests

Over 95 percent of the prefire aboveground N and C pools and about 55 percent of the prefire aboveground P were lost by combustion in a severe fire in neotropical dry forests of northeastern Brazil (Kauffman et al. 1993). These losses amounted to over 500 kg/ha of N and 20 kg/ha of P. It would require over 100 years of fallow to replace these losses, although the fallow system used allows only 15 years between successive burns. At the other extreme, productive coniferous forest stands in western Oregon have lost in excess of 1.1 Mg/ha of N during stand-replacing fires, and the soils still maintained a productive forest stand (McNabb and Cromack 1990). ■

Site productivity can be related to the proportion of total N lost. To illustrate this point, the left side of Figure 6.3 represents large quantities of N on a site of high productivity. Moving to the right, both N capital and productivity decrease. This decrease is nonlinear, because there are likely to be greater losses in productivity per unit loss of N capital on sites having lower productivity (right side of Figure 6.3) than on sites having higher site productivity (left side of Figure 6.3). Therefore, losses in site productivity per unit N loss (a_h to b_h) from sites of high productivity (l_h) are less than losses in site productivity per unit N loss (a_i to b_i) from sites having low productivity (l_i). One implication of this relationship is that attention must be given to replen-

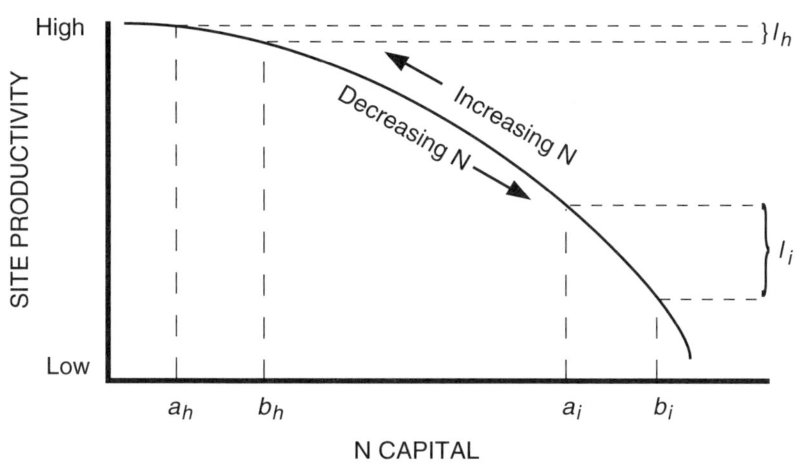

FIGURE 6.3 Relationship between nutrient loss and site productivity, with N capital increasing from right to left. (Adapted from Barnett 1989.)

ishing N lost as a result of fire on low-productivity sites. Similar relationships likely exist for P, S, and organic matter, although these relationship have not been studied. Cations are usually not limiting nutrients in natural ecosystems.

NITROGEN REPLACEMENT

Mechanisms that replace N in the soil must be considered in postfire management, particularly on sites where N is limiting. Nitrogen additions to the soil come from several sources including:

- Nitrogen present in precipitation and dust
- Fixation of atmosphere N_2 gas into usable forms by soil- and root-inhabiting microorganisms (see Chapter 7)
- Mineral and organic fertilizers

Atmospheric inputs are usually small in relation to other inputs, ranging from 1 to 4 kg/ha/yr in forests of the Pacific Northwest (Barnett 1989). Nitrogen inputs by precipitation in areas of high air pollution, such as the Los Angeles basin in southern California, can reach 8.2 kg/ha/yr (Riggan et al. 1985).

Phosphorus and Sulfur Cycling and Losses

Phosphorus is limited in some natural ecosystems. Deficiencies of P are most likely to appear in P-fixing soils (Vlamis et al. 1955) or in conjunction with N fertilization applications (Heilman and Gessel 1963). Phosphorus uptake and availability to plants appear dependent on the relationship between mycorrhizae and organic matter rather than involving a simple absorption from the soil solution (Trappe and Bollen 1979). Phosphorus responds differently to temperature increases during soil heating than N, with only about 60 percent of the total P being lost by nonparticulate transfer when organic matter is totally combusted (Raison et al. 1985a). As a result, relatively large amounts of highly available P can be found in the ash and on the soil surface immediately following fire. The highly available P can quickly be immobilized in insoluble compounds if calcareous materials are present in the ash.

The role of S in ecosystem productivity is unclear, although its fluctuations in the soil appear to parallel those of inorganic N. Sulfur is the second most limiting nutrient in some coastal forest soils of the Pacific Northwest, particularly when forest stands are fertilized with N (Barnett 1989). Percent loss of S by volatilization is intermediate to N and P (Tiedemann 1987), and burning has been reported to remove 20 to 40 percent of the S in aboveground biomass (Barnett 1989). Most important, however, neither S nor P (unlike N) can be fixed by biological processes, but instead they are added primarily by burning fossil fuels (a source of acid rain), as fallout from volcanic eruptions, or by weathering of rocks during soil development.

Estimating Nitrogen and Phosphorus Losses

A critical step in assessing the effect of fire on the soil chemical system is estimating losses of plant nutrients, particularly N. The three-step procedure for assessing fire effects described in Chapter 4 can be used to illustrate a procedure for estimating N and P losses from a stand of chaparral shrubs in central Arizona (Example 6.4). This procedure requires information on N and P for plant parts, litter, and soil, along with soil temperatures occurring at different fire severities. This procedure can also be used to estimate nutrient change for other nutrients and ecosystems, if information is available on the nutrient content of different organic components and the temperatures or severities of burning. A body of information describing the chemical composition of different plants and litter is found in the literature for a wide range of ecosystems. This information can be used to approximate the amounts of N or other nutrients that can be found in different plant and litter materials.

Example 6.4: Estimating the N and P Loss from a Chaparral Ecosystem in Central Arizona (DeBano 1990)

Biomass and soil nutrient data were collected as part of a research–application program. The distribution of biomass, N, and P contained in aboveground vegetation and soil (to 20-cm depth) are presented in the table presented below. Data on amount of biomass and nutrients (N and P) in the table below are used along with information on soil temperatures (see Figure 5.3); organic matter losses from the canopy, litter, and soil at different severities (see Chapter 5); and temperatures at which N and P are lost to develop the following scenario describing N and P losses.

Biomass and Nutrient Distribution in a Shrub Live Oak–Mixed Shrub Community at Battle Flat on the Prescott National Forest in Arizona

		Nutrient			
		N		*P*	
Ecosystem Component	***Biomass (Mg/ha)***	*(%)*	*(kg/ha)*	*(%)*	*(kg/ha)*
Leaves	4.06	1.82	74	0.15	6
Annual growth	1.21	1.05	13	0.12	1
Large and small stems	23.79	0.53	126	0.03	7
Litter	34.00	0.75	255	0.10	34
Soil					
–0–10 cm		0.11	13,000	0.009	250
10–20 cm		0.09	10,000	0.010	250

Source: Adapted from DeBano (1990).

- A low-severity fire would be expected to cause about a 40 percent reduction in canopy biomass and result in only an irregular spotty burning of surface litter, with perhaps 10 to 15 percent of the litter lost (see Chapter 3). If leaves and the current growth are totally combusted and large and small stems only partially combusted, the total loss of 11.63 Mg/ha in canopy biomass would be distributed as 4.06 Mg/ha in leaves, 1.21 Mg/ha in current growth, and 6.35 Mg/ha in large and small stems (see table below).

Loss of Biomass, N, and P from Vegetation, Litter, and Soil During Low-, Moderate-, and High-Severity Fires in Arizona Chaparral

Ecosystem Component	Biomass Loss (Mg/ha)			Nutrient Loss (kg/ha)					
				N			P		
	Low	Moderate	High	Low	Moderate	High	Low	Moderate	High
Vegetation	11.63	17.44	26.16	121	151	188	5	8	9
Litter	5.10	17.00	30.60	38	128	230	3	10	18
Soil				0	65	130	0	0	2
Total				159	344	548	8	18	29

Source: Adapted from DeBano (1990).

Associated with the combustion of 11.63 Mg/ha of canopy biomass would be about 121 kg/ha of N loss, assuming N loss is in direct proportion to total biomass combustion, and 5 kg/ha of P loss, assuming only a 60 percent loss of P upon total combustion. The 100 and 60 percent losses of N and P, respectively, upon total combustion of biomass, were reported by Raison et al.(1985a). In addition to the N lost from standing plants, 15 percent of litter N (38 kg/ha) and about 9 percent of litter P (3 kg/ha) would be lost. Maximum soil temperatures would be about 225°C at the surface and 125°C at 2.5 cm in the mineral soil (see Figure 5.3A). Soil temperatures 5.0 cm deep in the soil would remain at about ambient temperature. No additional losses of N and P would be expected from the soil at these temperatures.

- Most of the litter is likely to be charred during a moderate-severity burn, but not ashed. Between 40 and 80 percent of the plant canopy is consumed by the fire, and the remaining charred twigs would be greater than 0.6 to 1.3 cm in diameter. Under these conditions, it is assumed that 60 percent (17.44 Mg/ha) of the canopy and 50 percent (17.00 Mg/ha) of the litter would burn. Associated with the 60 percent combustion of the canopy would be a loss of about 151 kg/ha of N and 8 kg/ha of P. An additional 128 kg/ha of N and 10 kg/ha of P would be lost during combustion of 50 percent of the surface litter. Soil temperatures would reach about 430°C at the surface and 200°C at 2.5 cm below the soil surface (see Figure 5.3B). Moderately sensitive soil properties, including organic matter, N, S, soil structure, and soil wettability, would all be affected near the soil surface. Relatively large losses of N from the soil organic matter could also occur. It is estimated that 5 percent of the 1300 kg/ha of N in the upper soil layer (or 65 kg/ha) would be lost, mostly from the upper centimeter of soil. Soil P would not be lost, because the maximum surface temperature (425°C) is below the threshold temperature for loss of P (774°C). Only microbiological properties would be affected at 2.5 cm below the soil surface. At 5 cm in the soil, temperatures would reach only 70°C and, therefore, would have no effect on soil chemical and physical properties and a minimal effect on biological properties.

- Only ash remains on the soil surface after a high-severity fire, and the only plant stems remaining are 1.3 cm or greater in diameter. Only charred remains of large plant stems remain after fire in many cases. Under these intense burning conditions, it is assumed that at least 90 percent of both the plant canopy and surface litter would be consumed. Large losses of N (188 kg/ha) and P (9 kg/ha) from the plant biomass would occur, with greater losses from the litter (230 kg/ha of N and 18 kg/ha of P). Soil heating would be severe, reaching 700°C at the soil surface, 250°C at 2.5 cm, and 100°C at 5.0 cm (see Figure 5.3C). Near the soil surface, severe damage could occur to all soil properties except K, P, Ca, Mn, and soil clays. Large amounts of N, perhaps 10 percent (130 kg/ha), and small amounts of P could be lost. The soil surface would most likely be sterilized, with no viable microorganisms or seeds surviving in the upper 5 to 6 cm of soil.

It is estimated that the total N lost from plants and litter during low-, moderate-, and high-severity burns would be about 159, 344, and 548 kg/ha, respectively, while P losses from plants and litter would be 8, 18, and 29 kg/ha under low-, moderate-, and high-severity burns, respectively. These estimates generally agree with losses measured during fires in southern California chaparral, where 149 kg/ha of N was lost during a prescribed burn when about 30 percent of the plants were consumed (a low severity burn) (DeBano and Conrad 1978). The amounts of N and P lost during burning will vary depending on the amount of aboveground biomass, litter, and soil organic matter combusted during the fire.

The calculations made in this example represent only changes under the plant canopy and, therefore, must be reduced in direct proportion to the amount of interspace area between shrub plants when estimating total nutrient losses on an area basis. However, the calculations of N and P losses illustrate the importance of burning chaparral as coolly as possible to reduce consumption of surface litter and soil heating. It is also important to be aware of these nutrient losses to ensure that N restoration to a site is included as part of postfire management. ■

NUTRIENT AVAILABILITY

Major processes affecting nutrient availability that are changed by soil heating are (Raison et al. 1993):

- Transformation of nutrients from organic to inorganic forms (mineralization)
- Volatilization losses to the atmosphere in smoke (as particulate and gaseous forms)
- Erosion of ash and nutrient-rich surface soil
- Changes in N-fixing systems
- Modification of the decomposition process of litter and soil organic matter

Translocation of some nutrients downward in the soil along temperature gradients can also occur (DeBano et al. 1976). While short-term effects of fire on nutrient availability are reasonably well known, less is understood about longer-term fire effects.

Heating the underlying mineral soil directly affects the availability of nutrients contained in the soil organic matter (*in situ* changes). Little change is expected to occur in nutrient availability deeper than 4 to 5 cm below the soil surface, unless a severe fire occurs.

Nitrogen

Although large amounts of N are lost during the combustion of plants, litter, and duff, available NH_4-N is usually higher in the underlying soil (DeBano et al. 1979). Increases in NH_4-N immediately following a fire appear related to the soil temperatures reached. For example, most of the soil N is probably volatilized by high-severity fires, particularly on, or near, the soil surface, and only small amounts are transferred downward in the soil. Conversely, large amounts of NH_4-N can be found in the ash and underlying soil after low-severity fires (DeBano 1991). Therefore, concentrations of NH_4-N can increase, decrease, or remain the same depending on the severity and duration of a fire.

Nitrogen availability can also be increased by the translocation of N compounds downward into the soil. The translocation can happen simultaneously with the movement of organic compounds responsible for water repellency (see Chapter 5). This occurs because large temperature gradients are created in the upper soil layers during combustion of litter and duff on the soil surface. Therefore, some of the vaporized organic matter and NH_4-rich nitrogenous compounds that are released during combustion are transferred downward, where they condense on cooler underlying soil particles (DeBano et al. 1976). Release of NH_4-N involves the decomposition of proteins and other nitrogen-rich organic matter (Example 6.5).

Example 6.5: Formation of Available NH_4-N During Thermal Decomposition of Organic Matter

One product from the combustion of organic matter—one that affects post-fire fertility—is highly available NH_4-N (Christensen 1973, Jurgensen et al. 1981, Kovacic et al. 1986, Kutiel and Naveh 1987, Marion et al. 1991, Knoepp and Swank 1993). Ammonium-N appears to be formed during decomposition of secondary amide groups and amino acids. Russell et al. (1974) showed that secondary amide groups in protein-like components of a soil clay–organic complex decomposed above 100°C to yield NH_4-N. Other volatile nitrogenous decomposition products were released at temperatures above 400°C. Clay surfaces appeared to catalyze decomposition, because the thermal decomposition of amino acids was found to be enhanced by combining a mixture of amino acids with a clay loam soil (Juste and Dureau 1967). Mixing of these amino acids with pure sand did not increase decomposition. ■

The effect of repeated burning on N availability is variable and depends largely on the ecosystem, its inherent fertility, and its ability to replenish N. The appropriate frequency of burning, when prescribed, is determined from the amount of organic matter destroyed and the rates of replacing N volatilization losses. The effect of repeated low-intensity burning in Australian eucalypt forests suggests that fire-free periods of about 10 years or more are required to allow natural processes to replace N lost during burning, if volatilization of N is limited to approximately 50 percent of that in the fuel (Raison et al. 1993). Ponderosa pine forests in northern Arizona that were burned repeatedly at 1-, 2-, and 4-year intervals showed no significant differences in total N, but available N (NH_4-N and nitrate (NO_3)-N) was higher on sites repeatedly burned, in comparison to unburned controls (Covington and Sackett 1986). It was concluded, therefore, that frequent periodic burning can be used to enhance N availability in southwestern ponderosa pine stands.

Annual burning of tallgrass prairies in the Great Plains of the central United States resulted in greater inputs of lower quality plant residues, causing a significant reduction in soil organic N, lower microbial biomass, lower N availability, and higher C:N ratios in soil organic matter (Ojima et al. 1994). Addition of available N, however, might not be desirable in all ecosystems. Burned lowland fynbos in South Africa showed that twofold increases in soil nutrient concentrations can be detrimental to survival of indigenous species adapted to a nutrient-impoverished environment (Musil and Midgley 1990)

Phosphorus

Phosphorus is not usually translocated downward in a soil profile as readily as N compounds. Instead, increases of available P after a fire are found mainly in the ash and on, or near, the soil surface (DeBano and Klopatek 1988). The fate of P following fire is also mediated by biogeochemical processes (inorganic adsorption) rather than biochemical (nitrification).

Prefire P is contained in organic matter and inorganic minerals. Organic P is slowly converted during decomposition into forms available for plant and microbial growth. The phosphate released during decomposition is mediated by the phosphatase enzyme (Saa et al. 1993). Fire interrupts these slow decomposition processes and rapidly releases P in the form of a highly available phosphate ion. Fire not only destroys the organic matter, but it also inactivates the phosphatase enzyme at 180°C.

Availability of P released during a fire is complicated by the direct effects of fire and its continued availability depends on postfire environmental conditions. The continued availability of P following fire is highly dependent on soil pH and the presence of soluble Ca, Fe, and Al compounds. Phosphate is most available in neutral and slightly acidic soils, where it is associated with Fe and Al hydroxides. Phosphorus can be "strongly" combined with Ca to form insoluble apatite-like precipitates in alkaline soils having high concentrations of Ca (Singer and Munns 1996).

The most reliable source of P is the organic P that is released slowly during decomposition. Therefore, management of P resources on a site involves minimizing loss of litter and soil organic matter during a fire if possible by burning at lower severities. Phosphorus management is an important component in the long-term productivity of burned Arizona chaparral stands (Wienhold and Klemmedson 1992).

CATIONS

Ash deposited on the soil surface contains high concentrations of available cations. Fire increases the availability of all cations, including NH_4-N (Marion et al. 1991). The amount of NH_4-N released by burning depends on the fuel load and quantity of fuel burned (Tomkins et al. 1991). Some of the cations can be lost through particulate transfer in the smoke (Clayton 1976). Monovalent cations, such as Na and K, are present largely as chlorides and carbonates that are readily mobilized (Soto and Diaz-Fierros 1993). Divalent ions, including Ca and Mg, are less mobile and present as oxides and carbonates. Calcium carbonate can play an important postfire role in limiting the availability of P, as discussed earlier.

Major cations (for example, Ca, Mg, and K) are not usually deficient in most wildland soils (DeBano 1991). Also, these cations have high temperature thresholds and, therefore, are not easily volatilized (see Table 4.1) and lost from burned sites. As a consequence, large quantities of the available cations remain on the soil surface following fire. Although these readily available cations probably do not enhance plant growth directly, their amount and composition determine base saturation, which plays an important role in controlling pH regimes in soils as described next.

OTHER CHEMICAL PROPERTIES

Colloidal organic matter in soil affects soil chemical properties, such as cation exchange capacity and buffer capacity (Tate 1987). Destruction of humus and soil organic matter during combustion reduces cation exchange capacity and buffering capacity of the organic component in a soil. Reductions in exchange capacity are proportional to that amount of the total cation capacity that is provided by the organic component.

Soil pH usually increases as a result of burning because basic cations released during combustion are deposited on the soil surface. Therefore, soil pH is usually raised temporarily depending on the original soil pH, amount of ash released, chemical composition of the ash, and wetness of the climate (Wells et al. 1979).

Major cations are not normally deficient in most soils in natural ecosystems (DeBano 1991). These cations have high temperature thresholds and, therefore, are not easily volatilized (see Table 4.1) and lost from the burned site.

ASH BED EFFECT

The combined response of physical, chemical, and biological soil properties to fire is sometimes referred to as the *ash bed effect* (Hatch 1960, Pryor 1963, Humphreys and Lambert 1965, Renbuss et al. 1972). As such, ash formation and deposition need to be considered within the context of physical, chemical, and biological responses. Soil responses can occur as a result of a fire itself (soil heating), the residual effect of the ash deposited on the soil surface, or a combination of both (Raison 1979). Many of the nutrient loss and availability responses discussed in this chapter result from the formation and deposition of ash during a fire.

The amount and type of ash remaining after fire depend largely on the nature of fuels combusted (see Chapter 3), fuel densities (packing ratios), fuel moisture content, total amount of the fuel load consumed, and severity of burning (Gillon et al. 1995). Ash remaining after a fire can range from small amounts of charred dark-colored fuel residues to thick layers of white ash several centimeters thick (Example 6.6). Complete combustion of large amounts of densely packed fuels can produce large amounts of residual white ash in one place on the soil surface following burning. Soil mineral particles will become reddish in color in extreme cases of soil heating and can even be fused together because of extremely high temperatures that have been maintained for long periods of time on the soil surface. The ash bed effect frequently occurs on sites where accumulations of fuels resulting from piling slash after timber harvesting have been burned.

Example 6.6: Deposition of Ash and Nutrients by Burning in Forest Ecosystems

The amount of ash and elements in the ash deposited on the soil surface after fire can vary widely in different forest ecosystems as shown in the table below

Mass of Ash and Elements in Ash Deposited on the Soil Surface After Fire in Different Forest Ecosystems

	Ash	*N*	*P*	*K*	*Ca*	*Mg*
Forest and Fire Type[a]	***(Mg/ha)***			***(kg/ha)***		
1. Tropical rainforest, slashburn	—	—	26	751	1582	190
2. Tropical rainforest, slashburn	7.5[b]	75	19	89	131	68
3. Mixed eucalypt rain forest, slashburn	12.2	—	46.6	251	712	89
4. Young tropical rain forest, slashburn	6.7	96	15.6	190	555	106
5. *Pinus radiata* windrow, slashburn	12.9[b]	87	20	79	272	68
6. Mixed coniferous forest, wildfire	2.9	23	—	70	314	54
7. *Eucalyptus pauciflora*, low-intensity fire	1.9[b]	20	2.6	6	35	9
8. *Eucalyptus delegatensis*, low-intensity fire	3.4[b]	26	2.9	7	40	9

[a]Cited in Raison et al. (1990): (1) Nye and Greeland 1960; (2) Alegre et al. 1988; (3) Harwood and Jackson 1975; (4) Ewel et al. 1981; (5) Flinn et al. 1979; (6) Grier 1975; (7) and (8) Raison et al. 1985b.

[b]Contains some partly combusted material.

Source: Adapted from R. J. Raison, H. Keith, and P. K. Khanna. 1990. Effects of fire on the nutrient supplying capacity of forest sails. In: W. J. Dyck and C. A. Mees, editors. *Impact of Intensive Harvesting on Forest Site Productivity.* FRI Bulletin No. 159, Rotorua, New Zealand, pp. 39–54. Copyright © 1990, Forest Research Institute.

(Raison et al. 1990). Ash deposits are usually greatest after the burning of concentrated fuels (piled slash and windrows) and least after low-intensity fires. Calcium is usually the dominant cation in the ash. Most of the cations are leached into the soil, where they are retained on the cation exchange sites located on clay or humus particles. ■

Effects of ash, soil heating, and the ash–heat interaction on soil respiration were studied in two Australian soils (Raison and McGarity 1980). Large amounts of ash slightly decreased respiration and small amounts had no effect. Additions of ash to sterilized soil produced no effect, indicating that ash acts via its influence on active soil biological populations. The chemical nature of ash affected soil respiration through its effect on:

- Increasing pH
- Changing the solubility in water of organic matter and associated minerals
- Addition of available nutrients for microbial populations

SUMMARY

Soil organic matter is an important soil chemical property that is affected by fire. Combustion of organic matter during soil heating affects both total nutrient pools and the cycling of nutrients and their eventual availability. Nitrogen and phosphorous are key nutrients in natural ecosystems that are affected by fire. Once you have read this chapter, you should be able to:

1. Describe the ecological importance of organic matter in ecosystems and its response to fire.
2. Develop a procedure for estimating site-specific losses of N using information on the amount of N in different ecosystem components and severity of burning.
3. Evaluate the effects of total N loss on low- and high-productivity ecosystems.
4. Explain how N concentrations can appear to increase, decrease, or remain the same after fire.
5. Describe how nutrients are made more available as a result of burning.
6. Explain why large N losses from some forest sites during a fire may not affect long-term productivity.
7. Explain the response of P to burning.

REFERENCES

Alegre, J. C., D. K. Cassel, and D. E. Bandy. 1988. Effect of land clearing method on chemical properties of an ultisol in the Amazon. *Soil Science Society of America Journal* 52:1283–1288.

Almendros, G., F. J. Gonzalez-Vila, and F. Martin. 1990. Fire-induced transformation of soil organic matter from an oak forest: an experimental approach to the effects of fire on humic substances. *Soil Science* 149:158–168.

Anderson, J. M. 1991. The effects of climate change on decomposition processes in grassland and coniferous forest. *Ecological Applications* 1:326–347.

Barnett, D. 1989. *Fire Effects on Coast Range Soils of Oregon and Washington and Management Implications.* USDA Forest Service, Soils R-6 Technical Report, Portland, Oregon.

Boerner, R. E. J. 1982. Fire and nutrient cycling in a temperate ecosystem. *Bioscience* 32:187–192.

Brown, G. W. 1980. *Forestry and Water Quality.* Oregon State University Bookstores, Corvallis, Oregon.

Carballas, M., M. J. Acea, A. Cabaneiro, C. Trasar, M. C. Villar, M. Diaz-Ravina, I. Fernandez, A. Prieto, A. Saa, F. J. Vazquez, R. Zehner, and T. Carballas. 1993. Organic matter, nitrogen, phosphorus and microbial population evolution in forest humiferous acid soils after wildfires. In: Trabaud, L., and P. Prodon, editors. *Fire in Mediterranean Ecosystems.* Commission of the European Countries, Ecosystem Research Report 5, Brussels, Belgium, pp. 379–385.

Chandler, C., P. Cheney, P. Thomas, L. Trabaud, and D. Williams. 1983. *Fire in Forestry: Volume I. Forest Fire Behavior and Effects.* John Wiley & Sons, New York.

Christensen, N. L. 1973. Fire and the nitrogen cycle in California chaparral. *Science* 181:66–68.

Clayton, J. L. 1976. Nutrient gains to adjacent ecosystems during a forest fire: an evaluation. *Forest Science* 22:162–166.

Covington, W. W., and S. S. Sackett. 1986. Effect of periodic burning on soil nitrogen concentrations in ponderosa pine. *Soil Science Society of America Journal* 50:452–457.

DeBano, L. F. 1990. Effects of fire on the soil resource in Arizona chaparral. In: Krammes, J. S., technical coordinator. *Effects of Fire Management of Southwestern Natural Resources.* USDA Forest Service, General Technical Report RM-191, pp. 65–77.

DeBano, L. F. 1991. The effect of fire on soil properties. In: Harvey, A. E., and L. F. Neuenschwander, compilers. *Proceedings: Management and Productivity of Western-Montane Forest Soils.* USDA Forest Service, General Technical Report INT-280. pp. 151–156.

DeBano, L. F., and C. E. Conrad. 1976. Nutrients lost in debris and runoff water from a burned chaparral watershed. *Federal Inter-agency Sedimentation Conference Proceedings* 3:13–27.

DeBano, L. F., and C. E. Conrad. 1978. The effect of fire on nutrients in a chaparral ecosystem. *Ecology* 59:489–497.

DeBano, L. F., and J. M. Klopatek. 1988. Phosphorus dynamics of pinyon-juniper soils following simulated burning. *Soil Science Society of America Journal* 52:271–277.

DeBano, L. F., S. M. Savage, and D. A. Hamilton. 1976. The transfer of heat and hydrophobic substances during burning. *Soil Science Society of America Journal* 40:779–782.

DeBano, L. F., G. E. Eberlein, and P. H. Dunn. 1979. Effects of burning on chaparral soils. I. Soil nitrogen. *Soil Science Society of America Journal* 43:504–509.

DeBell, D. S., and C. W. Ralston. 1970. Release of nitrogen by burning light forest fuels. *Soil Science Society of America Proceedings* 34:936–938.

Ewel, J., C. Berish, B. Brown, N. Price, and J. Raich. 1981. Slash and burn impacts on a Costa Rican wet forest site. *Ecology* 62:816–829.

Flinn, D. W., P. Hopmans, P. W. Farrell, and J. M. James. 1979. Nutrient loss from burning of *Pinus radiata* logging residue. *Australian Forest Research* 9:17–23.

Gillon, D., V. Gomendy, C. Houssard, J. Marechal, and J. C. Valette. 1995. Combustion and nutrient losses during laboratory burns. *International Journal of Wildland Fire* 5:1–12.

Grier, C. C. 1975. Wildfire effects on nutrient distribution and leaching in a coniferous ecosystem. *Canadian Journal of Forestry Research* 5:599–607.

Harvey, A. E. 1982. The importance of residual organic debris in site preparation and amelioration for reforestation. In: Baumgartner, D. M., editor. *Site Preparation and Fuels Management on Steep Terrain: Symposium Proceedings.* Washington State University, Spokane, Washington, pp. 75–85.

Harvey, A. E., M. F. Jurgensen, M. J. Larsen, and R. T. Graham. 1987. *Decaying Organic Materials and Soil Quality in the Inland Northwest: A Management Opportunity.* USDA Forest Service, General Technical Report INT-225.

Harwood, C. E., and W. D. Jackson. 1975. Atmospheric losses of four plant nutrients during a forest fire. *Australian Forestry* 38:92–99.

Hatch, A. B. 1960. Ash-bed effects in western Australian forest soils. Forestry Department Western Australia, Bulletin 64.

Heilman, P. E., and S. P. Gessel. 1963. Nitrogen requirements and the biological cycling of nitrogen in Douglas-fir stands in relation to effects of nitrogen fertilization. *Plant and Soil* 18:386–402.

Hosking, J. S. 1938. The ignition at low temperatures of the organic matter in soils. *Journal of Agricultural Science* 28:393–400.

Humphreys, F. R., and M. J. Lambert. 1965. An examination of a forest site which has exhibited the ash-bed effect. *Australian Journal of Soil Research* 3:81–94.

Jurgensen, M. F., A. E. Harvey, and M. J. Larsen. 1981. Effects of prescribed fire on soil N levels in a cutover Douglas-fir/western larch forest. USDA Forest Service, Research Paper INT-275.

Juste, C., and P. Dureau. 1967. Production of ammonia nitrogen by thermal decomposition of amino acids with a clay-loam soil. *Comptes Rendus Hebdomadaires des Seances de l'Academie des Sciences Serie D: Sciences Naturelles* 265:1167–1169.

Kauffman, J. B., R. L. Sanford, Jr., D. L. Cummings, I. H. Salcedo, and E. V. S. B. Sampaia. 1993. Biomass and nutrient dynamics associated with slash fires in neotropical dry forests. *Ecology* 74:140–151.

King, N. K., D. R. Packham, and R. G. Vines. 1977. On the loss of selenium and other elements from burning forest litter. *Australian Forestry Research* 7:265–268.

Knight, H. 1966. Loss of nitrogen from the forest floor by burning. *Forestry Chronicle* 42:149–152.

Knoepp, J. D., and W. T. Swank. 1993. Site preparation burning to improve southern Appalachian pine-hardwood stands: nitrogen responses in soil, water, and streams. *Canadian Journal of Forest Research* 23:2263–2270.

Kovacic, D. A., D. M. Swift, J. E. Ellis, and T. E. Hakonson. 1986. Immediate effects of prescribed burning on mineral soil nitrogen in ponderosa pine of New Mexico. *Soil Science* 141:71–76.

Kutiel, P., and Z. Naveh. 1987. Soil properties beneath *Pinus halepensis* and *Quercus calliprinos* trees on burned and unburned mixed forest on Mt. Carmel, Israel. *Forest Ecology and Management* 20:11–24.

Maars, R. H., R. D. Roberts, R. A. Skeffinton, and A. D. Bradshaw. 1983. Nitrogen in the development of ecosystems. In: Lee, J. A., S. McNeill, and I. H. Rorison, editors. *Nitrogen as an Ecological Factor.* Blackwell Science Publishing, Oxford, England, pp. 131–137.

Marion, G. M., J. M. Moreno, and W. C. Oechel. 1991. Fire severity, ash deposition, and clipping effects on soil nutrients in chaparral. *Soil Science Society of America Journal* 55:235–240.

McGill, W. B., and C. V. Cole. 1981. Comparative aspects of cycling of organic C, N, S, and P through soil organic matter. *Geoderma* 26:267–286.

McNabb, D. H., and K. Cromack, Jr. 1990. Effects of prescribed fire on nutrients and soil productivity. In: Walstad, J. D., S. R. Radosevich, and D. V. Sandberg, editors. *Natural and Prescribed Fire in Pacific Northwest Forests.* Oregon State University Press, Corvallis, Oregon, pp. 125–142.

Musil, C. F., and G. F. Midgley. 1990. The relative impact of invasive Australian acacias, fire and season on the soil chemical status of a sand plain lowland fynbos community. *South African Journal of Botany* 56:419–427.

Nye, P. H., and D. J. Greeland. 1960. The soil under shifting cultivation. Commonwealth Bureau of Soils Technical Communication No. 51, Farnham Royal, United Kingdom.

Ojima, D. S., D. S. Schimel, W. J. Parton, and C. E. Owensby. 1994. Long- and short-term effects of fire on nitrogen cycling in tallgrass prairie. *Biogeochemistry* 24:67–84.

Pryor, L. D. 1963. Ash-bed growth response as a key to plantation establishment on poor sites. *Australian Forestry* 27:48–51.

Raison, R. J. 1979. Modification of the soil environment by vegetation fires, with particular reference to nitrogen transformations: a review. *Plant and Soil* 51:73–108.

Raison, R. J., and J. W. McGarity. 1980. Effects of ash, heat, and the ash–heating interaction on biological activities in two contrasting soils. I. Respiration rate. *Plant and Soil* 55:363–376.

Raison, R. J., P. K. Khanna, and P. V. Woods. 1985a. Mechanisms of element transfer to the atmosphere during vegetation fires. *Canadian Journal of Forest Research* 15:132–140.

Raison, R. J., P. K. Khanna, and P. V. Woods. 1985b. Transfer of elements to the atmosphere during low-intensity prescribed fires in three Australian sub-alpine eucalypt forests. *Canadian Journal of Forest Research* 15:657–664.

Raison, R. J., H. Keith, and P. K. Khanna. 1990. Effects of fire on the nutrient supplying capacity of forest soils. In: Dyck, W. J., and C. A. Mees, editors. *Impact of Intensive Harvesting on Forest Site Productivity.* FRI Bulletin No. 159, Rotorua, New Zealand, pp. 39–54.

Raison, R. J., A. M. O'Connell, P. K. Khanna, and H. Keith. 1993. Effects of repeated fires on nitrogen and phosphorus budgets and cycling processes in forest ecosystems. In: Trabaud, L., and P. Prodon, editors. *Fire in Mediterranean Ecosystems.* Commission of the European Countries, Ecosystem Research Report 5, Brussels, Belgium, pp. 347–363.

Renbuss, M. A., G. A. Chilvers, and L. D. Pryor. 1972. Microbiology of an ashbed. *Proceedings of the Linnean Society of New South Wales* 97:302–310.

Riggan, P. J., R. N. Lockwood, and E. N. Lopez. 1985. Deposition and processing of airborne nitrogen pollutants in Mediterranean-type ecosystems of southern California. *Environmental Science and Technology* 19:781–789.

Russell, J. D., A. R. Fraser, J. R. Watson, and J. W. Parsons. 1974. Thermal decomposition of protein in soil organic matter. *Geoderma* 11:63–66.

Saa, A., M. C. Trasar-Cepeda, F. Gil-Sotres, and T. Carballas. 1993. Changes in soil phosphorus and acid phosphatase activity immediately following forest fires. *Soil Biology and Biochemistry* 25:1223–1230.

St. John, T. V., and P. W. Rundel. 1976. The role of fire as a mineralizing agent in a sierran coniferous forest. *Oecologia* 25:35–45.

Schnitzer, M., and I. Hoffman. 1964. Pyrolysis of soil organic matter. *Soil Science Society of America Proceedings* 28:520–525.

Singer, M. J., and D. N. Munns. 1996. *Soils: An Introduction.* Prentice Hall, Upper Saddle River, New Jersey.

Soto, B., and F. Diaz-Fierros. 1993. Interactions between plant ash leachates and soil. *International Journal of Wildland Fire* 3:207–216.

Tate, R. L. III. 1987. *Soil Organic Matter: Biological and Ecological Effects.* John Wiley & Sons, New York.

Tiedemann, A. R. 1987. Combustion losses of sulfur and forest foliage and litter. *Forest Science* 33:216–223.

Tomkins, I. B., J. D. Kellas, K. G. Tolhurst, and D. A. Oswin. 1991. Effects of fire intensity on soil chemistry in a eucalypt forest. *Australian Journal Soil Research* 29:25–47.

Trappe, J. M., and W. B. Bollen. 1979. Forest soil biology. In: Heilman, P. E., H. A. Anderson, and D. M. Baumgartner, compilers. *Forest Soils of the Douglas-Fir Region.* Washington State University Cooperative Extension Service, Pullman, Washington, pp. 145–151.

Turner, J. 1977. Effect of nitrogen availability on nitrogen cycling in a Douglas-fir stand. *Forest Science* 23:307–316.

Vlamis, J., H. H. Biswell, and A. M. Schultz. 1955. Effects of prescribed burning on soil fertility in second growth ponderosa pine. *Journal of Forestry* 53:905–909.

Wells, C. G., R. E. Campbell, L. F. DeBano, C. E. Lewis, R. L. Fredricksen, E. C. Franklin, R. C. Froelich, and P. H. Dunn. 1979. *Effects of Fire on Soil: A State-of-Knowledge Review.* USDA Forest Service, General Technical Report WO-7.

White, E. M., W. W. Thompson, and F. R. Gartner. 1973. Heat effects on nutrient release from soils under ponderosa pine. *Journal of Range Management* 26:22–24.

Wienhold, B. J., and J. O. Klemmedson. 1992. Effect of prescribed fire on nitrogen and phosphorus in Arizona chaparral soil-plant systems. *Arid Soil Research and Rehabilitation* 6:285–296.

Chapter

7

Biological Soil System

INTRODUCTION

Soils are alive with biological organisms. Large populations of microorganisms, roots and mycorrhizae, invertebrates, and burrowing animals inhabit the upper part of the soil profile (Singer and Munns 1996). Here, they collectively contribute to soil productivity by enhancing decomposition, N cycling, humus formation, soil physical properties (aggregation and soil structure), plant promulgation, and plant nutrition and stability.

This chapter presents background information on the kinds of living organisms present in the soil and the specific roles that soil microorganisms play in soil biological processes. This information is used as the basis for discussing fire effects on specific soil microorganisms. Effects of fire on other biological organisms living in the soil are also discussed, including effects on roots, mycorrhizae, cryptogamic crusts, invertebrates, and small (mostly burrowing) mammals.

ROLE OF BIOLOGICAL ORGANISMS

Soil organisms range from microbiota (such as algae, protozoa, fungi, bacteria, and cyanobacteria) to macrobiota (including insects, earthworms, and plant roots) (Figure 7.1). Microbiota are less than about 100 μm in body width, mesobiota are between 100 μm and 2 mm, and macrobiota are between 2 and 20 mm (Verhoef and Brussaard 1990). The respective sizes of

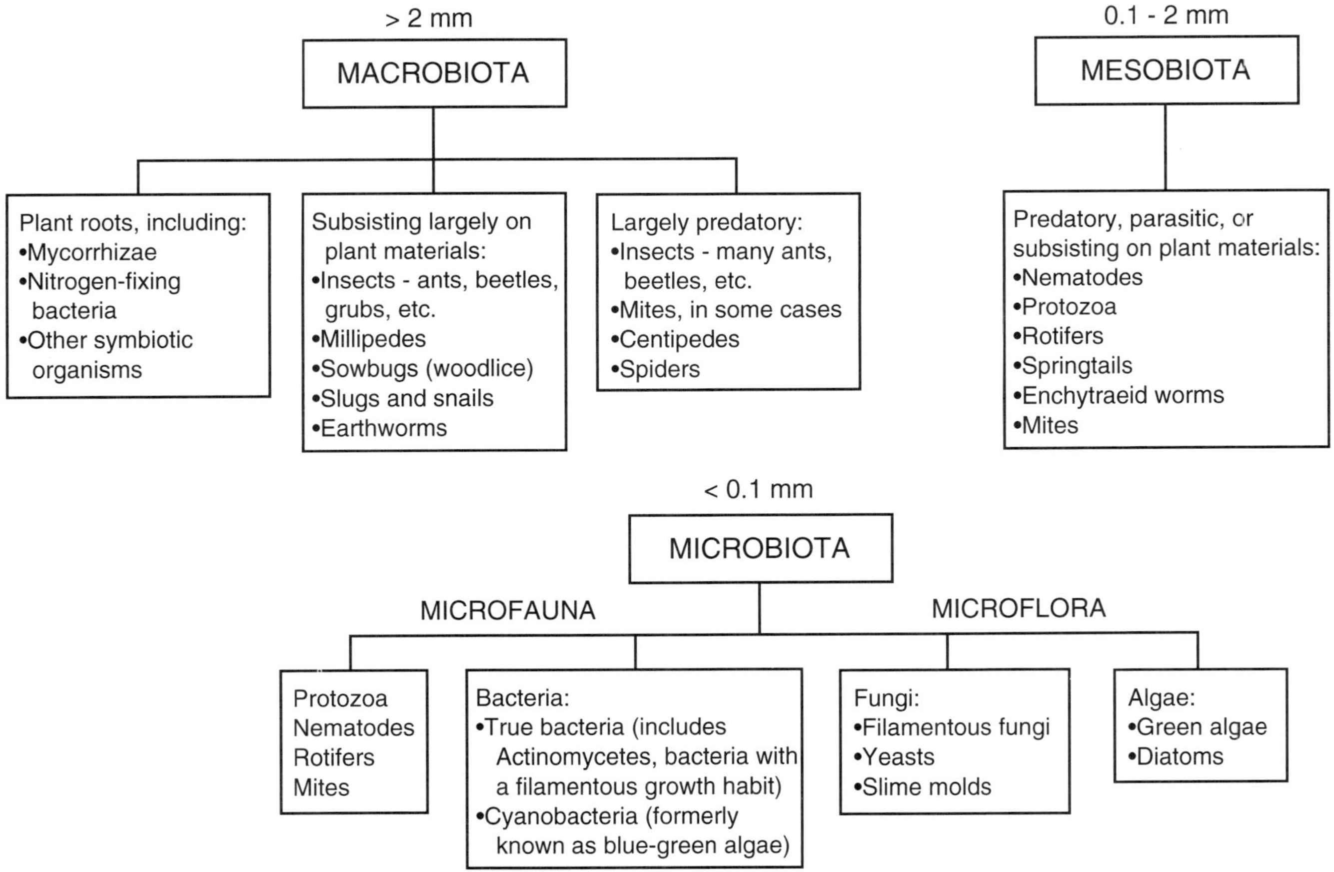

FIGURE 7.1. Types and sizes of soil organisms that can be affected by fire. (Adapted from Wallwork 1970).

biotic organisms have major roles in enhancing soil structure and nutrient cycling, with different groups having specialized functions (Table 7.1). Microbiota contribute large quantities of biomass to the organic matter component of soils. Bacteria and fungi alone can contribute nearly 5.0 Mg/ha of biomass to some forest soils (Bollen 1974).

Soil organisms inhabit a complex matrix of mineral soil particles, organic matter, air, and water (Borchers and Perry 1990). Clay particles and organic matter store nutrients (see Chapter 6) and are sites for root and microbial activity.

Organic matter, clays, and microorganisms are distributed unevenly below the soil surface. Generally, surface horizons of soils in natural ecosystems have a greater abundance and diversity of soil organisms, more organic matter, greater acidity, and a more aggregated structure than agricultural soils. Although soil organisms respond directly to variations in their environment, they also modify their environment physically and chemically and are a major soil-forming factor (Jenny 1980).

TABLE 7.1 Activities of Different Soil Biota During Nutrient Cycling and Formation of Soil Structure

	Nutrient Cycling	*Soil Structure*
Microbiota	Catabolize organic matter Mineralize and immobilize nutrients Regulate bacterial and fungal populations Alter nutrient turnover	Produce organic compounds that bind aggregates Hyphae entangle particles onto aggregates May affect aggregate structure through interactions between microfauna and microflora
Mesobiota	Regulate fungal and microfaunal populations Alter nutrient turnover Fragment plant residue	Produce fecal pellets Create biopores Promote humification
Macrobiota	Fragment plant residues Stimulate microbial activity	Mix organic and mineral particles Redistribute organic matter and microorganisms Create biopores Promote humification Produce fecal pellets

Source: (Adapted from P. F. Hendrix, D. A. Crossley, Jr., J. M. Blair, and D. C. Coleman. 1990. Soil biota as components of sustainable agroecosystems. In: C. A. Edwards, R. Lal, P. Madden, R. H. Miller, and G. House, editors. *Sustainable Agricultural Systems.* Copyright © 1990 Soil and Water Conservation Society, Ankeny, Iowa, pp. 637–654.)

BIOLOGICAL RESPONSES TO FIRE

Fire affects many biological organisms that interact with plants, animals, and the below-ground environment (Ahlgren and Ahlgren 1960, Borchers and Perry 1990). Soil microorganisms are concentrated in the surface *L* layer and *D* layer (see Chapter 4), because these soil layers contain large amounts of organic matter (see Chapter 6) and are active sites for decomposition and other microbial processes. Surface plant roots are frequently found throughout the *H* layer and upper part of the *A* horizon. Likewise, invertebrates are generally concentrated in the upper soil horizons.

Living organisms and nonliving organic matter in soils are located on, or near, the soil surface and, therefore, are exposed to heat radiated by flaming surface fuels and smoldering *L* and *D* layers. As a result, the thickness and the extent of combustion in the *D* layer are important factors affecting the survival of living organisms (see Chapter 2). It follows that their location in the soil and the susceptibility of biological organisms to heat damage are important considerations when assessing fire-related changes in biological systems. Depending on fire severity, organisms in the *L* layer and *D* layer can be killed outright, while those in the deeper soil horizons, or in isolated unburned locations, might survive. Even low-severity fires can damage organisms that are on or near the soil surface, because biological organisms are damaged at lower temperatures than those that cause changes in physical and chemical soil properties during fire (see Table 4.1).

Heat penetration into the soil during a fire affects the welfare of biological organisms located below the soil surface. Heat transfer into soil varies widely (see Chapter 2). Dry soils are relatively effective insulators to heat transfer and, as a consequence, protect organisms under the soil surface. Conversely, moist soils conduct heat rapidly until the maximum temperatures reach the boiling point of water (100°C), a temperature that is lethal to most organisms.

Fire affects soil organisms directly by killing or injuring the organisms, and indirectly by its effect on plant succession, soil organic matter transformations, and microclimate.

Direct Effects

Fire directly affects a wide range of microorganisms that carry out nutrient cycling processes. An important component of the soil biological system that can be affected by soil heating is the *rhizosphere,* a focal point of intense biological activity at the interface between plant roots and the surrounding soil. The energy stored in C-rich exudates and secretions produced by the rhizosphere supports populations of bacteria that fix N and release enzymes, growth hormones, antibiotics, or chelating compounds. Chelating compounds are organic molecules that bring into solution insoluble forms of certain nutrients (such as Fe), which might not otherwise be available to plants. The rhizosphere also includes *mycorrhizae*, which enhance nutrient uptake by plants and contribute directly to the productivity of terrestrial ecosystems.

Soil heating kills soil organisms, particularly microbiota, directly or by altering their reproductive capacity (Covington and DeBano 1990). Direct effects on microorganisms are also related to changes in the soil environment. For example, heterotrophic bacteria can be impacted by loss of their energy source during combustion of organic matter (Barbour et al. 1980). *Heterotrophs* are organisms that are able to derive C and energy for growth and cell synthesis by utilizing organic compounds (Soil Science Society of America 1997). Soil acidity can also decrease after fire in vegetative communities, affecting organisms that are sensitive to those changes. However, little change is likely to occur in neutral pH soils (Covington and DeBano 1990).

Indirect Effects

Indirect effects of fire on soil organisms are more complex than direct effects and can alter ecosystem processes for years. Disturbances that kill or injure living plants impact organisms that depend on plant products for energy, nutrients, and habitat, especially mycorrhizal fungi and organisms present in the rhizosphere (Borchers and Perry 1990). Similarly, combustion of coarse woody debris, forest floor duff, and soil organic matter can have long-term effects on site productivity, soil aggregation, and water relations in the soil.

NUTRIENT CYCLING PROCESSES

Soil organisms play a major role in nutrient cycling, where they participate in several processes, including:

- Decomposition of organic matter
- Nitrogen cycling, including mineralization, denitrification, and N fixation
- Nutrient uptake, for example, mycorrhizae in symbiosis with plant roots facilitating uptake of P and other nutrients

Decomposition of Organic Matter

The decomposition of organic matter is essential to the functioning and productivity of terrestrial ecosystems throughout the world. *Decomposition*, the breakdown of organic matter, results in catabolism of organic matter into smaller organic materials (Coleman and Crossley 1996). It is especially important because it contributes to humus formation in soils. Carbon is recirculated to the atmosphere as CO_2 during the decomposition of plant and animal residues, and N is stored in microbial biomass until it is released by mineralization. During decomposition, some of the C and N is immobilized in microbial tissue, and part is microbially converted to resistant humic substances (humification), which make up the largest percentage of the organic matter. Soil humus in natural ecosystems undergoes both continuous decomposition and mineralization, so that the total soil organic matter and N content may remain in a steady-state condition. Factors affecting decomposition are the composition of the decomposing litter (substrate quality), environmental factors (moisture and temperature), types of microorganisms involved, and other soil factors including soil pH and deficiencies of inorganic nutrients (P, K, and micronutrients).

Fire affects decomposition in two ways. First, fire can kill the biological organisms that decompose organic matter. A wide range of microbiota and invertebrates play an active role in the conversion of leaves and other larger organic debris into humus. Microorganisms affected by fire include bacteria and fungi, which are numerically the most abundant organisms in terrestrial ecosystems and the primary decomposers of organic matter in soil (Van Veen and Kuikman 1990). A second, and more dramatic, way in which fire affects decomposition is the rapid decomposition of organic matter during combustion (see Chapters 3 and 6).

Nitrogen Cycling

Because N is one of the most limiting nutrients in natural ecosystems, attention should be given to the various processes by which N can be affected by fire. Fire can volatilize large amounts of N, which need to be replaced following fire (see Chapter 6). Important N-cycling processes affected by fire are mineralization (ammonification and nitrification), N fixation, and denitrification.

MINERALIZATION

Mineralization is conversion of an element from an organic form to an inorganic state as the result of microbial activity (Soil Science Society of America 1997). Mineralization includes the transformation of organic N compounds (such as proteins and amino acids) into NH_4 *(ammonification)* and, subsequently, into nitrite (NO_2) and NO_3 *(nitrification)*, and the conversion of organic C into CO_2. Ammonification and nitrification are processes involved in the mineralization of N that are affected by fire. Nitrification is carried out mainly by autotrophic microorganisms, which derive their energy solely from the oxidation of NH_4 and NO_3 (Haynes 1986). Several genera of autotrophic bacteria are able to oxidize NH_4-N to NO_2, including *Nitrosomonas, Nitrosolobus,* and *Nitrospira.* Oxidation of NO_2 to NO_3 is done almost exclusively by *Nitrobacter* in natural systems (Haynes 1986). Up to 30 percent of the N nitrified in natural ecosystems is provided by faunal populations (Verhoef and Brussaard 1990).

Both *Nitrosomonas* and *Nitrobacter* were found to be sensitive to soil heating during fire in southern California chaparral shrublands. *Nitrosomonas* bacteria can be killed in dry soil at temperatures of 120 to 140°C, and in moist soil at 75 to 80°C (Dunn and DeBano 1977, Dunn et al. 1985). The more sensitive *Nitrobacter* groups of bacteria are killed at 100°C in dry soil and at 50°C in a moist soil. The sensitivity of ammonifying and nitrifying bacteria to soil heating has important implications concerning plant nutrition, because N is frequently a limiting nutrient in wildland soils (Maars et al. 1983). In unburned chaparral stands, high levels of total N are present as organic N, while relatively low levels of inorganic mineral N (NH_4-N and NO_3-N) are present.

Christensen (1973) hypothesized that suppression of mineralization in chaparral soils occurred because heterotrophic microorganisms responsible for mineralization were inhibited by allelopathic substances present, or because high lignin content of chaparral plant leaves resisted decomposition and subsequent mineralization of N. Allelopathic substances in forest types can decrease N mineralization rates (White 1991, Everett et al. 1995).

Higher concentrations of NH_4-N and NO_3-N are generally present after a fire (Christensen and Mueller 1975, DeBano et al. 1979). It has been hypothesized that NH_4-N and NO_3-N are formed by different processes in response to burning. Apparently, large amounts of NH_4-N are produced chemically by soil heating during a fire (see Chapter 6) and also microbially following a fire. Conversely, NO_3 is not produced directly by heating during a fire but is formed during subsequent nitrification of excess NH_4-N produced during burning. However, postfire nitrification does not appear to be carried out by the classical nitrifying bacteria (*Nitrosomonas* and *Nitrobacter*), because these bacteria are sensitive to soil heating and other disturbances and, consequently, are absent (or at extremely low levels) for several months following burning (Dunn et al. 1979). This absence suggests that nitrification can involve heterotrophic nitrification, possibly by fungi. Dormant forms of fungi are stimulated by mild heating (Dunn et al. 1985), which might have contributed to the fungal growth that paralleled NO_3 production (Dunn et al. 1979).

Increased levels of NH_4-N and NO_3-N following a fire are relatively short-lived, returning to prefire levels within 2 years after burning. Following prescribed burning in tropical forests in Costa Rica, the increase in available N returned to background levels in 6 months (Matson et al. 1987). Increased NH_4-N produced during a fire in chaparral ecosystems in central Arizona remained at an elevated level for about 6 months and then began decreasing, at which time NO_3-N began increasing and both NH_4-N and NO_3-N levels returned to prefire levels after about a year (Figure 7.2). Nitrogen mineralization was found to increase and remain elevated for 1 year in *Agropyron spicatum* and *Stipa comata* grasslands and for 2 years in *Artemisia tridentata*

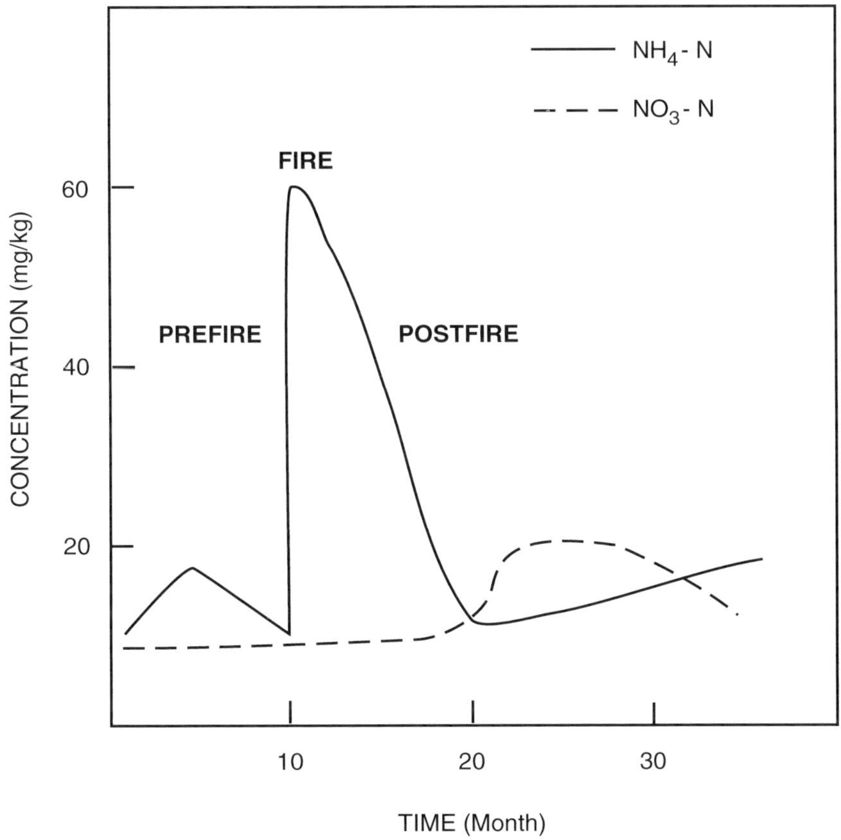

FIGURE 7.2. Sequence of NH_4-N and NO_3-N increases and decreases preceding and following a prescribed fire in a chaparral shrubland ecosystem in central Arizona.

and *Purshia tridentata* mountain shrublands following prescribed burning (Hobbs and Schimel 1984).

DENITRIFICATION

Denitrification is the process of reducing NO_3 to N_2 and N_2O by biological means. Little research has been done on the biological losses of N in relation to fire, however, partly because of the overwhelming losses that occur chemically by volatilization (see Chapter 6).

NITROGEN FIXATION

Nitrogen is unique among the soil nutrients because it is present in the soil almost entirely as part of organic compounds. No inorganic N reserve is normally available to replace the soil N lost by volatilization during a fire

(Harvey et al. 1976). Therefore, N is recovered from the atmosphere by N fixation and atmospheric deposition (see Chapter 6). *Nitrogen fixation* is the conversion of molecular N (N_2) to ammonia and, subsequently, to organic combinations or forms utilizable in biological processes (Soil Science Society of America 1997).

The atmosphere supplies N to soil in natural ecosystems mainly through organisms that fix inert N_2 into forms useful to plants. Although there is some N added by atmospheric deposition in precipitation (see Chapter 6), and by the adsorption of NH_3 from the air, most of N additions to the soil are by N fixation instead of being produced by rock weathering (Harvey et al. 1989). Nitrogen additions to the soil by N-fixing organisms counterbalance the volatilization losses during combustion, losses due to biological denitrification, and losses from leaching through the soil. Nitrogen fixation is mainly accomplished by two groups of organisms that fix N by two microbiological processes, symbiotic and nonsymbiotic (Example 7.1). Atmospheric N fixed by microorganisms ranges from 32 to 20 kg/ha/yr in dense stands of red alder in the northern Rocky Mountains of the United States, to 0.1 to 1.3 kg/ha/yr by free-living N-fixers (Jurgensen et al. 1979).

Living protoplasm of free-living, N-fixing microorganisms is likely to be damaged by soil heating, as are other microorganisms (bacteria and fungi) found in the surface organic layers. Symbiotic N-fixing microorganisms can be affected in two ways. Loss of the host plant could affect the symbiotic relationship by removing the source of energy. Direct damage to symbiotic microorganisms would be expected for those roots in the *D* layer and upper *A* horizon. Conversely, little or no direct damage would occur to deeper roots, which are far removed from the soil surface where lethal temperatures occur.

Example 7.1: Nitrogen-Fixing Organisms in Soil

- Symbiotic N fixation is carried out by microorganisms that are associated with the roots of higher plants (symbiotic), with energy required for fixing N being supplied by the host plant. The most common symbiotic relationships are those formed by rhizobia or actinomycetes with plant roots. Symbiosis between *Rhizobium* and the roots of leguminous plants is widespread. There are approximately 700 different genera of the Leguminosae family and over 14,000 individual species (Haynes 1986). Locust trees are nodulated and enhance the N status of forest soils (Klemmedson 1994). Postfire herbaceous species (*Lotus* sp. and *Lupinus* sp.) occupying southern California chaparral stands form symbiotic relationships with *Rhizobium* and are known to fix N, with the amount depending on postfire moisture and temperature regimes.

  Nitrogen fixation by actinomycetes is widespread. One example of forest trees that form these symbiotic root nodules belongs to the genus *Alnus* (Haynes 1986). It has been estimated in studies in Alaska that alder enriched the soil with about 156 kg of N per year (Van Cleve et al. 1971). Bitterbrush *(Purshia tridentata)* and mountain mahogany (*Cercocarpus* spp.) also form symbiotic relationships that fix N.

- Nonsymbiotic N fixation is performed by microbes that are free-living (nonsymbiotic), with energy for N fixation being supplied through the breakdown (decay) of organic matter (Harvey et al. 1989). Three main groups of free-living bacteria that fix N are heterotrophs, chemoautotrophs, and photoautotrophs (Haynes 1986). Chemoautotrophs utilize energy from chemical reactions, while photoautotrophs utilize light energy for N fixation. The ability of heterotrophs to fix N is generally low.

■

The role of both symbiotic and nonsymbiotic organisms in postfire N fixation and replenishment is unclear at this time. Results of a study in forests of the northwestern United States showed that soil N additions from symbiotic N fixation are not as large as previously assumed (Harvey et al. 1989). Exceptions were found in alder riparian ecosystems and relatively dense stands of snowbrush *(Ceanothus velutinus)* (Jurgensen et al. 1979). Therefore, storage and fixation of N in these forests are more likely to depend on nonsymbiotic processes, which receive energy from the biological oxidation of organic matter. More than one-third of the N-fixing capacity of some forest soils can be provided by microorganisms responsible for decaying wood on the surface and in the soil profile (Harvey et al. 1989). As a consequence, management of woody residues (coarse woody debris) within a fire prescription can be an important consideration for N management of forest soils after a fire.

Coarse Woody Debris

Coarse woody debris is made up of tree limbs, boles, and roots in various stages of decay that are greater than 7.5 cm in diameter (Graham et al. 1994). It is an important component of organic matter affected by fire, one that plays a special role in the productivity of forest soils because of the N-fixing capabilities associated with it. This woody debris also enhances the soil physical, chemical, and biological systems, in addition to providing a favorable environment for tree seedling establishment and growth. It not only affects soil properties but physically protects the soil surface from erosion and mechanical disturbance and new seedlings from livestock damage.

Coarse woody debris can be found on the soil surface or covered by soil and humus layers, where it is slowly incorporated into the soil during decomposition. *Soil wood,* consisting of buried or partially buried woody debris, can comprise over 50 percent of all surface organic matter in old-growth forests in the Inland Northwest of the United States (Page-Dumroese et al. 1991), amounting to between 37 and 50 Mg/ha. Soil wood found in these forests is a product of brown-rot decay and is made up of heartwood from pine and Douglas-fir trees. It is relatively resistant to decay and, as a result, can remain in the soil for hundreds of years (Jurgensen et al. 1997).

NUTRIENT STORAGE

Coarse woody debris stores plant nutrients and particularly N, P, and S, which are released slowly when coarse woody debris decays or rapidly when it burns. The debris provides sites for N fixation by nonsymbiotic N-fixers (Larsen et al. 1978, Jurgensen et al. 1992). Organic materials, particularly humus and buried residue in advanced stages of decay, also provide environments for ectomycorrhizal fungi. Although these fungi only make up a small portion of the soil, they help woody plants take up water and nutrients. Their fruiting bodies are part of the food chain of small rodents and larger predators.

Removal of large debris during prescribed burning and intensive timber harvesting in the forests of Idaho and Montana has reduced nonsymbiotic N fixation (Jurgensen et al. 1992). Relatively large reductions in N fixation have occurred on sites that were either broadcast burned or had all trees and woody residue greater than 2.5 cm removed by harvesting (Jurgensen et al. 1990). Nitrogen fixation was reduced 36 percent on Engelmann spruce *(Picea engelmannii)*/subalpine fir *(Abies lasiocarpa)* sites in Montana and 61 percent on western red cedar *(Thuja plicata)*/ western hemlock *(Tsuga heterophylla)* sites in Idaho after timber harvesting and preparation of the sites for subsequent regeneration. These reductions were attributed to the burning and removal of large amounts of coarse woody debris. It was estimated it would take between 10 and 275 years to replace the N lost from these sites, depending on the severity of the site treatments, the presence or absence of N-fixing plants, and the amounts of N contributed by atmospheric deposition (Jurgensen et al. 1997).

Leaving enough woody residue after timber harvesting to increase biodiversity and maintain long-term productivity is important to forest managers. To maintain the role of coarse woody debris in forest productivity, variable amounts of residue need to be left after silvicultural treatments, including prescribed fire (Example 7.2). The amounts necessary to maintain long-term productivity depend on the inherent productivity of individual forest stands.

Example 7.2: Recommended Amounts of Residual Woody Debris Necessary to Maintain Forest Productivity

Recommendations on the amounts of residual woody material to leave after silvicultural treatments vary with the forest type. A minimum of 22 to 36 Mg/ha of residual woody material should be left in moist habitat types of northwestern Montana and northern Idaho (Harvey et al. 1987). In contrast, less than 10 Mg/ha can be adequate to maintain long-term productivity on drier sites occupied by ponderosa pine forests (Graham et al. 1994).

Recommendations for the amounts of coarse woody debris to be left after any management activity have been proposed for several forest types in Arizona, Idaho, and Montana (see figure below). These recommendations are based on a minimum level thought to be necessary for maintaining long-term forest productivity (Graham et al. 1994). These recommendations will not immediately replace the present forest floor but will ensure its replacement

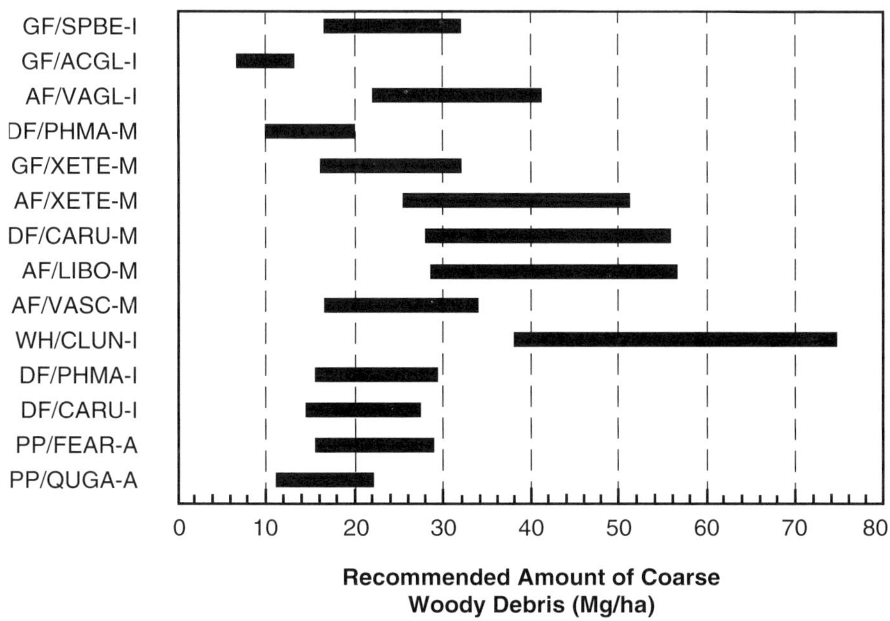

Legend	Habitat type
GF/SPBE	Grand fir/snowberry (*Abies grandis/ Spiraea betulifolia*)
GF/ACGL	Grand fir/mountain maple (*Abies grandis/Acer glabrum*)
AF/VAGL	Subalpine/huckleberry (*Abies lasiocarpa/ Vaccinium globulare*)
DF/PHMA	Douglas-fir/ninebark (*Pseudotsuga menziesii/ Physocarpus malvaceus*)
GF/XETE	Grand fir/beargrass (*Abies grandis/ Xerophyllum tenax*)
AF/XETE	Subalpine fir/beargrass (*Abies lasiocarpa/Xerophyllum tenax*)
DF/CARU	Douglas-fir/pinegrass (*Pseudotsuga menziesii/ Calamagrostis rubescens*)
AF/LIBO	Subalpine fir/twinflower (*Abies lasiocarpa/ LInnaea borealis*)
AF/VASC	Subalpine fir/ whortleberry (*Abies lasiocarpa/ Vaccinium scoparium*)
WH/CLUN	Western hemlock/queencup beadlily (*Tsuga heterophylla/ Clintonia uniflora*)
DF/PHMA	Douglas-fir/ninebark (*Pseudotsuga menziesii/ Physocarpus malvaceus*)
DF/CARU	Douglas-fir/pinegrass (*Pseudotsuga menziesii/ Calamagrostis rubescens*)
PP/FEAR	Ponderosa pine/Arizona fescue (*Pinus ponderosa/ Festuca arizonica*)
PP/QUGA	Ponderosa pine/Gambel oak (*Pinus ponderosa/ Quercus gambelii*)

Recommended amounts of coarse woody debris to leave after timber harvesting to maintain forest productivity in Arizona (A), Idaho (I), and Montana (M). (Adapted from Graham et al. 1994.)

over about 100 years. Maintaining these levels of coarse organic debris may temporarily increase the risk of wildfire or favor the activity of some insects and disease fungi (Jurgensen et al. 1997). However, prescribed burns and mechanical site preparation can be used on most sites with a relatively low impact on the soil organic level, while attaining management objectives concerned with forest productivity and sustainability. ■

ROLE OF FIRE

There must be a balance between organic matter buildup, including coarse woody debris, and decomposition. Fire can be implemented to better establish this balance. Soil organic matter, insect and disease activities, and fire incidence are interrelated (Jurgensen et al. 1997). In many cases, organic matter in the western United States is produced at a higher rate than it can be cycled by decay. Forest health concerns have evolved because large accumulations of live and dead organic matter have occurred, along with high stand densities (dog-hair thickets) and changes in species composition. These accumulations of fuels have resulted in nutrient deficiencies on some sites. The buildup of organic matter has resulted from fire suppression and tree mortality due to pine blister rust in the Inland Northwest (Covington et al. 1994, Harvey 1994). The accumulation of woody biomass may increase the incidence of severe stand-replacing fires. Consequently, regulating fuel buildups on some sites may be as important to forest health and productivity as maintaining adequate organic reserves (Jurgensen et al. 1997).

BACTERIA AND FUNGI

Complex interrelationships exist between soil heating and the well-being of bacteria and fungi. Duration of heating, maximum temperatures, and soil water content are the most important factors affecting the response of microbes and other biological organisms to soil heating (Dunn and DeBano 1977). The type of microbial population also influences the response to heat. A study in southern California chaparral soils showed fungi were more sensitive to soil heating than NO_2 oxidizers, which (in turn) were more sensitive than heterotrophic bacteria (Dunn et al. 1985). Heating increased germination of dormant forms, yielding higher fungi counts than those in unheated soil.

Physiologically active populations in moist soil were often more sensitive to temperature increases than dormant populations in dry soil. Survival of fungi was reduced to 1 percent at a temperature of about 60°C in moist soil (20 percent water), in comparison to about 80°C in dry soil (3 percent water). Nitrate oxidizers were less sensitive and were reduced to 1 percent at a temperature of about 80°C and 90°C in moist and dry soil, respectively.

Heterotrophic bacteria were most resistant. Survival was reduced to 1 percent at about 100°C in the moist soil and was above 1 percent at a temperature of 120°C in the dry soil.

Effects of soil water on microbial inactivation during fire are unclear. It is known that heat directly affects survival, because cellular components such as proteins, membrane lipids, and nucleic acids are unstable at temperatures of 60 to 70°C. Soil water also affects the transfer and storage of heat energy in the soil (see Chapter 2) and, therefore, indirectly affects microbial survival. The effect of soil water on lethal temperatures for bacteria and fungi has practical implications. If the upper 5-cm layer of soil is moist (10 percent water) and heated to 70°C, it would kill about 95 percent of the heterotrophic bacteria and a greater proportion of the fungi (Dunn et al. 1985). Conversely, this same temperature (70°C) would kill less than 25 percent of the soil heterotrophic bacteria if the soil is dry.

CRYPTOGAMIC CRUSTS

Cryptogamic crusts are communities of lichens, blue-green bacteria (cyanobacteria), fungi, mosses, and algae that are found on the surfaces of rocks and soil in dryland regions throughout the world (Isichei 1990, Johansen 1993, Loftin and White 1996). Cryptogamic crust development is most common in semiarid forest and shrublands, where pedicled crusts such as those found in pinyon-juniper and sagebrush communities are widespread (Johansen 1993). These crusts can account for as much as 70 percent of all living ground cover in arid Western Hemisphere ecosystems.

Composition and Functions

Lichens colonize rocks, tree trunks, and exposed soil surfaces. Lichens result from a symbiosis among fungi, algae, and photosynthetic blue-green bacteria. The blue-green bacteria photosynthesize and fix N, while the fungi provide water and mineral nutrients. Pedicled algal and algal–lichen crusts can form pedicels of up to 6 cm in height. They are typically found on erodible soils with some topographic relief, or in shallow soil pockets in slickrock habitats (Johansen 1993). Cryptogamic crusts can benefit soils in a number of ways (Example 7.3).

Example 7.3: Benefits of Cryptogamic Crusts in Soils

- *Retarding erosion on desert and steppe rangelands by binding individual soil particles together* (particularly in sandy soils). The filaments pro-

duced by mosses, algae, and other microflora components of these cryptogams proliferate on the soil surface and form a crust that stabilizes the soil surface. The aggregation of soil particles is hypothesized to occur as a result of the production of extracellular polysaccharides (Lynch and Bragg 1985).
- *Fixing gaseous N.* The N fixation is believed to be a result of free-living and lichenized blue-green bacterial components of the crusts. These cyanobacteria can fix up to 25 kg/ha of N annually (West and Skujins 1977). The amount of N fixed is largely dependent on the abundance and activity of the crusts and favorable climatic conditions (Loftin and White 1996).
- *Promoting organic matter buildup.* Cryptogamic crusts in semiarid areas of southwestern United States can accumulate 6 to 23 kg/ha of C annually (Jeffries et al. 1993).
- *Increasing P levels by retaining fine soil particles* (Kleiner and Harper 1977).
- *Enhancing seedling establishment.* Although germination can be inhibited by allelopathic substances produced by the crust, the seedling establishment is benefited overall (St. Clair et al. 1984). ■

Because of their effect on N fixation, C accumulation, and P capture, cryptogamic crusts enhance nutrient cycling in soils, particularly during the early successional stages. Crusts can increase or decrease infiltration rates, although their effect on conserving soil moisture is not well-defined (Johansen 1993).

Fire Effects

Once fire destroys cryptogamic crusts, it can take several years for their populations to redevelop to prefire levels. High-severity fire in the dry summer months causes the greatest damage to cryptogamic crusts. Recovery following disturbance can be fast or slow, depending on the type of crust (algal or lichenous), soil conditions, and climate. Crusts can take much longer to develop in hotter and more arid shrublands. A site in a blackbrush *(Coleogyne ramosissima)* community in southern Utah did not show any cryptogamic development following a severe range fire after 37 years (Callison et al. 1985). Soils in this area are less likely to support cryptogamic crusts even in the absence of disturbance.

Annual fires for 7 years completely destroyed cryptogamic crusts on degraded semiarid woodlands in Australia (Greene et al. 1990). Conversely, recovery after a single fire only took 4 years to reach prefire levels. Components making up the crptogamic crust (algae, cyanobacteria, or lichens) also affect the rate of recovery after disturbance (Example 7.4).

Example 7.4: Rate of Recovery for Cryptogamic Crusts After Disturbance Depends on the Composition of the Crusts

Algae generally recover from disturbance most rapidly, attaining undisturbed densities within 1 to 5 years (Johansen 1993). Algal cells of many species are usually able to survive even the most severe disturbance, so dispersal and recolonization are faster. The first organisms to colonize soil on burned English heaths are algae, especially simple green algae such as *Cryptococcus*, and species with a gelatinous thallus such as *Gloeocystis* (Warcup 1981). Blue-green bacteria are also favored by the higher pH after fire. Species with wind-blown propagules, such as coccoid chlorophytes and xanthophytes, are able to recolonize disturbed areas rapidly. Filamentous cyanophytes are less likely to recolonize by wind dispersal because of their size.

Mosses and lichens are slower to reestablish themselves than algae. Acrocarpic mosses can reoccupy burned sites in the Mediterranean area 9 to 15 months after fire (De Las Heras et al. 1993). Lichens that produce vegetative diaspores are able to move into disturbed areas more quickly than lichens that do not produce these propagules (Johansen et al. 1984). Lichens can take 10 to 20 years to develop diverse and abundant communities (Anderson et al. 1982). Epiphytic cryptogams have also rapidly established on rock surfaces following fire and contribute to soil formation (Garty 1992). ■

MYCORRHIZAE AND PLANT ROOTS

Mycorrhizal fungi form symbiotic relationships with plant roots in the rhizosphere. Mycorrhizal fungi are dependent on their hosts to complete their life cycle (Borchers and Perry 1990). This mutualistic root–fungal association occurs at a cost of 5 to 30 percent of the total photosynthate translocated below ground but contributes significantly to the welfare of the plants by obtaining inorganic nutrients and water over a much wider range of the soil volume than the roots alone.

Mycorrhizae are easily damaged by soil heating during fire, which, in turn, affects nutrient cycling and uptake by plants. Mycorrhizae and roots frequently occupy the *L* layer and *D* layer of soil and, in these locations, can be subjected to lethal soil temperatures during a fire. Intense fires, particularly those produced during duff ignition and burnout (see Chapter 2), easily produce temperatures that are lethal to mycorrhizae and roots.

Mycorrhizae

Two types of mycorrhizae are found in soil, *endomycorrhizae* (arbuscular mycorrhizae) and *ectomycorrhizae,* and both can be affected by fire (Example 7.5). Effects of fire on mycrorrhizae depend on the severity of the fire, the reproductive structure exposed to fire (spores or hyphae), and the type of mycorrhizal fungi present (ecto- or endomycorrhizae).

Example 7.5: Types and Functions of Mycorrhizae in Soils

Endomycorrhizae are characterized by structures within plant roots called arbuscules, because they develop into tree-like structures within the cell (Coleman and Crossley 1996). Ectomycorrhizae differ in their physiology and ecology. These fungi, principally basidiomycetes, proliferate between cells, not inside them as endomycorrhizae do. Ectomycorrhizae can send out hyphae for several meters into the soil. Therefore, these hyphae enhance the uptake of nutrients and water to the host plant. These mycorrhizal associations and other rhizosphere microbial populations, and the general soil microbial flora, contribute importantly to the mineralization and assimilation of P by plants (Tate 1987).

The hyphae of ectomycorrhizae constitute a significant proportion of C allocated to below-ground net primary productivity in coniferous forests, with the reproductive structure often present as mushrooms on the forest floor (Read 1991). The ecotomycorrhizae are found at shallow depths in the soil profile, tending to concentrate in organic matter, that is, in woody material during dry seasons and in the *H* layer during moist conditions. Ectomycorrhizae are also important in that they have decomposing abilities and, as a result, can obtain a portion of their reduced C from the decomposing litter layer. Because the ectomycorrhizae are located near the soil surface, both the resting stages and the hyphae are readily subjected to soil heating during fire. ■

ECTOMYCORRHIZAL FUNGI

The occurrence of ectotrophic mycorrhizae formed by basidiomycetes on the roots of forest trees generally decreases on burned areas, and, after fires of high severity, the mycorrhizae were found deeper in the soil (Wright and Tarrant 1958). Because tree seedlings often thrive on the burned area, it has been concluded that the decrease in mycorrhizae was compensated for by increased nutrition from the ash.

Ectomycorrhizal fungi have been reported to both increase and decrease with fire intensity during prescribed burns in pine forests in Canada (Herr et al. 1994). Fire intensity was positively correlated with percent ectomycorrhizal roots for white pine *(P. strobus)* but not for red pine *(P. resinosa)*. Two hypotheses were proposed to explain the increased ectomycorrhizae with fire intensity in white pine. One hypothesis is that the higher-intensity fires would be expected to volatilize large amounts of N and P in the soil, creating a more nutrient-poor root environment. As a consequence, the lower concentrations of available nutrients might be more conducive to greater ectomycorrhizal formation (Trofymow and van den Driessche 1991). A second hypothesis is that the more intense fires may volatilize phytotoxic substances, which can inhibit the growth and germination of ectomycorrhizal fungal propagules.

ENDOMYCORRHIZAL FUNGI

Fire severity has an important effect on recovery of endomycorrhizae following a fire (Vilarino and Arines 1991, Dhillion and Anderson 1993). Ecosystems that have high fuel loading are more likely to produce heavier losses in mycorrhizal fungi than areas that have lower fuel loads and experience less intense burning situations. For example, a reduction of up to 80 percent in vesicular arbuscular mycorrhizal spores occurred after high-severity fires under pinyon-juniper trees, in comparison to 47 percent on unburned interspace areas (Klopatek et al. 1991). Maximum soil temperatures of 94°C almost completely destroyed vesicular arbuscular mycorrhizal colonization by propagules in dry soil (Klopatek et al. 1988). The magnitude and duration of soil temperatures were believed to be principal factors causing damage to vesicular arbuscular mycorrhizal fungi.

Vesicular arbuscular mycorrhizae may be less affected by disturbances involving destruction of aerial biomass (including fire) than are ectomycorrhizal fungi, because they may form symbiotic relationships with more plant species (Puppi and Tartaglini 1991). Ectomycorrhizae are more abundant in the litter layer compared to vesicular arbuscular mycorrhizae, which concentrate in the lower mineral soil horizons (Reddell and Malajczuk 1984). Fire that removes only the litter layer would, therefore, favor vesicular arbuscular mycorrhizae.

Plant Roots

Plant roots are sensitive to both duration and temperature of heating. Temperatures of 60°C for 1 minute are sufficient to coagulate protein (Precht et al. 1973). Plant roots are sensitive to soil heating and lethal temperatures can be below the coagulation temperature of protein. The roots of loblolly pine have been killed when immersed in water at 54°C for 5 minutes, 50°C for 30 minutes, or 48°C for 2 hours (Hare 1961). Individual tree species vary in their response to heating. Slash *(P. caribaea)* and sand pine *(P. clausa)* have been killed by heating to 52°C for 17 minutes, while 49°C was lethal only to sand pine but not to slash pine. Neither species was affected by temperatures of 47°C for 25 minutes. The temperature at which death of the plant tissue occurs is largely dependent on the tissue moisture content. Those tissues with higher moisture contents are killed at lower temperatures and shorter time intervals (Zwolinski 1990). Although protein coagulation defines the lethal temperature of individual cells, the differences in internal heat effects are due mostly to differences in insulation from heat sources.

The better insulated the plant roots are by the soil, the less their chance of being subjected to lethal temperatures during a fire. Two factors affecting the insulation of roots against soil heating are their depth in the soil and the soil water content. Generally, the deeper the plant roots are located in the soil, the greater will be the survival rate (Flinn and Wein 1977). Low-severity fire might combust only the *L* layer and kill aboveground plant parts, while high-severity fire can consume all of the surface organic matter and heat the mineral soil above the lethal temperature for roots. Plant species that have their reproductive structures in the litter probably will not survive even a low-severity fire, while those with reproductive organs in mineral soil will survive the deep and long-lasting soil heating produced during most severe fires.

Soil heating and heat transfer processes are more complicated in moist soils compared to dry soils, a fact that complicates determination of the effect of heating on plant root mortality. Dry soil is a poor conductor of heat, and, as a result, heat does not penetrate deeply in the soil if the residence time of the flaming front is short (see Chapter 2). The soil surface of dry soil can easily exceed the lethal temperature of living tissue of roots, while ambient daily soil temperatures can prevail 2 cm or more downward in the soil, with little damage occurring to the roots. Therefore, when the roots are in dry soil below 2 cm, they will likely not be damaged by soil heating unless the residence time of the flaming front is long in duration. Conversely, those plant structures located on or near the soil surface can easily be damaged.

INVERTEBRATES

Effects of fire on invertebrates, as with other living organisms, depend largely on the severity of the fire. Two situations should be recognized in this regard—uncontrolled wildfire (high fire severity) and low-intensity prescribed fire (low fire severity) (Campbell and Tanton 1981). In either case, the effect of fire on invertebrates can be transitory or longer lasting (Lyon et al. 1978). Invertebrates generally decrease during a fire because the organisms or their eggs are killed by the flames or lethal soil temperatures, or because their food supply and shelter are destroyed. Flying insects are especially vulnerable, because they can be attracted to fire by heat or smoke. Conversely, some populations of invertebrates can be increased after a fire because trees that are damaged or killed by the fire provide a more suitable habitat for those invertebrates that survive.

Invertebrates occupying fire-affected ecosystems can be placed in two broad classes—those that spend all or most of their time in the upper soil layers, and those that reside in these habitats only temporarily or not at all. Invertebrates residing more permanently in the upper soil layers (*L* layer and *D* layer) are most likely affected when these soil layers are heated to lethal temperatures. Macroinvertebrates dwelling exclusively in litter were found to be particularly vulnerable to wildfire in Greece that destroys surface fuels and litter (Sgardelis et al. 1995). However, invertebrates that permanently occupy deeper soil horizons are protected from even high-severity fire. Some macroinvertebrates have been found to move deeper into the soil during the summer and, as a result, they are insulated from lethal soil temperatures during fire. A study in South Africa showed that most of the invertebrates in the top 2.5 cm of soil survived relatively cool burning forest fires (Coults 1945). But, in Australia, a reduction in litter quantity by fire decreased both the number of species and species density of soil and litter invertebrates (Springett 1976).

No invertebrates appear to be adapted to respond solely to fire. Invertebrate orders least affected by fire show high mobility or adaptations to conserve water and resist high ambient temperatures in seasonally dry habitats. These adaptations by invertebrate fauna appear not to be related to fire environments but instead to characteristics of specific invertebrates. The varying balance of invertebrate groups within prefire communities, selective modification of this balance by burning, and postfire community structure, which develops as the survivors accommodate to changes in litter structure and distribution, make it difficult to predict what will happen after a fire (Campbell and Tanton 1981). Nevertheless, several case studies of individual invertebrate responses to fire are presented in Example 7.6.

Example 7.6: Invertebrate Response to Fire: Case Studies

ANTS

Fire can increase ant populations (Anderson et al. 1989). An Australian study of sclerophyllous vegetation showed that ants consumed many of the seeds produced by early successional plants following fire (Andersen 1988). The seed-removing activity of ants continued for 2 years following fire, resulting in selective postfire recruitment of plants.

GRASSHOPPERS

Burning of Illinois prairies reduced both soil fauna and surface insect populations, including grasshoppers (Rice 1932). Combined adult and nymph grasshopper densities in southern Arizona declined over 60 percent on burned plots compared to unburned plots during the first year after a fire (Bock and Bock 1991). These differences disappeared by the second year. Similarly, grasshopper biomass was significantly greater on an unburned sand prairie site compared with an adjacent burned site in another study (Anderson et al. 1989).

EARTHWORMS

Earthworms have long been recognized as an important invertebrate component of soil systems (Coleman and Crossley 1996). They act as biological agents in the decomposition of litter and C mineralization in soil and litter (Zhang and Hendrix 1995). The effect of soil heating on earthworms is not well understood, however. One study in tallgrass prairie showed that indirect effects of fire are probably more important than direct heating of earthworm populations (James 1982). The increased earthworm activity resulted from differences in plant productivity following the fire, with the soil on the burned sites more thoroughly penetrated with roots and rhizomes. Earthworms are frequently found 10 to 20 cm below the soil surface, so they are probably protected from the direct effects of fuel combustion and soil heating, except in the case of severe long duration fires that might occur under piles of slash and logs. ■

AMPHIBIANS AND REPTILES

The direct effects of fire on amphibians and reptiles are generally minimal, because these animals seem able to avoid direct heat by either moving away from it or burrowing in the soil (Mushinsky 1985). The indirect effect of fire on amphibians and reptiles is less clearly understood, however, although patchy burns with high structural heterogeneity seemed to favor some species in southern California chaparral ecosystems (Lillywhite 1977). Likewise, although a variable response to fire was reported for different lizard species in Australia, some species were benefited by a low-severity patch fire interspersed with severely burned spots (Braithwaite 1987).

SMALL MAMMALS

Fire occasionally kills small mammals. More commonly, fire modifies their habitat conditions. Among small mammals affected by fire are voles, mice, shrews, squirrels, chipmunks, hares, and cottontails. (Effects of fire on the habitats of larger mammals are described in Chapter 9.)

Coarse woody debris on the soil surface before and after a fire is important for the welfare of small mammals in that it:

- Provides shelter for animals under the woody fraction where they can survive, even though the litter layer around the woody material is completely consumed in the fire. The number of animals under a log appears generally to be related to the dimensions of the log, its state of decay, and the depth of accumulated litter about the log prior to burning (Campbell and Tanton 1981).
- Lowers the temperature under the woody fraction compared to that under the normal leaf litter.
- Provides a less fluctuating moisture regime under the wood.

Mortality

Small mammals can be killed directly by burning, or they can die of suffocation (Wright and Bailey 1982). Their survival within a burn depends on several factors, namely:

- The uniformity, intensity and severity, size, and duration of the burn
- The mobility of the animal
- The position of the animal relative to the soil surface during passage of a fire

Uniformly severe wildfire over large areas is the most damaging to small animals that cannot get underground to avoid the direct effects of the fire. Lethal temperatures for small mammals are about 63°C, although when animals are killed in fire, it is usually a result of suffocation rather than high temperatures (Lyon et al. 1978). Mice tolerate temperatures up to 63°C for short periods of time if the relative humidity is below 22 percent, but they will die quickly when exposed to 49°C if the relative humidity is above 60 percent (Wright and Bailey 1982). Survival of small animals is a direct function of temperature decrease with increasing depth of mineral soil, and, in most cases, temperatures will decrease enough for survival within 2 to 3 cm of the soil surface (Lyon et al. 1978). Consequently, low-severity prescribed fire, which results in a mosaic of burned and unburned areas, is likely to be least damaging to small animals.

Habitat Modification

Regardless of the level of survival of small mammals in relation to the direct effect of the fire, other effects such as the temporary loss of shelter and food, exposure of surface runways and burrow openings, and increased predation will decrease the number and diversity of small mammals for 1 to 3 years following a fire (Ream 1981). Individual species vary in their ability to adapt to postfire habitats (see Example 7.7).

Example 7.7: Response of Small Mammals to Habitat Modification by Fire: Case Studies

SQUIRRELS

Ground squirrels prefer open areas, where they can feed on plants that grow in open habitats (Ream 1981). Burrowing systems provide protection from predators, so cover is not as important as it is for chipmunks. As a result, ground squirrels often increase in density following fire (Wright and Bailey 1982). Conversely, tree squirrels (such as red and flying squirrels) require trees for dens. These squirrels, therefore, are displaced when fire eliminates these trees. However, cavities in fire-killed trees can be used as dens for red squirrels, if these trees are surrounded by living conifer trees that provide food. Flying squirrels prefer fungi and can forage in fire-created openings (Ream 1981).

POCKET GOPHERS

Pocket gophers are favored by fire disturbances that remove the forest canopy, scarify the soil, and result in the production of herbaceous vegetation as a food source (Ream 1981). Protective burrow systems allow these animals to use large open burned areas that lack protective cover.

DEER MICE

Deer mice are generally found in the early successional stages of vegetation development and are adaptable to xeric environments. Although fire might initially decrease the number of deer mice, these small mammals are adapted to the postfire environment and, commonly, increase severalfold in burned areas compared to unburned sites within 2 years of a fire (Wright and Bailey 1982). As a consequence, deer mice are often the most abundant small mammals found on severely burned sites. Their primary food sources are insects, wind-dispersed seeds, and seeds that remain in the soil after burning (Ream 1981).

VOLES

Voles are found in organic layers of the soil surface, where important components of their habitats are a mat of ground cover, palatable herbaceous plants, and moisture (Ream 1981). However, a severe fire that destroys surface organic layers can eliminate voles. Voles in mesic habitats are particularly vulnerable to postfire predation for several years following fire (Wright and Bailey 1982). ■

SUMMARY

Soil organisms consist of an assemblage of animals and plants ranging from microscopic bacteria and fungi to small mammals and reptiles that remain below-ground during all or part of their life cycle. The surface horizons of soils in natural ecosystems generally have an abundance and diversity of soil organisms, with more organic matter, greater acidity, and a more aggregated structure than are found in agricultural soils. The soil organisms are important for nutrient cycling (including the processes of decomposition, mineralization, N fixation, and facilitating nutrient uptake) and, as such, they are essential for maintaining site productivity. At this point in the book, you should be able to:

1. Explain why biological systems are vulnerable to fire.
2. Describe the importance of coarse woody debris in nutrient cycling and in other biological responses to fire.
3. Explain why N-fixing organisms are important to the productivity of ecosystems that burn regularly.
4. Discuss the behavior of a fire that has the greatest effect on the biological soil system.
5. Describe how fire directly and indirectly affects invertebrates and small mammals.
6. Discuss the soil microorganisms that are most likely to be affected by fire, and what their role is relative to ecosystem productivity.

REFERENCES

Ahlgren, I. F., and C. E. Ahlgren. 1960. Ecological effects of forest fires. *Botanical Review* 26:483–533.

Anderson, A. N. 1988. Immediate and longer-term effects of fire on seed predation by ants in sclerophyllous vegetation in south-eastern Australia. *Australian Journal of Ecology* 13:285–293.

Anderson, D. C., K. T. Harper, and S. R. Rushforth. 1982. Recovery of cryptogamic soil crusts from grazing on Utah winter ranges. *Journal of Range Management* 35:355–359.

Anderson, R. C., T. Leahy, and S. Dhillion. 1989. Numbers and biomass of selected insect groups on burned and unburned sand prairie. *American Midland Naturalist* 122:151–162.

Barbour, M. G., J. H. Burk, and W. D. Pitt. 1980. *Terrestrial Plant Ecology.* Benjamin/Cummings Publishing Co., Menlo Park, California.

Bock, C. E., and J. H. Bock. 1991. Response of grasshoppers *(Orthoptera acrididae)* to wildfire in a southeastern Arizona grassland. *American Midland Naturalist* 1254:162–167.

Bollen, W. B. 1974. Soil microbes. In: Cramer, O., organizer and technical editor. *Environmental Effects of Forest Residues Management in the Pacific Northwest. A State-of-Knowledge Compendium.* USDA Forest Service, General Technical Report PNW-24, pp. B-1–B-30.

Borchers, J. G., and D. A. Perry. 1990. Effects of prescribed fire on soil organisms. In: Walstad, J. D., S. R. Radosevich, and D. V. Sandberg, editors. *Natural and Prescribed Fire in Pacific Northwest Forests.* Oregon State University Press, Corvallis, Oregon, pp. 143–157.

Braithwaite, R. W. 1987. Effects of fire on lizards in the wet-dry tropics of Australia. *Journal of Tropical Ecology* 3:265–275.

Callison, J., J. D. Brotherson, and J. E. Browns. 1985. The effects of fire on the blackbrush *(Coleogyne ramosissima)* community of southwestern Utah. *Journal of Range Management* 38:535–538.

Campbell, A. J., and M. T. Tanton. 1981. Effects of fire on the invertebrate fauna of soil and litter of a eucalypt forest. In: Gill, A. M., B. H. Groves, and I. R. Noble, editors. *Fire and the Australian Biota.* Australian Academy of Science, Canberra, Australia, pp. 215–241.

Christensen, N. L. 1973. Fire and the nitrogen cycle in California chaparral. *Science* 181:66–68.

Christensen, N. L., and C. H. Mueller. 1975. Effects of fire on factors controlling plant growth in *Adenostoma* chaparral. *Ecological Monographs* 45:29–55.

Coleman, D. C., and D. A. Crossley, Jr. 1996. *Fundamentals of Soil Ecology.* Academic Press, New York.

Coults, J. R. H. 1945. Effect of veld burning on the base exchange capacity of a soil. *South Africa Journal of Science* 41:218–224.

Covington, W. W., and L. F. DeBano. 1990. Effects of fire on pinyon-juniper soils. In: Krammes, J. S., technical coordinator. *Effects of Fire Management of Southwestern Natural Resources.* USDA Forest Service, General Technical Report RM-191, pp. 78–86.

Covington, W. W., R. L. Everett, R. Steele, L. L. Irwin, T. A. Daer, and A. N. D. Auclair. 1994. Historical and anticipated changes in forest ecosystems in the Inland West of the United States. *Journal of Sustainable Forestry* 2:13–63.

DeBano, L. F., G. E. Eberlein, and P. H. Dunn. 1979. Effects of burning on chaparral soils: I. Soil nitrogen. *Soil Science Society of America Journal* 43:504–509.

De Las Heras, J., J. M. Herranz, and J. J. Martinez. 1993. Influence of bryophyte pioneer communities on edaphic changes in soils of Mediterranean ecosystems damaged by fire (SE Spain). In: Trabaud, L., and P. Prodon, editors. *Fire in Mediterranean Ecosystems.* Commission of the European Countries, Ecosystem Research Report 5, Brussels, Belgium, pp. 387–393.

Dhillion, S. S., and R. C. Anderson. 1993. Seasonal dynamics of dominant species of arbuscular mycorrhizae in burned and unburned sand prairies. *Canadian Journal of Botany* 71:1625–1630.

Dunn, P. J., and L. F. DeBano. 1977. Fire's effect on biological and chemical properties of chaparral soils. In Mooney, H. A., and C. E. Conrad, technical coordinators. *Proceedings of the Symposium on the Environmental Consequences of Fire and Fuel Management in Mediterranean Ecosystems.* USDA Forest Service, General Technical Report WO-3, pp. 75–84.

Dunn, P. H., L. F. DeBano, and G. E. Eberlein. 1979. Effects of burning on chaparral soils: II. Soil microbes and nitrogen mineralization. *Soil Science Society of America Journal* 43:509–514.

Dunn, P. H., S. C. Barro, and M. Poth. 1985. Soil moisture affects survival of microorganisms in heated chaparral fire. *Soil Biology and Biochemistry* 17:143–148.

Everett, R. L., B. J. Java-Sharpe, G. R. Scherer, F. M. Wilt, and R. D. Ottmar. 1995. Co-occurrence of hydrophobicity and allelopathy in sand pits under burned slash. *Soil Science Society of America Journal* 59:1176–1183.

Flinn, M. A., and R. W. Wein. 1977. Depth of underground plant organs and theoretical survival during fire. *Canadian Journal of Botany* 55:2550–2554.

Garty, J. 1992. The postfire recovery of rock-inhabiting algae, microfungi, and lichens. *Canadian Journal of Botany* 70:301–312.

Graham, R. T., A. E. Harvey, M. F. Jurgensen, T. B. Jain, J. R. Tonn, and D. S. Page-Dumroese. 1994. Managing coarse woody debris in forests of the Rocky Mountains. USDA Forest Service, Research Paper INT-RP-477.

Greene, R. S. B., C. J. Chartres, and K. C. Hodgkinson. 1990. The effects of fire on the soil in a degraded semi-arid woodland: I. Cryptogam cover and physical and micromorphological properties. *Australian Journal of Soil Research* 28:755–777.

Hare, R. C. 1961. Heat effects on living plants. USDA Forest Service, Southern Forest Experiment Station, Occasional Paper 183, New Orleans, Louisiana.

Harvey, A. E. 1994. Integrated roles for insects, diseases and decomposers in fire dominated forests of the Inland Western United States: past, present and future forest health. *Journal of Sustainable Forestry* 2:211–220.

Harvey, A. E., M. F. Jurgensen, and M. J. Larsen. 1976. *Intensive Fiber Utilization and Prescribed Fire: Effects on the Microbial Ecology of Forests.* USDA Forest Service, General Technical Report INT-28.

Harvey, A. E., M. F. Jurgensen, M. J. Larsen, and R. T. Graham. 1987. *Decaying Organic Materials and Soil Quality in the Inland Northwest: A Management Opportunity.* USDA Forest Service, General Technical Report INT-225.

Harvey, A. E., M. F. Jurgensen, and R. T. Graham. 1989. Fire–soil interactions governing site productivity in the northern Rocky Mountains. In: Baumgartner, D. M., D. W. Bruer, B. A. Zamora, L. F. Neuenschwander, and R. H. Wakimoto, compilers and editors. *Prescribed Fire in the Intermountain Region: Forest Site Preparation and Range Improvement: Symposium Proceedings.* Cooperative Extension Service, Washington State University, Pullman, Washington, pp. 9–18.

Haynes, R. J. 1986. Origin, distribution, and cycling of nitrogen in terrestrial ecosystems. In: Hayes, R. J., editor. *Mineral in the Plant–Soils System.* Academic Press, New York, pp. 1–51.

Hendrix, P. F., D. A. Crossley, Jr., J. M. Blair, and D. C. Coleman, 1990. Soil biota as components of sustainable agroecosystems. In: C. A. Edwards, R. Lal, P. Madden, R. H. Miller, and G. House, editors. *Sustainable Agricultural Systems.* Soil and Water Conservation Society, Ankeny, Iowa, pp. 637–654.

Herr, D. G., L. C. Duchesne, R. Tellier, R. S. McAlpine, and R. L. Peterson. 1994. Effect of prescribed burning on ectomycorrhizal infectivity of a forest soil. *International Journal of Wildland Fire* 42:95–102.

Hobbs, N. T., and D. S. Schimel. 1984. Fire effects on nitrogen mineralization and fixation in mountain shrub and grassland communities. *Journal of Range Management* 37:402–404.

Isichei, A. O. 1990. The role of algae and cyanobacteria in arid lands: a review. *Arid Soil Research and Rehabilitation* 4:1–17.

James, S. W. 1982. Effects of fire and soil type on earthworm populations in a tallgrass prairie. *Pedobiologia* 24:37–40.

Jeffries, D. L., S. O. Link, and J. M. Klopatek. 1993. CO_2 fluxes of cryptogamic crusts: I. Response to resaturation. *The New Phytologist* 125:163–173.

Jenny, H. 1980. *The Soil Resource: Origin and Behavior.* Springer-Verlag, New York.

Johansen, J. R. 1993. Cryptogamic crusts of semiarid and arid lands of North America. *Journal of Phycology* 29:140–147.

Johansen, J. R., L. L. St. Clair, B. L. Webb, and G. T. Nebeker. 1984. Recovery patterns of cryptogamic soil crusts in desert rangelands following fire disturbance. *Bryologist* 87:238–243.

Jurgensen, M. F., S. F. Arno, A. E. Harvey, M. J. Larsen, and R. D. Pfister. 1979.

Symbiotic and nonsymbiotic nitrogen fixation in northern Rocky Mountain forest ecosystems. In: Gordon, J. C., C. R. Wheeler, and D. A. Perry, editors. *Symposium Proceedings on Symbiotic Nitrogen Fixation in the Management of Temperate Forests.* Oregon State University, Corvallis, Oregon, pp. 294–308.

Jurgensen, M. F., A. E. Harvey, R. T. Graham, M. J. Larsen, J. R. Tonn, and D. S. Page-Dumroese. 1990. Soil organic matter, timber harvesting and forest productivity in the Inland Northwest. In: Gessel, S. P., D. S. Lacate, G. F. Weetman, and R. F. Powers, editors. *Sustained Productivity of Forest Soils.* Seventh North American Forest Soils Conference, University of British Columbia, Vancouver, British Columbia, Canada, pp. 392–415.

Jurgensen, M. F., R. T. Graham, M. J. Larsen, and A. E. Harvey. 1992. Clear-cutting, woody residue removal, and nonsymbiotic nitrogen fixation in forest soils of the Inland Pacific Northwest. *Canadian Journal of Forestry Research* 22:1172–1178.

Jurgensen, M. F., A. E. Harvey, and T. B. Jain. 1997. Impacts of timber harvesting on soil organic matter, nitrogen, productivity, and health of Inland Northwest Forests. *Forest Science* 43:234–251.

Kleiner, E. F., and K. T. Harper. 1977. Soil properties in relation to cryptogamic ground cover in Canyonlands National Park. *Journal of Range Management* 30:202–205.

Klemmedson, J. O. 1994. New Mexico locust and parent material: influence on macronutrients of forest floor and soil. *Soil Science Society of America Journal* 58:974–980.

Klopatek, C. C., L. F. DeBano, and J. M. Klopatek. 1988. Effects of simulated fire on vesicular–arbuscular mycorrhizae in pinyon-juniper woodland soil. *Plant and Soil* 109:245–249.

Klopatek, C. C., C. Friese, M. F. Allen, L. F. DeBano, and J. M. Klopatek. 1991. The effect of high intensity fire on the patch dynamics of VA mycorrhizae in pinyon-juniper woodlands. In: Nodvin, S. C., and T. A. Waldrop, editors. *Fire and the Environment: Ecological and Cultural Perspectives: Proceedings of an International Symposium.* USDA Forest Service, General Technical Report SE-69, pp. 123–128.

Larsen, M. J., M. F. Jurgensen, and A. E. Harvey. 1978. N_2 fixation associated with wood decayed by some common fungi in western Montana. *Canadian Journal of Forestry Research* 8:341–345.

Lillywhite, H. B. 1977. Effects of chaparral conversion on small vertebrates in southern California. *Biological Conservation* 11:171–184.

Loftin, S. R., and C. S. White. 1996. Potential nitrogen contribution of soil cryptogams to post-disturbance forest ecosystems in Bandelier National Monument, NM. In: Allen, C. D., technical editor. *Fire Effects in Southwestern Forests: Proceedings of the Second La Mesa Fire Symposium.* USDA Forest Service, General Technical Report RM-GTR-286, pp. 140–148.

Lynch, J. M., and F. Bragg. 1985. Microorganisms and soil aggregation stability. *Advances in Soil Science* 2:133–171.

Lyon, J. L., H. S. Crawford, E. Czuhai, R. L. Fredriksen, R. F. Harlow, L. J. Metz, and H. A. Pearson. 1978. *Effects of Fire on Fauna.* USDA Forest Service, General Technical Report WO-6.

Maars, R. H., R. D. Roberts, R. A. Skeffinton, and A. D. Bradshaw. 1983. Nitrogen in the development of ecosystems. In: Lee, J. A., S. McNeill, and I. H. Rorison, editors. *Nitrogen as an Ecological Factor.* Blackwell Science Publishing, Oxford, England, pp. 131–137.

Matson, P. A., P. M. Vitousek, J. J. Ewel, M. J. Mazzarino, and G. P. Robertson. 1987. Nitrogen transformations following tropical forest felling and burning on a volcanic soil. *Ecology* 68:491–502.

Mushinsky, H. R. 1985. Fire and the Florida sandhill herpetofaunal community: with special attention to responses of *Cnemidophorus sexlineatus. Herpetologica* 41:333–342.

Page-Dumroese, D. S., A. E. Harvey, M. F. Jurgensen, and R. T. Graham. 1991. Organic matter function in the Inland Northwest soil system. In: Harvey, A. E., and L. F. Neuenschwander, compilers. *Proceedings: Management and Productivity of Western Montane Forest Soils.* USDA Forest Service, General Technical Report INT-280, pp. 95–100.

Precht, J., J. Chrisphersen, H. Hensel, and W. Larcher. 1973. *Temperature and Life.* Springer-Verlag, New York.

Puppi, G., and N. Tartaglini. 1991. Mycorrhizal types in three Mediterranean communities affected by fire to different extents. *Acta Oecologica* 12:295–304.

Read, D. J. 1991. Mycorrhizae in ecosystems. *Experientia* 47:376–391.

Ream, C. H. 1981. *The Effects of Fire and Other Disturbances on Small Mammals and Their Predators: An Annotated Bibliography.* USDA Forest Service, General Technical Report INT-106.

Reddell, P., and N. Malajczuk. 1984. Formation of mycorrhizae by jarrah (*Eucalyptus marginata* Dom ex Smith) in litter and soil. *Australian Journal of Botany* 32:511–520.

Rice, L. A. 1932. The effect of fire on the prairie animal communities. *Ecology* 13:392–401.

St. Clair, L. L., B. L. Webb, J. R. Johansen, and G. T. Nebeker. 1984. Cryptogamic soil crusts: enhancement of seedling establishment in disturbed and undisturbed areas. *Reclamation Revegetation Research* 3:129–136.

Sgardelis, S. P., J. D. Pantis, M. D. Argyropoulou, and G. P. Stamou. 1995. Effects of fire on soil macroinvertebrates in a Mediterranean Phryganic ecosystem. *International Journal of Wildland Fire* 5:113–121.

Singer, M. J., and D. N. Munns. 1996. *Soils: An Introduction.* Prentice Hall, Upper Saddle River, New Jersey.

Soil Science Society of America. 1997. *Glossary of Soil Science Terms.* Soil Science Society of America, Madison, Wisconsin.

Springett, J. A. 1976. The effect of prescribed burning on the soil fauna and on litter decomposition in western Australian forests. *Australian Journal of Ecology* 1:77–82.

Tate, R. L. III. 1987. *Soil Organic Matter: Biological and Ecological Effects.* John Wiley & Sons, New York.

Trofymow, J. A., and R. van den Driessche. 1991. Mycorrhizae. In: van den Driessche, R., editor. *Mineral Nutrition of Conifer Seedlings.* CRC Press, Boston, Massachusetts, pp. 183–227.

Van Cleve, K., L. A. Viereck, and R. L. Schlentner. 1971. Accumulation of nitrogen in alder *(Alnus)* ecosystems near Fairbanks, Alaska. *Arctic and Alpine Research* 3:101–114.

Van Veen, J. A., and P. J. Kuikman. 1990. Soil structural aspects of decomposition of organic matter by micro-organisms. *Biogeochemistry* 11:213–233.

Verhoef, H. A., and L. Brussaard. 1990. Decomposition and nitrogen mineralization in natural and agroecosystems: the contribution of soil animals. *Biogeochemistry* 11:175–211.

Vilarino, A., and J. Arines. 1991. Numbers and viability of vesicular-arbuscular fungal propagules in field soil samples after wildfire. *Soil Biology and Biochemistry* 23:1083–1087.

Wallwork, J. A. 1970. *Ecology of Soil Animals.* McGraw-Hill, Maidenhead, Berkshire, England.

Warcup, H. H. 1981. Effect of fire on the soil microflora and other non-vascular plants. In: Gill, A. M., B. H. Groves, and I. R. Noble, editors. *Fire and the Australian Biota.* Australian Academy of Science, Canberra, Australia, pp. 203–214.

West, N. E., and J. Skujins. 1977. The nitrogen cycle in North American cold-winter semidesert ecosystems. *Oecologia Plant* 12:45–53.

White, C. S. 1991. The role of monoterpenes in soil nitrogen cycling processes in ponderosa pine. *Biogeochemistry* 12:43–68.

Wright, E., and R. F. Tarrant. 1958. Occurrence of mycorrhizae after logging and slash burning in the Douglas fir forest type. USDA Forest Service, Research Note PNW-160.

Wright, H. A., and A. W. Bailey. 1982. *Fire Ecology—United States and Southern Canada.* John Wiley & Sons, New York.

Zhang, Q. L., and P. F. Hendrix. 1995. Earthworm *(Lumbricus rubellus* and *Aporrectodea caliginosa)* effects on carbon flux in soil. *Soil Science Society of America Journal* 59:816–823.

Zwolinski, M. J. 1990. Fire effects on vegetation and succession. In: Krammes, J. S., technical coordinator. *Proceedings of a Symposium on the Effects of Fire Management of Southwestern Natural Resources.* USDA Forest Service, General Technical Report RM-191, pp. 18–24.

PART

III

Responses of Other Resources

The soil responses to fire that were described in Part II also directly or indirectly affect the way in which many other ecosystem resources are affected by fire. Topics of Part III are the responses of these other resources to fire. Effects of fire on components of the hydrologic cycle, streamflow regimes, soil erosion and sedimentation, and water quality are discussed in Chapter 8. How fire affects plants, succession, and vegetative resources, including timber, forage, and wildlife habitat components, is the subject of Chapter 9. The emphasis is placed on the manner in which fire affects ecosystem resources in wetlands and riparian ecosystems in Chapter 10 because of the unique settings of these often fragile environments.

The impacts of fire on the atmosphere must be understood to better meet air quality regulations. Chapter 11 addresses this general topic by considering the chemical products of combustion, effects of smoke on the atmosphere, and practices needed to meet air quality regulations in managing fuels and developing fire prescriptions. Chapter 12 considers fire effects on the cultural resources of ecosystems from an archeological perspective, focusing largely on how fires having different severities affect ceramics, grinding stone, architectural materials, and other artifacts.

Chapter

8

Water

INTRODUCTION

Effects of a fire on water resources are determined largely by the severity of the fire, decisions made relative to suppression of the fire, and the immediate postfire precipitation regime. Unfortunately, it is not possible to adequately describe the possible impacts of fire in all conceivable situations, because information is scarce for portions of the spectrum of conditions in which a fire might occur. As a consequence, understanding the nature of hydrologic processes is a key to interpretation of fire effects on them. Hydrologic processes to be considered in this chapter include those represented by components of the hydrologic cycle, streamflow regimes, soil erosion and sedimentation, and those processes affecting water quality. The concept of watershed condition helps to place this chapter in perspective.

WATERSHED CONDITION

Watershed condition is a subjective term to indicate the health (status) of a watershed in terms of its *hydrologic function* and *soil productivity.* The hydrologic function of a watershed relates to its ability to receive and process precipitation into streamflow without ecosystem deterioration. Soil productivity reflects the capabilities of a watershed for supporting sustained plant growth and plant communities, or the natural sequences of plant communities. In general, on a watershed in *good condition:*

- Precipitation infiltrates into the soil.
- Precipitation does not contribute excessively to erosion, since the resultant overland flow (when a pathway of flow on the watershed) does not dislodge and move soil particles.
- Streamflow response to precipitation is relatively slow.
- Baseflow (when a pathway of flow) is sustained between storms.

At the other end of the spectrum, on a watershed in *poor condition:*

- Precipitation flows on the soil surface.
- Excessive erosion occurs during precipitation events.
- Streamflow response to precipitation is rapid.
- There is little or no baseflow between storms.

While different procedures have been devised to assess watershed condition, embedded criteria for most of these schemes are based largely on factors that can affect the hydrologic function and soil productivity of the watershed. These criteria usually include some measure of:

- Vegetative cover
- Stream channel dynamics
- Degree of surface and gully erosion
- On-site soil mass stability
- Sedimentation regimes
- Soil productivity in relation to plant growth

Fire can be a determinant of watershed condition. The role that fire assumes in this regard depends largely on the severity, extent, and timing of the burn, and on the decisions made in relation to its containment and suppression. How fire can affect the hydrologic processes on a watershed and, as a consequence, watershed condition from both a beneficial and detrimental viewpoint is the subject of this chapter.

HYDROLOGIC CYCLE

The hydrologic cycle represents processes and pathways by which water is circulated from land and water bodies to the atmosphere, and back again. While the hydrologic cycle is complex and dynamic, it can be simplified as a system of water-storage components, and the solid, liquid, or gaseous flows of water within and between storage points (Figure 8.1). Precipitation inputs to a watershed are affected little by burning, although chemistry of the precipitation can be altered by the resulting smoke. However, interception, evapotranspiration, infiltration, soil moisture storage, and runoff can be af-

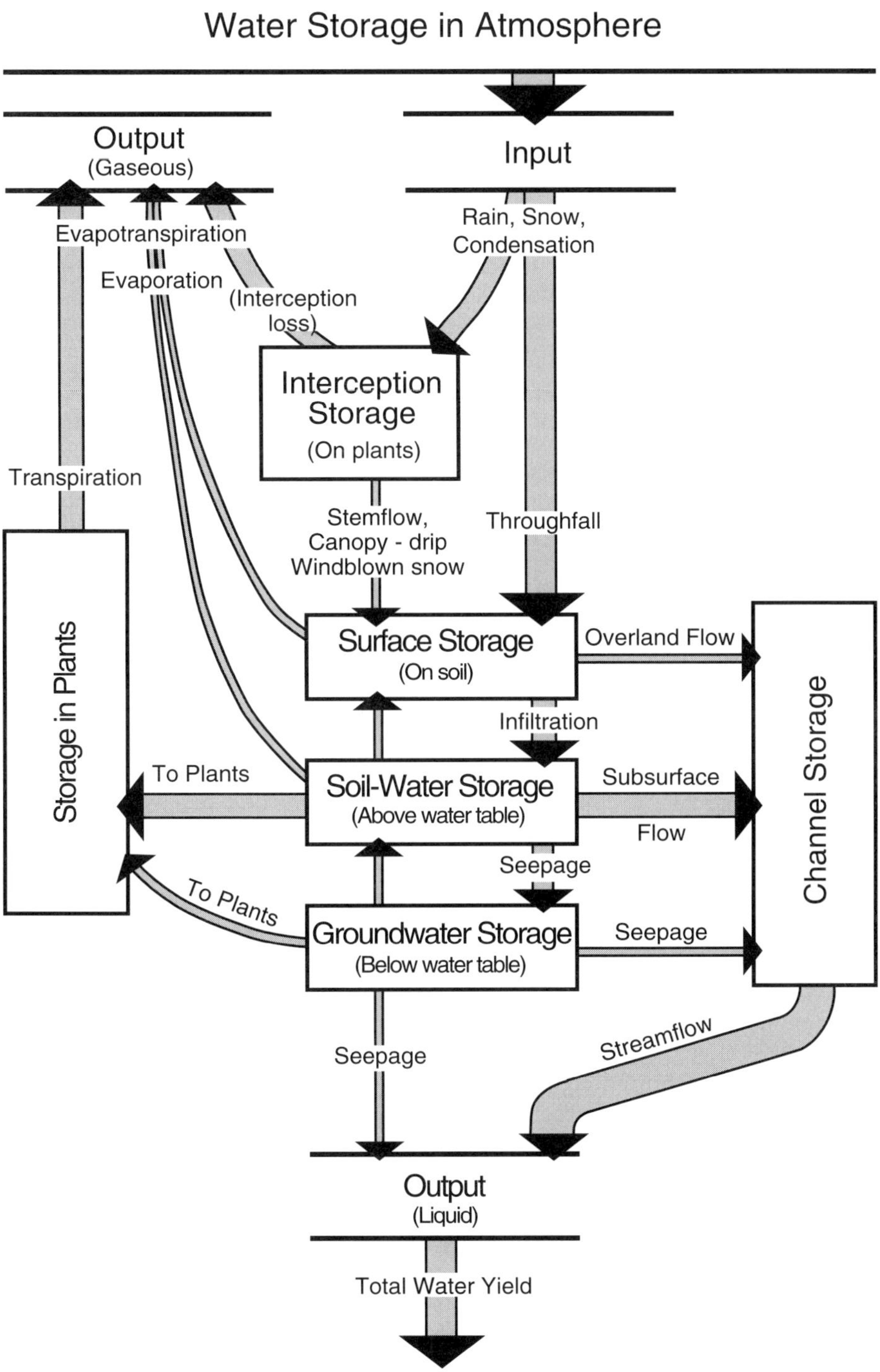

FIGURE 8.1 Hydrologic cycle of a watershed. (Adapted from Anderson et al. 1976.)

fected by fire. These components of the hydrologic cycle are closely interrelated, and, therefore, it is difficult to isolate impacts of fire on one component alone.

Interception

Interception is the process in which vegetative canopies and litter accumulations and other decomposed organic matter on the soil surface interrupt the fall of precipitation (rain or snow) to the soil surface. Interception plays a hydrologic role of protecting the soil surface from the energy of falling raindrops. Without this dissipation of energy, the mineral soil surface can become compacted or dislodged by raindrop splash, which then affects the infiltration characteristics of the soil surface and the processes of runoff and erosion.

HYDROLOGIC SIGNIFICANCE

Much of the precipitation that is intercepted returns to the atmosphere by evaporation and, therefore, becomes a loss of water from a watershed. As a consequence, interception is a storage term that is subtracted from the *gross precipitation* input to a watershed in water budget studies. Not all of the precipitation intercepted by a vegetative canopy or litter layer is returned through vaporization to the atmosphere, however. Much of the water intercepted by a vegetative canopy, especially in large storms of long durations, drips off the foliage or flows down the stems to the soil surface (Brooks et al. 1997). A portion of the water that is intercepted by litter layers also drains to the soil surface.

There is considerable variability in the magnitude of rainfall interception throughout the world (Helvey 1971, Ffolliott and Brooks 1996). Interception losses in temperate forests of North America, Europe, and Australia range from less than 5 to over 35 percent of the annual rainfall. The proportion of rainfall that is intercepted by humid and subtropical forests is generally greater than that of forests in temperate regions. For example, interception of rainfall in the humid tropical forests of Thailand, Peninsular Malaysia, and the Philippines often reaches 50 to 70 percent of the annual amount of rain. Interception losses in the sparsely stocked woodlands, shrublands, and grasslands of the dry tropics and temperate regions are typically less than 10 percent of the annual rainfall. Regardless of the region, however, interception of rainfall represents a transient form of water storage in the vegetative cover of a watershed.

Interception of snowfall on foliage of forest overstories is more difficult to quantify than interception of rainfall, largely because neither the initial amount of snowfall nor the amount of water in the snow that accumulates on foliage can be measured adequately. In many situations, much of the intercepted snow is ultimately deposited on the ground (Hoover and Leaf 1967, Tennyson et al. 1974).

Interception of precipitation (either rain or snow) by vegetative canopies in a storm is generally a function of:

- The form (rain or snow), intensity, and duration of the precipitation
- The wind velocity, water vapor gradient, and other storm characteristics
- The type of vegetation (broad- or needle-leaved), number of vegetative layers, and amount of surface leaf area

Interception of precipitation by litter and other decomposed organic matter on the soil surface ranges, in general, from 5 to 35 percent of the total annual precipitation input. Storage of intercepted water in litter layers can represent a large proportion of small (25-mm and less) rainfall events. Interception and the storage of water in litter on the soil surface are related to the depth and relative development of the layers.

EFFECT OF FIRE

One hydrologic consequence of fire destroying vegetative canopies, reducing litter accumulations, or both is its effect on interception losses. Most of the vegetative canopy and litter can be lost completely in severe fires and, as a result, comparatively little interception of precipitation occurs. The effect of fire on interception in this case is a likely increase in net precipitation reaching the soil surface. However, when only small quantities of a vegetative cover are consumed in a fire of low severity, the effect of fire on the interception process is less pronounced.

Persistence of prefire levels of litter and decomposed organic matter covering the soil surface is important in protecting the soil surface in those situations where vegetation is destroyed by fire. Increased soil erosion is often a consequence when both protective layers (vegetation and litter) are lost to fire.

Evapotranspiration

Evaporation from soils, plant surfaces, and water bodies, together with the water losses from transpiring plants, are collectively called *evapotranspiration* (ET). Therefore, interception that evaporates from plant foliage is part of the total ET component of the hydrologic cycle. ET is usually a high percentage of the annual precipitation in a water budget, approaching 100 percent on some forested watersheds. The ET component interests hydrologists and watershed managers, because this component of the hydrologic cycle largely determines what proportion of the precipitation input to a watershed is likely to become streamflow runoff and groundwater recharge. ET also represents the component of the hydrologic cycle that is influenced the most by vegetative change.

HYDROLOGIC SIGNIFICANCE

While the ET process largely controls the hydrologic response of a watershed to rainfall and snowmelt inputs, hydrologists and watershed managers still understand little about the process itself, and the feedback mechanisms that control the ET process in natural environments (Morton 1990). It is known, however, that the type, density, and structure of vegetation influence transpiration losses through time. Differences in transpiration rates among plant communities and individual plant species are attributed largely to:

- Differences in rooting characteristics
- Stomatal response
- Albedo of leaf surfaces
- Length of growing season

What is known about the magnitude of ET losses has been obtained mostly from small plot studies, the use of tracers, or measurements of water vapor flux. While estimates of ET obtained from these methods are site-specific and technique-dependent, they are reflective of the range of ET losses in different regions of the world.

Estimated ET values in temperate forests of North America and Europe range from 40 to over 80 percent of the annual precipitation (Ffolliott and Brooks 1996). ET losses in humid and tropical forests of Southeast Asia are estimated at 50 and 80 percent of the annual precipitation, respectively. These estimates of ET vary greatly with different compositions and structures of forest overstories. ET represents the largest loss of water in the hydrologic cycle of dryland regions. In some situations, soil water storage following the growing season is nil, regardless of the type of vegetative cover, indicating that large quantities of precipitation are lost through the ET process.

On a watershed scale, it has been estimated that 80 to 95 percent of the annual precipitation is evaporated from land surfaces or transpired by plants on many forested watersheds in the southwestern United States, leaving only 5 to 20 percent available for runoff (Ffolliott and Thorud 1977). While runoff approaches 50 percent of the rainfall and larger snowmelt inputs on high mountain watersheds in Colorado, the ET component of the hydrologic cycle is still large and potentially subject to modification.

EFFECT OF FIRE

Watershed management studies throughout the world have demonstrated that the runoff component of the hydrologic cycle can increase following a vegetative change that reduces ET losses (Bosch and Hewlett 1982, Whitehead and Robinson 1993). That is, following the vegetative change, less precipitation is converted into vapor as a result of the ET process and, as a consequence, more water is available for runoff. One cause of such a vegetative change is that resulting from either *vegetation-modifying* or *vegetation-replacing* fire.

Fire that modifies the composition and structure of vegetation by removing foliar volume will result in less ET losses from a watershed. Fire that results in replacement of deep-rooted, high-profile trees and shrubs by shallow-rooted, low-profile herbaceous plants is also likely to reduce ET losses. In either instance, less ET loss following a fire often translates into increased runoff. Moreover, a relatively small change in ET losses because of fire can result in a significant increase in runoff (Example 8.1). The reduced ET losses are due to changes in structure and composition of the vegetation.

Example 8.1: A Small Percent Change in ET Losses Can Result in a Large Percent Increase in Runoff: Hypothetical Case (Ffolliott and Thorud 1977)

A hypothetical, but realistic, water budget for a forested watershed in the southwestern United States illustrates the above concept. Prior to a change in vegetative composition and structure, such as might occur as a result of a fire, annual precipitation and ET were 635 and 550 mm, respectively, leaving 85 mm for the runoff component. After the burn, annual runoff was increased by 25 mm due to a reduction in ET of 25 mm attributed to the impact of the fire on vegetation. It has been assumed that the precipitation input is not affected by the fire. Runoff increased by 30 percent in this example, while ET losses decreased by only 5 percent. The percent decrease in the ET component is likely to be small no matter what the vegetative change on a watershed. ■

Infiltration

Precipitation that reaches the soil surface, that is, *net precipitation,* moves into the soil, forms puddles of water on the soil surface, or flows over the surface of the soil (Figure 8.2). The process of water entering the soil is *infiltration.* The maximum rate at which water can enter the soil is called the infiltration *capacity.* Water that infiltrates into the soil moves either downward and laterally to a stream channel by *interflow* or downward to a groundwater aquifer. When more water is supplied to a site than can infiltrate, the excess water flows off the surface as *overland flow* to a stream channel.

HYDROLOGIC SIGNIFICANCE

The relative proportions of net precipitation that infiltrate into the soil and move toward the groundwater aquifer, or run off to a stream channel, deter-

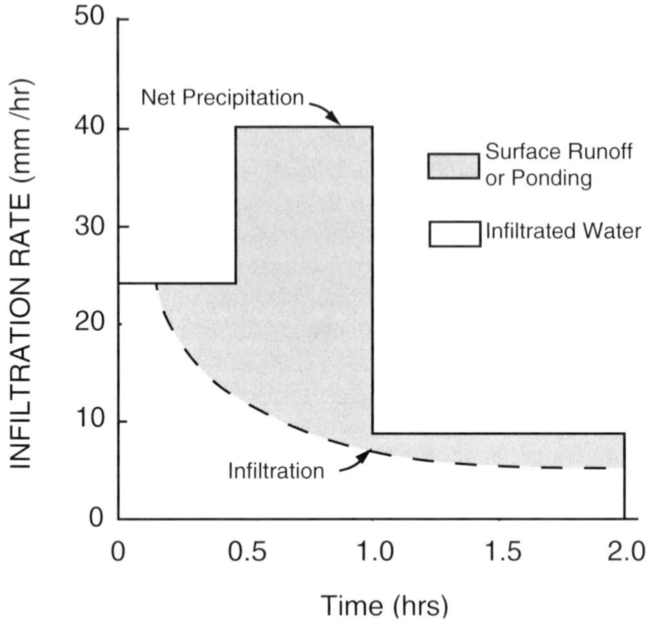

FIGURE 8.2 Relationship between rainfall and infiltration resulting in puddling of water on the soil surface, or flowing of water over the surface of the soil. (Adapted from Brooks et al. 1997.)

mine the amount and timing of streamflow that ultimately occurs. Infiltrometer measurements have been the basis of most investigations of infiltration. Therefore, the infiltration capacity values reported are essentially "point estimates" in that they were generated from measurements taken on relatively small plots representing site-specific conditions, often limiting their extrapolation to a larger watershed scale (Ffolliott and Brooks 1996).

Infiltrometer measurements indicate that undisturbed forest soils have high infiltration capacities compared to other types of soil. This high rate of infiltration is a major factor contributing to the popularly held idea that forests have a moderating effect on streamflow. Because of organic accumulations on the soil surface, the high organic content of forest soils, and the large number of macropores that typically occur as a result of worm and burrowing animal activities, forest soils are generally porous and open on the surface.

Infiltration into humid and subtropical soils is generally higher than that observed in temperate regions, largely because of the more permeable structure of the soil and the greater stability of aggregates. Infiltration capacities of soils in dryland regions are often higher on sites dominated by shrub species than on grass-dominated sites or sites heavily grazed by livestock. Regardless of region or condition of the watershed, however, variables generally affecting infiltration include:

- Soil texture and porosity
- Litter accumulations and other decomposed organic matter on the soil surface
- Vegetative cover
- Land use and vegetative changes, which influence infiltration capacities primarily by altering soil water storage

EFFECT OF FIRE

When fire destroys vegetation and litter layers on a watershed, infiltration properties of soils are often altered. When burning exposes bare soil, infiltration can be reduced due to:

- A collapse of the soil structure and a subsequent increase in bulk density of the soil because of the removal of the organic matter that serves as a binding material
- A consequent reduction in soil porosity
- The impact of raindrops on the soil surface causing compaction and a further loss of soil porosity
- The kinetic forces of raindrop impact displacing surface soil particles and causing a sealing of surface pores
- Ash and charcoal residues clogging soil pores

A number of the variables that affect both cumulative infiltration in time and infiltration capacity can be affected by fire to varying degrees, often adversely, resulting in decreased infiltration (Zwolinski 1971, Biswell 1973, McNabb et al. 1989) and increased runoff and, on many watersheds, streamflow discharge. These variables include:

- The vegetative cover type
- The portion of soil surface covered by litter accumulations and other decomposed organic matter
- The weight (depth) of the litter accumulations
- Soil texture, porosity, bulk density, and so on

Another soil property that can influence infiltration is the *wettability* of the soil. Soils in some vegetative types can develop a characteristic of water repellency following fire, which can reduce infiltration capacities (Example 8.2). Although occurrence of these hydrophobic soils is common, the causes of this condition are not always well known. Most hydrophobic soils repel water as a result of organic, long-chained hydrocarbon substances coating soil particles (see Chapter 5). As a consequence, water "beads up" on the soil and will not penetrate readily, and accelerated overland flow and increased surface erosion can occur, particularly on steep slopes.

Example 8.2: Hydrologic Importance of Hydrophobic Soils (DeBano 1981, Dunn et al. 1988)

Hydrophobic soils are common in the chaparral shrublands found in southern California. Fires, which occur frequently in this region, intensify the hydrophobic condition and, apparently, volatilize organic substances that accumulate in the litter layer between fires. The resulting water-repellent layer is then driven deeper into the soil profile. This layering arrangement allows rainfall to infiltrate only to a limited depth before the wetting front reaches the water-repellent layer, often causing increases in overland flow. In addition, the soil layer above this water-repellent layer is easily eroded and, therefore, affects sedimentation and debris flow production after fire. ■

Fire can also influence the microclimate of a site by causing greater air and soil temperature extremes (Fowler and Helvey 1978). In cooler temperate regions, these temperature changes can increase potentials for concrete-type soil frost to form, which can cause a reduction in infiltration capacity indirectly related to burning (Bullard 1954).

Soil Water Storage

The maximum amount of water that a soil mantle retains against the force of gravity is the *field capacity* of the soil. When water is added to a soil that is already charged to field capacity, the excess water either flows overland or drains from the soil. Soil is normally charged to, or near to, field capacity in periods of high precipitation and at the start of the growing season.

However, much of the water that is stored in the soil is diminished in periods of low precipitation and by ET losses as the growing season progresses. The soil water deficit at the end of the growing season must be satisfied, therefore, when high precipitation amounts occur once again.

HYDROLOGIC SIGNIFICANCE

The amount of stored soil water that is lost to ET is largely dependent on the vegetative type occupying the site. Trees and shrubs have roots that can penetrate deep into the soil and, as a consequence, are able to extract water throughout much of the soil mantle. On the other hand, grasses and forbs have relatively shallow root systems and, therefore, are only able to use water in the upper meter or so of the soil.

Water at the soil surface that infiltrates is stored in the soil profile, percolates through the soil profile, or both. Vegetative change has a generally lesser effect on subsoil properties that influence soil water storage than on those impacting on infiltration. It follows, therefore, that effects of vegetative change on subsurface soil properties that influence soil water storage are not likely to be controlling factors in the hydrologic cycle of a watershed. However, a vegetative change that affects ET processes and infiltration can influence soil water storage.

EFFECT OF FIRE

The effect of fire on soil water storage largely results from the loss of vegetation by the burn, which lowers the ET losses. Lower ET losses, in turn, leave more water in the soil at the end of the growing season than would be present if the vegetation had not been burned (Tiedemann et al. 1979, Wells et al. 1979). As a consequence of this increased soil water storage, runoff and, ultimately, streamflow become more responsive to subsequent precipitation events (Example 8.3). It is likely that soil water deficits at the end of the growing season will return to prefire levels in time if the vegetative cover also recovers to conditions that characterized the watershed before burning.

Example 8.3: Soil Water Trends Following a Wildfire (Klock and Helvey 1976)

A wildfire that occurred on a 564-ha watershed in eastern Washington effectively changed the magnitude of the autumnal soil water deficit on the water-

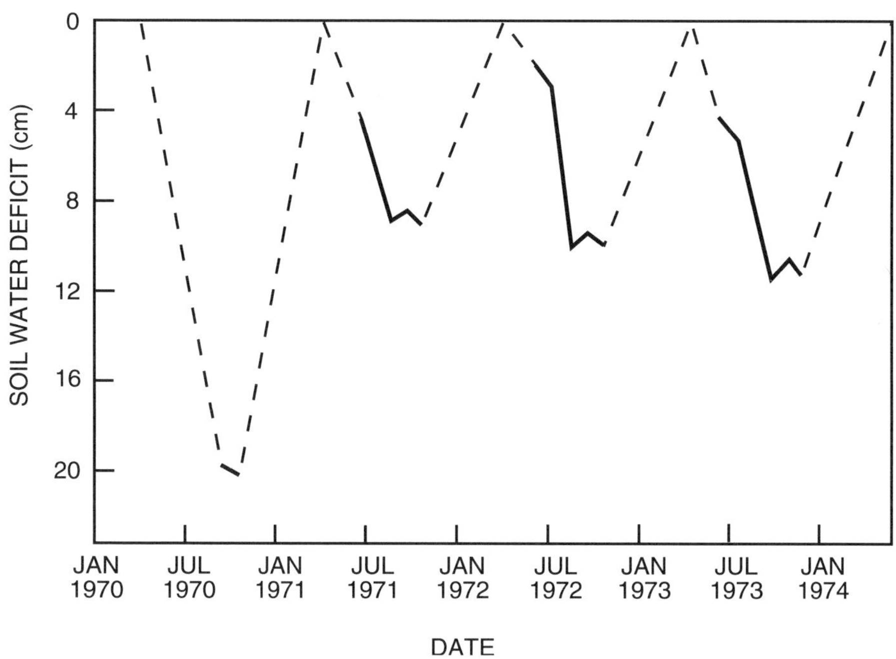

Trends in the autumnal soil water deficit in the upper 120 cm of the soil profile for 3 years following the August 1970 wildfire. (Adapted from Klock and Helvey 1976. Copyright © 1976 Tall Timbers Research Station.)

shed (see figure). The mixed conifer forest vegetation apparently depleted all of the available soil water in the upper 120 cm of the soil profile immediately before the August 1970 fire. The difference between the soil water deficit from 1970 to 1971 was about 11.6 cm, which (it was concluded) contributed a significant part to the increased streamflow discharge reported by Helvey et al. (1976). The transpiration draft of large conifer trees had been removed in 1971, and the soil water deficit was an apparent result of surface evaporation and transpiration by the newly established vegetation.

Increased autumnal soil water deficit in 1972 and 1974 was caused by the greater ET demand by the increased vegetative regrowth (Tiedemann and Klock 1976). Trends for postfire years of 1971 to 1974 suggest that the minimum soil contents might reach prefire levels in about 5 years after the wildfire. ■

The situation described above does not hold true in all cases, however. For example, a reduction in water storage was observed in the upper 30 cm of the soil on a watershed in northern Arizona where the ponderosa pine forest had been severely burned, in comparison to an unburned watershed (Campbell et al. 1977). Greater runoff from the burned watershed was the factor underlying this difference. Water repellency of the soil and the increased drying of the more exposed soil contributed to this increased runoff.

Effects of fire on water storage in rangeland soils are more variable than those in forested soils, however. Some investigators have reported that soil water storage is higher on burned sites, others have found lower soil water storage on these sites, and still others observed no change (Wells et al. 1979). Varying severities of fire are often cited as the reason for the differences.

Snow Accumulation and Melt Patterns

Much of this chapter has focused on rainfall as the precipitation falling on a watershed. However, snow is also an important form of precipitation input in many regions. Melt of snowpacks that accumulate at higher latitudes and higher elevations is often a primary source of water to downstream users. It is not surprising, therefore, that hydrologists and watershed managers can be interested in snow accumulation and melt patterns, and effects of vegetative changes on these patterns.

HYDROLOGIC SIGNIFICANCE

The total snow on a watershed at any time throughout the winter is largely a function of the total snowfall (Baker 1990). However, greater snow accumulations tend to be found at higher elevations on a watershed than at lower elevations because of the generally greater snowfall and lower temperatures at higher elevations. More snow accumulates and is retained longer into the winter season on "cooler" than "warmer" sites because of lower solar radiation levels on the former. More snow also accumulates in sparsely stocked forests than in dense forests; and additional snowfall is deposited in small openings in a forest because of increased turbulence (Troendle 1983), or through the reduction in the amount of snow intercepted by the forest canopy (Troendle and Meiman 1984, Satterlund and Adams 1992, Brooks et al. 1997).

Once snowmelt is initiated in the spring, the rate of melt is often more rapid in the forest openings than under dense vegetative canopies, largely because of greater levels of solar radiation impinging on the open site (Brooks et al. 1997).

EFFECT OF FIRE

Fire can affect snow accumulation and melt patterns when the burn creates openings in formerly dense forest vegetative canopies. Not only is snowfall interception decreased after a fire has destroyed the canopies, but frequently additional snowfall is deposited into openings due to the disruptions in wind turbulence over the canopy surface. Characteristics of openings that are created by fire depend largely on the severity of the fire. A severe fire can destroy much of the forest cover, creating many large openings in the process, while only a few, relatively small openings are likely to be created by low-severity fires that consume only the surface fuels.

Charred trees and other black bodies protruding from a snowpack after a fire can change the albedo of the ground surface, inducing earlier and more rapid rates of snowmelt. Earlier snowmelt in the spring can also be attributed to a reduced soil water deficit on a burned site (see Example 8.3). Not as much water is needed to satisfy this deficit and, as a consequence, overland flow originating from snowmelt is initiated earlier in the season. Such changes in the timing of snowmelt can also alter the timing and, as a consequence, quantity of runoff on a watershed.

Runoff

There are various pathways by which *excess water* becomes streamflow. Excess water represents that portion of total precipitation that flows off the land surface plus that which drains from the soil and, therefore, is neither consumed by ET nor leaked into deep groundwater aquifers. Pathways of flow that contribute to streamflow include the following: interception of precipitation that falls directly into a stream channel, called *channel interception, overland flow,* also called *surface runoff,* which is waterflow that has not infiltrated into the mineral soil, and *subsurface flow,* or interflow, which is that part of the precipitation input that infiltrates into the soil and then flows to a stream channel in a time short enough to be part of the stormflow. *Stormflow* is the flow resulting from a precipitation or snowmelt event. The stormflow of an *intermittent stream* is the sum of channel interception, overland flow, and subsurface flow. A *perennial stream* is fed by groundwater, or *baseflow,* which sustains streamflow between precipitation or snowmelt. The streamflow regime of a watershed is the integrated result of the processes of the hydrologic cycle.

The relative contribution of overland flow to the streamflow regime of a watershed is variable, depending largely on how impervious the soil surface is.

HYDROLOGIC SIGNIFICANCE

Overland flow occurs from sites that are impervious, locally saturated, or where the infiltration capacity has been exceeded by net precipitation or rate of snowmelt. Influences that vegetation and soil exert on interception, ET, infiltration rates, and soil moisture content ultimately affect the flow of water overland.

EFFECT OF FIRE

When fire decreases infiltration (see earlier discussion), an increase in overland flow often results. Wildfire often consumes all or nearly all of the protective vegetative and litter layers over extensive watershed areas, producing a significant effect on overland flow and, as discussed next, on streamflow from a watershed (Tiedemann et al. 1979, Baker 1990). Widespread formation of hydrophobic soils following fire also reduces infiltration, increases overland flow, and speeds delivery of the increased overland flow to stream channels (Hibbert et al., 1974, Rice 1974, Scott and Van Wyk 1990). Persistence of increased overland flow following fire often relates to the rate at which burned sites become revegetated. Prescribed burning often has its greatest hydrologic influence on infiltration processes and, as a consequence, on the potential for increased overland flow.

STREAMFLOW REGIME

It is impossible to separate pathways of flow in most investigations of streamflow responses to the effects of fire on components of the hydrologic cycles (Brooks et al. 1997). Therefore, responses of streamflow are evaluated by separating the stormflow component of the *hydrograph*, a graphical relationship of streamflow discharge (m^3/s) to time, from the baseflow (Figure 8.3), and then studying the two flow components separately. Responses of a streamflow regime to fire are reflected by changes in streamflow discharge, peak flow, timing of flow, and baseflow.

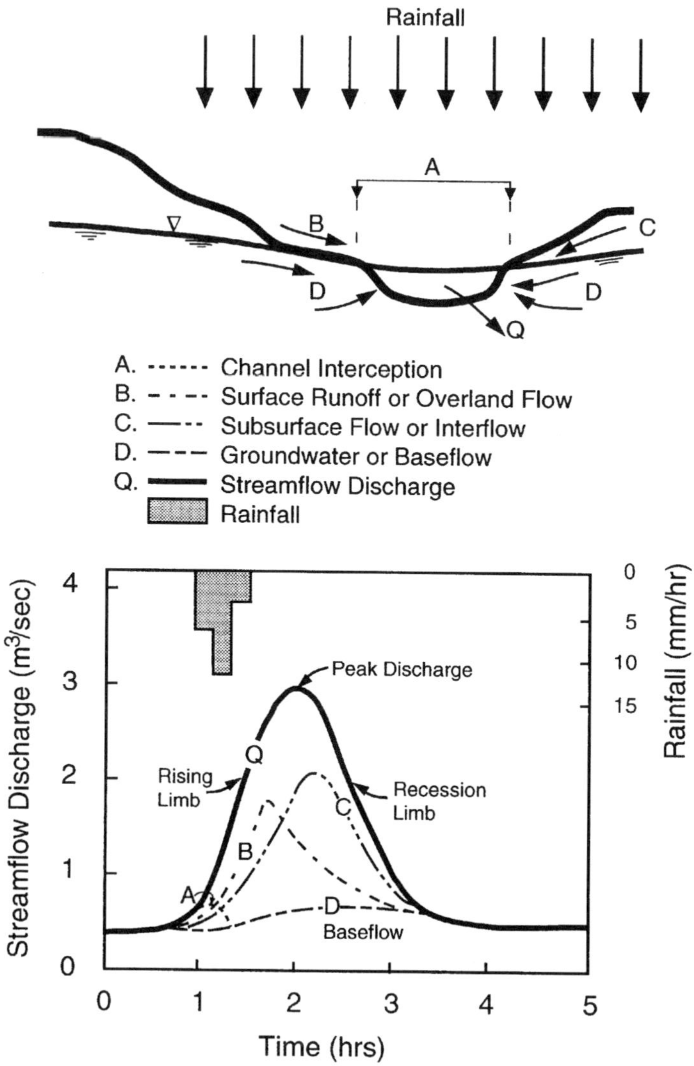

FIGURE 8.3 Relationship between pathways of flow from a watershed and the resultant streamflow hydrograph. (Adapted from Brooks et al. 1997.)

Streamflow Discharge

When vegetation and litter accumulations and other decomposed organic matter on the soil surface are destroyed by fire, reductions in interception and ET can result, along with decreases in infiltration and increases in overland flow and subsurface flow. Increases in runoff often translate into increases in streamflow discharge.

To determine the responses of streamflow discharge and, more generally, the responses of a streamflow regime to fire requires that on-site fire effects be determined initially, and that these effects then be evaluated within the context of the entire watershed. Determining effects of fire on a watershed scale is difficult, however, because of the routing and combining of all pathways of flow to the outlet of the watershed and other downstream points of interest (Brooks et al. 1997). Effects of fire on the magnitude and timing of streamflow discharge can be diminished and lengthened, respectively, as the postfire streamflow event moves downstream.

EFFECT OF WILDFIRE

There is little doubt that wildfire often has an influence on streamflow discharge. The combined effects of a loss of vegetative cover, a decrease in litter accumulations and other decomposed organic matter on the soil surface, and a frequent formation of water-repellent soils are the causative mechanisms for the increase in streamflow discharge (Tiedemann et al. 1979, Baker 1990). Increases in streamflow discharge are highly variable, although they are generally greater in regions with high ET losses, as illustrated by the following studies.

Annual streamflow discharge from a 564-ha watershed in the Cascade Range of eastern Washington, on which a wildfire killed nearly 100 percent of the mixed conifer forest vegetation, increased dramatically relative to a prefire streamflow relationship between the burned watershed and an unburned control (Figure 8.4). Differences between prefire and postfire streamflow discharge varied from 107 mm in a dry year (1977) to 47.7 cm in a wet year (1972). Soil water storage remained high for the period of record largely because of abnormally high precipitation (rain and snow) inputs (Helvey 1980). As a consequence, the burned and control watersheds became more sensitive to precipitation.

Following a wildfire in a ponderosa pine forest in northern Arizona, Campbell et al. (1977) observed a 3.5–times increase (in excess of 20 mm) in average annual stormflow discharge from a small (8.1-ha) severely burned watershed. Average annual stormflow discharge from a smaller (4-ha) moderately burned watershed increased to 2.3 times (almost 15 mm) in relation to an unburned (control) watershed. *Average runoff efficiencies* (ROE), the percentage of runoff to precipitation, increased from 0.8 percent on the unburned watershed to 3.6 and 2.8 percent on the severely burned and moderately burned watersheds, respectively. Compared to the moderately burned watershed, ROE on the severely burned watershed was 357 percent greater when the precipitation input was rain, and 51 percent less in snowmelt periods. It was speculated by the authors that the differences during rainfall events were largely due to the lower tree density, reduced litter cover, and

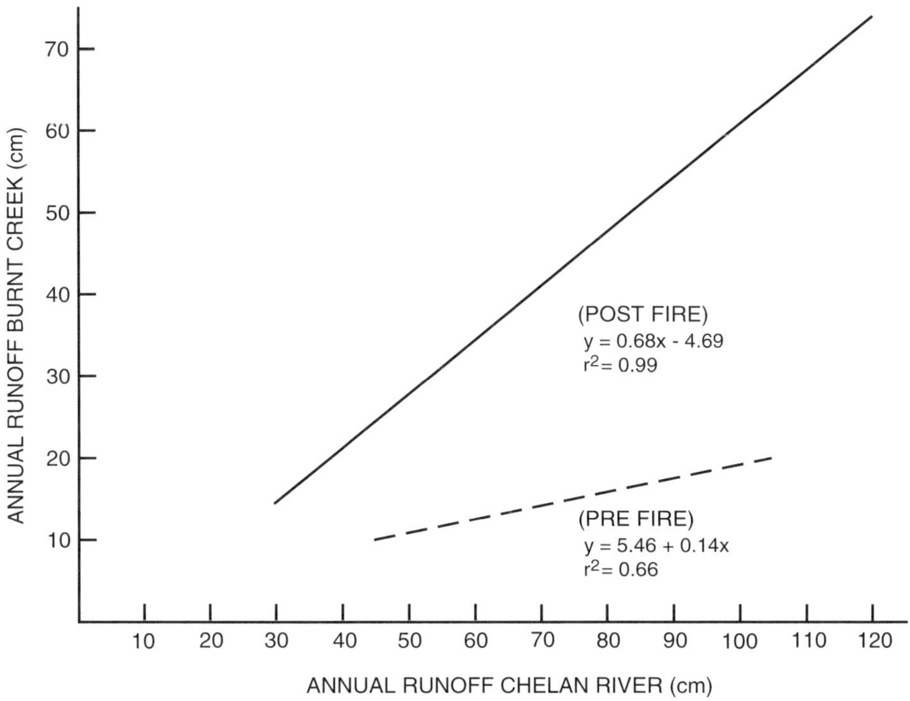

FIGURE 8.4 Annual streamflow from a burned watershed before and after the fire in relation to annual streamflow from a control. (Adapted from Helvey 1980. Copyright © 1980 American Water Resources Association.)

hydrophobic soil resulting in lower ET losses and more stormflow on the severely burned watershed than on the moderately burned watershed. In the snowmelt period, the lower tree density of the severely burned watershed allowed more of the snowpack to be lost to evaporation and, as a result, less stormflow occurred than on the more shaded, moderately burned watershed.

Scott (1993) reported on changes in streamflow discharge after the occurrence of wildfire on three watersheds in the Cape Region of South Africa. Average annual streamflow discharge increased by about 10 percent (120 mm) on a 245-ha watershed following a wildfire that consumed most of the indigenous fynbos (sclerophyllous) shrubs. This increase was attributed to the reductions in interception and ET losses.

Increases of 12 percent (70 mm) in average annual streamflow discharge and over 60 percent (almost 6.5 mm) in stormflow discharge were observed after a wildfire on a 201-ha watershed in the Cape Region that had been converted from fynbos shrubs to a *Pinus radiata* plantation (Scott 1993). While hydrologic effects of the fire were the consequence of changes on the watershed itself, reductions in interception and ET were the dominant factors

(Scott and Van Wyk 1990). There were no changes in annual streamflow or stormflow discharges after a wildfire swept across a watershed 132 ha in size. The fire killed most of the planted *Eucalyptus festigala* trees, consumed all of the surface fuels, and completely exposed the soil beneath dying trees. However, the unusually low rainfall that followed the burn could have limited postfire interpretations of the statistically weak calibration relationship with an unburned control watershed.

There was a 30 percent (nearly 60-mm) increase in streamflow discharge the first year after a wildfire occurred on a 146-ha watershed in southern France, near the Mediterranean Sea (Lavabre et al. 1993). Prefire vegetation on the watershed was a mixture of maquis, cork oak, and chestnut trees. The increase in streamflow discharge was largely related to the reduction in ET due to destruction of the vegetative cover by the fire.

Annual streamflow discharges from watersheds in the fire-prone chaparral shrublands of the southwestern United States have increased by varying magnitudes, at least temporarily, as a result of wildfire (Davis 1977, Hibbert 1984). The combined effects of loss of vegetative cover, decreased litter accumulations, and formation of water-repellent soils following burning are the presumed reasons for these streamflow increases (Rundel 1977).

EFFECT OF PRESCRIBED BURNING

Streamflow responses to prescribed fire are generally smaller in magnitude, or almost nonexistent in instances of low fire severity, in contrast to responses to wildfire. It is generally not the purpose of prescribed burning to completely consume extensive areas of litter accumulations and other decomposed organic matter on the soil surface, with resultant drastic alterations in streamflow regimes.

A burn that was prescribed to reduce fuel loads on a 180-ha watershed in the Cape Region of South Africa caused a 15 percent increase (80 mm) in average annual streamflow discharge (Scott 1993). The watershed was vegetated with fynbos shrubs, most of which were not damaged by the fire. Effectiveness of the prescribed fire was less than anticipated, however, because of the unseasonably high rainfall at the time of burning.

Annual streamflow discharge was not significantly changed relative to prefire levels in 6 years following a prescribed fire on 43 percent of a 471-ha watershed in northern Arizona supporting a ponderosa pine forest (Gottfried and DeBano 1990). The burning plan specifying a 70 percent reduction in fine fuels and a 40 percent reduction in heavy fuels was met satisfactorily. Damage to the residual stand of trees was minimal.

Burning of understory vegetation and logging residues in timber harvesting operations, burning of competing vegetation to prepare a site for planting, and burning of forests in the process of clearing land for agricultural production are common practices in many parts of the world. Depending on their intensity and extent, these prescribed burns can also cause changes in streamflow discharge from watersheds on which the treatments are conducted (Example 8.4). However, it is difficult, if not impossible, to isolate effects of these burning treatments from the accompanying hydrological impacts of timber harvesting operations, site preparation, and clearing of forest vegetation in analyzing the responses of streamflow discharge to fire.

Example 8.4: Changes in Streamflow Discharge After Burning with Other Management Activities

MALAYSIA

Clearing of rainforest vegetation before establishing plantations of fast-growing *Acacia mangium* trees was the main reason for increased streamflow discharge observed in a paired watershed near Sabah (Malmer 1992). However, the normal practice of burning the residual secondary vegetation as part of the clearing operation added an unknown increment to the increased streamflow. Even less vegetation was left to transpire following this prescribed burn.

SOUTHEASTERN UNITED STATES

Slash burning is a technique of site preparation that is done in converting stands of low-quality broad-leaved trees to pine stands in the southeastern United States. Robichaud and Waldrop (1994) reported that water flow from burned sites following slash burning increased with severity of the fire. Variability of observed flows was largely attributed to differences in the moisture content of litter accumulations prior to ignition. The hydrologic effects of tree removal are unknown.

CHILE

In a land management practice dating back to colonization, farmers in central Chile clear native trees and shrubs by felling and burning in the autumn; the cleared land is then planted to wheat once winter rains begin (Jones et al. 1975). The effect of this practice on streamflow discharge was studied on three small watersheds, 28.9, 7.6, and 3.4 ha, respectively. The larger watershed was planted to wheat following clearing, in the traditional land use, while the other two watersheds were burned and planted with *Pinus radiata* trees. Resultant stormflow discharges from the three watersheds did not differ. But, peak flows from the burned (control) watersheds increased up to 90 percent, in comparison to those on the unburned watershed. It was not possible to isolate the effect of burning on peak flows, however. Clearing of trees and shrubs beforehand was also a likely contributor to the altered streamflow regime. ■

Peak Flow

Destruction of vegetation and reductions in litter accumulations and other decomposed organic matter by fire can also increase peak flow. However, the magnitude of increased peak flow is generally variable, as indicated by the responses of peak flows following wildfire shown in Table 8.1.

Increases in peak flow as a result of fire appear generally related to the occurrence of intense, short-duration rainfall inputs, slope steepness on the watershed, and formation of soil water repellency after burning. Postfire streamflow events with excessively high peak flows are often characteristic of flooding regimes.

Timing of Flow

Another concern of hydrologists and watershed managers regarding effects of fire on a streamflow regime is the likelihood of changes in timing of flow. While information is limited, it has been observed that streamflow regimes from burned nearby watersheds often respond to rainfall inputs faster than watersheds supporting protective vegetative cover, producing streamflow events where time-to-peak is earlier (Brooks et al. 1997). Earlier time-to-peak, coupled with higher peak flow (see above), can increase the frequency of flooding.

Timing of snowmelt in the spring can also be advanced by fire. Earlier snowmelt can be initiated by lower snow albedo caused by blackened trees and increased surface exposure with elimination of vegetative cover (Helvey 1973).

TABLE 8.1 Effect of Wildfire on Peak Flows

Location	*Vegetation Type*	*Increase (percent)*	*References*
Eastern Oregon	Ponderosa pine	45	Anderson et al. (1976)
Central Arizona	Ponderosa pine	500–1500	Rich (1962)
Cape Region of South Africa	Monterey pine	290	Scott (1993)
Southwestern United States	Chaparral	200–45,000	Sinclair and Hamilton (1955), Glendening et al. (1961)

Baseflow

It has been observed that fire, often in combination with other management treatments, can increase baseflow. For example, while the hydrologic mechanisms involved are largely unclear, Berndt (1971) observed immediate increases in baseflow following a wildfire on a 564-ha watershed in eastern Washington. In addition, removal of riparian (streambank) vegetation by the fire eliminated diurnal fluctuations of flow. Increased baseflow persisted above prefire levels 3 years after the fire.

Crouse (1961) reported increased dry-season flows (baseflows) from burned watersheds on the San Dimas Experimental Forest in southern California. While these watersheds had been cleared of chaparral vegetation by burning, seeded to grass, and maintained in a grass cover with herbicides to induce higher streamflow discharges, the author and others (Dunn et al. 1988) felt that wildfire had made a significant contribution to the increased dry-season flow.

Following a wildfire in central Arizona, streamflow from watersheds dominated by chaparral shrub species became perennial at the watershed outlets after the shrubs were chemically treated to suppress resprouting, even though upstream channels were frequently dry (Hibbert et al. 1982). Streamflow was largely absent in the immediate prefire years, but it became intermittent shortly after the wildfire and remained so until the chemical treatment was imposed, at which time the streamflow regime became perennial.

SOIL EROSION AND SEDIMENTATION

Processes of the hydrologic cycle directly and indirectly affect the rate of soil erosion and, as a consequence, the transport and deposition of eroded soil as sediment in streamflow, and the physical and chemical characteristics that collectively determine the quality of water. Therefore, fire affects soil erosion, sedimentation, and water quality through changes that burning causes in the hydrologic cycle.

Soil Erosion

Soil erosion is the dislodgement and transport of soil particles and small aggregates of soil by the actions of water and wind. The rate at which soil erosion occurs on a watershed is generally affected by:

- The prevailing climate
- Soil characteristics
- Physiography of the landscape
- Extent of vegetative cover
- Land use

SURFACE EROSION

The process of *surface erosion* is caused by the actions of falling raindrops, thin films of water flowing on the soil surface, concentrated overland flow, or the erosive power of wind. Surface erosion from the action of water is affected largely by the force of gravity and, as a consequence, is most pronounced on sloping landscapes. Slope can also be a factor in surface erosion from the action of wind, particularly when eddying effects are induced.

Erosion is a natural, continuing process. However, erosive forces at the soil surface are intensified when fire denudes land by consuming the protective litter accumulations and other decomposed organic matter on the soil surface, exposing mineral soil (Figure 8.5). Soil erodibility can also be increased because of the volatilization of soil organic material and destruction of soil aggregates (Tiedemann et al. 1979). The effects of fire, therefore, trend toward an increase in the rate of surface erosion.

Postfire rates of surface erosion are often increased more on hydrophobic soils than on nonhydrophobic soils. The reduced infiltration on hydrophobic soils (see Chapter 5) and consequent accelerated overland flows combine to create conditions that are favorable for dislodgement and transport of soil particles. The water repellency produced by fire is usually confined to areas beneath plant canopies. Therefore, continuous, widespread water-repellent layers are found on sites where a dense vegetative cover has burned, while a

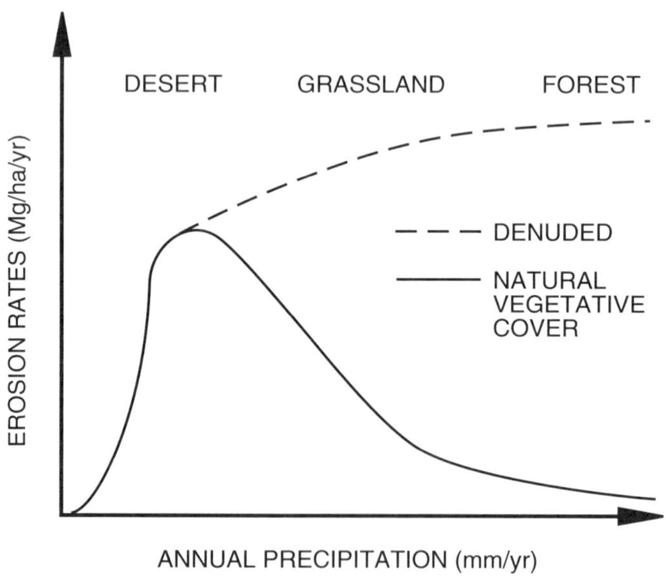

FIGURE 8.5 Relationship between erosion rates and annual precipitation for different cover types and vegetative cover conditions. (Adapted from Hudson 1981. Copyright © 1981, 1995 B. T. Batsford/Norman Hudson.)

weakened, discontinuous layer is likely to be formed where a sparse vegetative cover was only partially consumed by fire.

Debris flows following fire in the chaparral shrublands of southern California are often attributed to the increased rate of surface erosion on hydrophobic soils. A feature of frequent accelerated postfire surface erosion in these ecosystems is the rapid development of extensive rill networks on hillslopes (Wells 1987). These rill networks are linked to a layer of water-repellent soil that forms a few millimeters below the soil surface during a fire. The networks result from numerous, tiny debris flows that occur on the hillslopes during the early postfire storms. The rill networks form rapidly, often in minutes, and provide a means for transporting surface runoff to stream channels. This phenomenon helps to explain why the postfire debris flows occur during small storms and after short periods of rainfall.

Wildfire tends to accelerate surface erosion more than prescribed burning. These burnings can partially or completely remove litter, leaving the soil surface exposed to the erosive actions of raindrop splash and overland flow. Raindrop impact on mineral soil directly detaches and transports soil particles short distances (Baker 1990). In dispersing fine particles of soil, raindrop splash also tends to seal the soil surface, reducing infiltration and promoting overland flow, the causal processes that transport eroded soil materials over longer distances.

Limited information available suggests that a sufficient layer of litter and other organic matter usually remains after a prescribed burn to protect the site from accelerated surface erosion. It is not normally the intent of prescribed burning to consume extensive areas of litter accumulations and other decomposed organic matter on the soil surface. As a result, the mosaic patterns produced by restricting the spatial extent and degree of water repellency often restrict the amount of soil movement within a watershed (Baker 1990). Even when there is an increase in soil exposure and, therefore, surface erosion, much of the material that is eroded only moves a short distance downslope.

SOIL MASS MOVEMENT

Soil mass movement, the process where cohesive masses of soil are displaced, is the downslope movement, driven by the force of gravity, of soil, rock, and debris masses. This movement might be rapid (landslides) or relatively slow (creep). Causative mechanisms and processes of soil mass movement are complex in either case (Dunne and Leopold 1978, Brooks et al. 1997). However, the principal factors responsible for this form of erosion are elimination of stability provided by roots of vegetation and high soil pore water pressures resulting from removal of transpiration drafts.

Soil mass movement either develops or is accelerated in periods of abnormally high rainfall after a fire and is closely related to peak flows generated by these storms. It is a significant postfire phenomenon in southern California, the Pacific Northwest, and the Intermountain West in the western United States, and elsewhere on slopes at or near the angle of repose and where rainfall is heavy.

Sedimentation

Sediment is eroded soil that is transported from watershed surfaces to stream channels by overland flow, and then through stream systems in streamflow. Sediment, therefore, is the product of erosion. *Sedimentation* is the process of deposition of sediment in stream channels or downstream reservoirs. The amount of sediment outflow from a watershed in a stream, or the *sediment yield,* is dependent on the supply of soil particles to the channel, magnitude and rate of streamflow discharge, and physical characteristics of the sediment. Supply of soil particles to the channel and streamflow discharge, in turn, depend on the climate, physiography, vegetation, and land use.

Natural rates of sedimentation are generally lower in high-rainfall regions than in dryland regions (Brooks et al. 1997). Sedimentation in dryland environments is often viewed as a discontinuous (unsteady) process, which transports sediment from its source through a stream channel system with intermittent periods of storage (Wolman 1977, Baker 1990). This episodic sediment transport process is attributed to the major cause of erosion being the infrequent "big" storms that are characteristic of dryland environments. These storms move soil materials from a number of sources, including materials that are temporarily stored in channel systems. The disproportionate amount of sediment transported by these big storms makes it difficult to determine a "normal rate" of sedimentation on either undisturbed or burned watersheds (DeBano 1977).

Watersheds severely denuded by fire are often vulnerable to accelerated rates of soil erosion and, therefore, can yield large, but often variable, amounts of postfire sediment. Values presented in Table 8.2 are indicative of the wide variability in sediment transport response to fire. Although the magnitudes shown in this table are site-specific, changes in sediment transport presented reflect responses that can be expected from wildfire and prescribed burning.

While many fires increase sediment transport, at least to some degree, wildfire often produces more sediment than prescribed burning, at least temporarily. Large inputs of sediment into a stream can tax the transport capacity of the stream and, in some instances, lead to channel deposition (aggradation). Conversely, prescribed burns, by their design, do not normally consume extensive layers of litter. Therefore, sedimentation is generally less than that resulting from wildfire.

Sediment yields are often highest in the first year after burning, particularly if the watershed has been subjected to large, high-intensity rainfall events immediately after fire has exposed the soil surface. Postfire sediment

TABLE 8.2 Effect of Fire on Sediment Transport

			Sediment Transport (kg/ha/yr)		
Location	*Vegetation*	*Treatment*	*Prefire*	*Postfire*	*Reference*
Eastern Washington	Ponderosa pine Douglas-fir	Wildfire	28	695	Helvey (1980)
Northern Arizona	Ponderosa pine	Wildfire	0–3	1–1254	Campbell et al. (1977)
North Carolina	Southern woodlands	Prescribed burn	0	11,200	Copley et al. (1944)
Texas	Oak-juniper	Broadcast burn	0.02	28	Wright et al. (1976)
Central Arizona	Chaparral	Wildfire	175	204,000	Glendening et al. (1961)
Southern California	Chaparral	Wildfire	5530	55,300	Krammes (1960)

yields are largely indicative of the partial or complete consumption of litter and other decomposed organic matter on the soil surface and a reduction in infiltration, with consequent increase in overland flow (Brooks et al. 1997). Sedimentation usually declines in subsequent years as protective vegetation becomes established.

WATER QUALITY

Water quality refers to the physical, chemical, and biological characteristics of water in reference to a particular use. Among the physical characteristics of interest to hydrologists and watershed managers are sediment concentrations, turbidity, and water temperature. Dissolved chemical constituents of importance include N, P, Ca, Mg, and K. Some of these nutrients are adsorbed on organic and inorganic sediment particles. Bacteriological quality is also important if water is used for human consumption or recreation.

Physical Characteristics

Sediment and *turbidity,* an expression of the optical property of water that scatters light, are often the most dramatic of water quality responses to fire

(Tiedemann et al. 1979). Turbidity reduces the depth to which sunlight can penetrate into water and, therefore, influences the rate of photosynthesis. Sediment and turbidity result from erosion and overland flow, channel scouring because of increased streamflow discharge, creep accumulations in stream channels, or combinations of all three actions.

Less is known about the effect of fire on turbidity than on sedimentation (see earlier discussion). One problem contributing to this lack of information is that turbidity is difficult to measure, largely because it is a highly transient and variable parameter. Nevertheless, it has been observed that postfire turbidity levels can be affected by the steepness of the burned watershed. For example, the turbidity of overland flow from burned steep slopes in central Texas that had been converted from woodland to a herbaceous cover was higher than that of overland flow from burned slopes of lesser steepness (Wright et al. 1976). Turbidity increase after a fire is generally a result of the postfire suspension of ash and silt- to clay-sized soil particles in the water.

Water temperature, a critical water quality characteristic of many streams, controls survival of certain flora and fauna in the water. Removal of streambank (riparian) vegetation by burning can cause water temperature to rise (thermal pollution), which can increase biological activity in a stream (Brooks et al. 1997). Increases in biological activity, in turn, place a greater demand on the dissolved O_2, one of the more important water quality characteristics from a biological perspective.

Dissolved Chemical Constituents

Major sources of dissolved chemical constituents (nutrients) in streamflow from upland watersheds are geologic weathering, decompositions of photosynthetic products into inorganic substances, and storm events. Vegetative communities accumulate and cycle large quantities of nutrients in their role of linking soil, water, and atmosphere into a biological continuum (Tiedemann et al. 1979). Nutrients are cycled in a largely orderly and often predictable manner until a disturbance alters the form of their distribution. One such disturbance is fire.

Effects of fire on the nutrient status of vegetative communities on a watershed are manifested by a rapid mineralization and dispersion of plant nutrients from an intrabiotic to an extrabiotic state (Grier 1975, Tiedemann et al. 1979). Part of the plant- and litter-incorporated N, P, K, Ca, Mg, copper (Cu), Fe, Mn, and Zn are volatilized and, through this process, evacuated from the system. Metallic nutrients such as Ca, Mg, and K are converted into oxides and deposited as ash on the soil surface. Oxides are low in solubility until they react with CO_2 and water in the atmosphere and, as a result, are converted into bicarbonate salts. In this form, they are more soluble and vulnerable to loss through leaching or overland flow than as oxides, or they are incorporated into plant tissues or litter.

A postfire situation of reduced vegetative cover increases susceptibility to nutrient loss through erosion. Soil–plant cycling mechanisms reduce nutrient uptake opportunities with a reduced vegetative cover, further increasing the potential nutrient loss by leaching. Responses of some of the more important nutrients to burning are discussed in the following paragraphs.

BICARBONATE RESPONSE

Bicarbonate ions in soil solution and streamflow are often increased as a consequence of burning. The bicarbonate ion represents the principal anion in soil solution, the end product of root respiration and a product of oxide conversion after fire (McColl and Cole 1968, Tiedemann et al. 1979). Concomitant fluctuations of bicarbonate and cation concentrations indicate that bicarbonate is a main carrier of cations in the soil solution (Davis 1987).

NITROGEN RESPONSE

NO_3-N, NH_4-N, and organic-N are the N forms most commonly studied as indicators of disturbance. Most of the attention of hydrologists and watershed managers relative to water quality responses to fire focuses on NO_3-N, since it is highly mobile. The response of NO_3-N to burning is varied, however. Some investigators have found no change in the levels of NO_3-N, while others report increases of NO_3-N in either soil solution or streamflow (Hibbert et al. 1974, Tiedemann et al. 1979). NO_3-N increases in streams appear to be a result of accelerated nitrification in soil in response to a more favorable pH and increased content of electrolytes.

Increased water flow and increased concentrations of N in streamflow from burned areas have caused accelerated loss of N from watershed lands (Brown et al. 1973). While these losses are not generally referenced to the total site N capitals on a watershed, it is felt that the losses pose little threat to continued on-site productivity.

PHOSPHORUS RESPONSE

Phosphorus is present in two forms in soil solution and streamflow—orthophosphate (inorganic phosphate) and organic phosphate, the difference between total phosphate-P and orthophosphate-P. Total phosphate is reported as total P in most studies of phosphorus response to fire. Even though phosphate-P is an anion, it is not as readily leached as NO_3-N, because it complexes with organic compounds in the soil (Black 1968). Studies of soil leachates show increased levels of total Phosphorus due to burning, indicating accelerated mobilization of Phosphorus after burning (McColl and Grigal 1975, Knighton 1977). Phosphorus concentrations in overland flow can increase as a result of burning, although these increases are not always sufficient to alter the quality of streamflow (Gifford et al. 1976, Longstreth and Patten 1975).

CATION RESPONSE

Fire can substantially alter the form and distribution of cations, as mentioned earlier, placing them in a position vulnerable to removal by overland flow. Responses of cations to fire are difficult to interpret, however, due to different amounts of vegetative biomass and litter, different fire intensities, different exchange capacities of litter accumulations and soil, and different moisture fluxes and timing (Tiedemann et al. 1979, Baker 1990). Studies of soil solutions and overland flow following fire indicate increases in levels of cations such as Ca, Mg, K, sodium (Na), and Mn (Lewis 1974, Gifford et al. 1976, Knighton 1977, Sims et al. 1981).

Watershed-scale studies provide an integrated view of the effects of fire on cation concentrations and losses. A number of investigators have reported little effect of burning on ionic concentrations following burning of a watershed, concluding that the commonly reported increased streamflow resulting from fire-caused transpiration reductions likely masked concentration effects (Helvey et al. 1976, Tiedemann et al. 1978, Gottfried and DeBano 1990). Other investigators observed higher concentrations of some cations in streamflow from burned watersheds (Synder et al. 1975, Campbell et al. 1977). These elevated concentrations often return to prefire levels after the first flush of flow (Example 8.5). The variability of responses indicates that information is insufficient to adequately predict the effect of fire on concentrations of cations in streamflow.

Example 8.5: Effects of Fire on Ions and Cations in Streamflow

WASHINGTON

Maximum concentrations of NO_3-N increased from less than 0.016 to 0.56 mg/L in the 3 years after a wildfire destroyed the mixed conifer forest vegetation on a 564-ha watershed in eastern Washington (Tiedemann et al. 1978). These increases appeared to be a result of increased nitrification. Concentrations of total Phosphorus in streamflow from the burned watershed were 1.5 to 3 times greater than those from an undisturbed watershed. Despite the lack of prefire data, elevated levels of these magnitudes indicate that the wildfire significantly affected Phosphorus levels in streamflow, at least in the short-term. Combined concentrations of Ca, Mg, K, and Na in streamflow before the fire ranged from 12.0 to 14.9 mg/L. The concentrations declined to a range of 7.4 to 10.5 in the second postfire year because of the dilution caused by increased streamflow. Relative to the total capital of these nutrients, however, these losses were considered insignificant.

ARIZONA

Chemical quality of streamflow from watersheds that originally supported ponderosa pine forests in northern Arizona was not greatly affected by a wildfire (Campbell et al. 1977). While concentrations of Ca, Mg, and K increased with the first postfire streamflow events, concentrations of these cations decreased rapidly in subsequent flow events. Na concentrations were largely unaffected by the fire or by the observed changes in streamflow discharges. Combined organic–inorganic N concentrations also increased in the initial postfire streamflow event and then quickly decreased to the level that was found in streamflow from the unburned watershed.

SPAIN

Heathland communities of northern Spain are burned periodically by shepherds to improve pasture. Belillas and Roda (1993) studied effects of this burning treatment on streamflow discharge and dissolved nutrients. While burning increased streamflow discharge, fire had little impact on stream chemistry. However, dissolved nutrient losses in the streamflow increased by about 35 percent in the initial postfire years. It was anticipated that these nutrient losses would decline as streamflow returned to prefire levels. ■

Sediment Losses of Nutrients

Nutrient losses from burned watersheds result primarily from streamflow export. Dissolved chemical concentrations are commonly measured in determining these nutrient losses. However, a potentially important source of nutrient and heavy metal loss is often ignored—that of transport by sediment (Gifford and Busby 1973, Fisher and Minckley 1978, Gosz et al. 1980). Sediment has been reported to transport relatively high levels of nutrients and heavy metals (Angino et al. 1974, Potter et al. 1975). However, most of these studies have looked at large river basins consisting of numerous combinations of bedrock and vegetation.

Few studies have focused on the effects of fire on nutrient and heavy metal losses that are transported by sediment from smaller upland watersheds. It was reported in one study, however, that sediment losses of N, P, and cations in streamflow from burned watersheds in chaparral shrublands of southern California can substantially exceed those lost in solution (DeBano and Conrad 1978). Nitrogen and P losses of 15.1 and 3.4 kg/ha, respectively, were found in sediments, as compared to only trace levels in solution. Furthermore, losses of Ca, Mg, Na, and K in solution were only one-fourth of the losses in sediment. Sediment and solution losses of N, P, K, Mg, Ca, and Na were only a small fraction of the total prefire nutrient capital of plants, litter accumulations, and the upper 10 cm of soil, however.

Biological Quality

Litter accumulations and other decomposed organic matter on the soil surface often function as a filter that removes bacteria and other biological organisms from overland flow. It has been observed that rainfall runoff and snowmelt that percolates through a litter layer or strip of organic matter contain fewer bacteria than water that had not passed through the strip (Brooks et al. 1997). It follows, therefore, that the destruction of this layer or strip by burning might result in higher concentrations of bacterial and other biological organisms flowing overland to a stream channel.

CUMULATIVE EFFECTS OF FIRE ON WATER RESOURCES

Land-use activities, fire, and other disturbances affect environmental parameters on a watershed by changing the character of vegetation and soil; importing and removing fauna, water, and chemicals; and introducing pathogens and heat. As these environmental parameters change, hydrologic processes that are associated with the transport of water through the watershed change in response and alter the production and transport of water, sediment, chemicals, and heat (Reid 1993). How fire changes the environmental parameters that alter the generation and transport of these watershed products, thus modifying watershed condition and aquatic ecosystems, has been the subject of this chapter.

Cumulative Effects of Fire

Cumulative effects of fire on water resources can be viewed at the site of the disturbance or downstream from the watershed. *On-site fire effects* include impacts on vegetation, soil, and nutrient cycling. Changes resulting from these cumulative effects influence long-term productivity of the watershed. These changes can have major impacts on water production, timber management, rangeland use, agriculture, and recreational opportunities. *Downstream effects* take many forms, but each cumulative effect occurs largely because the environmental change alters the production or transport of water, sediment, organic matter, chemicals, and heat. Downstream fire effects occur when hydrologic processes are altered for a long enough time that the changes can accumulate through time, when responses from a number of sites are transported to the same site, or when a transported response interacts with an on-site change at another site.

Hydrologic processes that are altered by fire often modify the amount and time distribution of overland flow and, as a consequence, change streamflow discharge, peak flow, and baseflow, as discussed earlier in this chapter. These changes, in turn, can alter flood flow frequencies; aggrade, incise, or widen stream channels; and change the particle distribution of sediment flows. Alteration of flood frequencies and changed channel morphology often affect downstream floodplain land use, fisheries resources, and agricultural irrigation schedules. Changes in sedimentation can cause other, often similar, responses through different mechanisms. Changes in the input and transport of organic matter after a burn are important because accumulations of large organic debris affect channel morphology. Organic matter is also a food source for aquatic animals throughout the stream system. Impacts of altered streamflow chemistry from fire can be manifested in downstream water quality problems affecting water supplies and aquatic biota. Changes in streamflow temperature are also important because of their effect on biota.

Triggering Mechanisms

A full appreciation of the cumulative effects of fire on water resources requires an understanding of the mechanisms that trigger these impacts. Paraphrasing a review of cumulative watershed effects by Reid (1993), fire-related activities influence only a limited number of environmental parameters in most instances (Path A in Figure 8.6). Resultant changes in vegetative conditions, soil characteristics, chemicals, water flow, and pathogens can induce compensatory changes along one path (Path B) and hydrologic processes on another path (Path C). Changes in hydrologic processes arise from the role of a watershed as a concentrator of water flow and include the production and transport of streamflow, sediment, chemicals, organic matter, and heat. These changes can then influence off-site environmental parameters (Path D); they can also interact with each other (Path E). Changes in either on-site environmental parameters or hydrologic processes can generate on-site cumu-

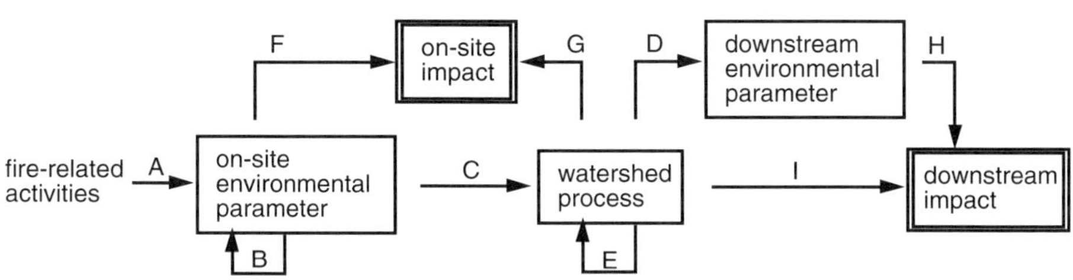

FIGURE 8.6 Influences of fire-related activities on on-site and downstream cumulative effects. Interaction of paths are described in the text. (Adapted from Reid 1993.)

lative effects (Paths F and G), but *only* changes in hydrologic processes can produce downstream effects (Paths H and I), though they might do so by changing off-site environmental parameters.

Any cumulative effect of fire on hydrologic processes, watershed condition, or aquatic ecosystems is a change that is influenced by multiple, generally progressive, or repeated hydrologic changes. Impacts of this change can accumulate through time or grow by contributions from multiple sources. Accumulations through time require that the triggering mechanism be persistent, or that the postfire recovery time be greater than the time period between fires. Spatially accumulative fire effects, in contrast, occur through immediate changes brought about by fire.

SUMMARY

Water resources are affected by fire largely through changes in the components of the hydrologic cycle. Processes represented by these components, in turn, directly and indirectly impact on soil erosion, transport and deposition of sediment, and the physical, chemical, and biological quality of water. This chapter has addressed these points and provided information so that you should be able to:

1. Discuss the concept of watershed condition in relation to how the occurrence of fire can be a determinant of this condition.
2. Understand the significance of the components of the hydrologic cycle and the ways in which fire affects the processes represented by these components.
3. Discuss the effects of wildfire and prescribed burning on surface erosion, soil mass movement, and sedimentation processes on watersheds.
4. Recognize the impacts of fire on the turbidity and temperature of water, dissolved chemical constituents of importance to ecosystem functioning, and the quality of water for human consumption and recreation.
5. Appreciate the importance of cumulative effects of fire in describing on-site and downstream postfire impacts.

REFERENCES

Anderson, H. W., M. D. Hoover, and K. G. Reinhart. 1976. *Forest and Water: Effects of Forest Management on Floods, Sedimentation, and Water Supply.* USDA Forest Service, General Technical Report PSW-18.

Angino, E. E., L. M. Magnuson, and T. C. Waugh. 1974. Mineralogy of suspended sediment and concentrations of Fe, Mn, Ni, Zn, Cu, and Pb in water and Fe, Mn, and Pb in suspended load of selected Kansas streams. *Water Resources Research* 10:1187–1191.

Baker, M. B. Jr. 1990. Hydrologic and water quality effects of fire. In: Krammes, J. S., technical coordinator. *Effects of Fire Management of Southwestern Natural Resources.* USDA Forest Service, General Technical Report RM-191, pp. 31–42.

Belillas, C. M., and F. Roda. 1993. The effects of fire on water quality, dissolved nutrient losses and the export of particulate matter from dry heathland catchments. *Journal of Hydrology* 150:1–17.

Berndt, H. W. 1971. Early effects of forest fire on streamflow characteristics. USDA Forest Service, Research Note PNW-148.

Biswell, H. H. 1973. Prescribed fire effects on water repellency, infiltration, and retention in mixed conifer litter. *University of California, Water Resources Center Report* 25:40.

Black, C. A. 1968. *Soil–Plant Relationships.* John Wiley & Sons, New York.

Bosch, J. M., and J. D. Hewlett. 1982. A review of catchment experiments to determine the effect of vegetation changes on water yield and evapotranspiration. *Journal of Hydrology* 55:3–23.

Brooks, K. N., P. F. Ffolliott, H. M. Gregersen, and L. F. DeBano. 1997. *Hydrology and the Management of Watersheds.* Iowa State University Press, Ames, Iowa.

Brown, G. W., A. R. Gahler, and R. B. Marston. 1973. Nutrient losses after clear-cut logging and slash burning in the Oregon Coast Range. *Water Resources Research* 9:1450–1453.

Bullard, W. E. 1954. *A Review of Soil Freezing by Snow Cover, Plant Cover, and Soil Conditions in Northwestern United States.* USDA Forest Service, Portland, Oregon.

Campbell, R. E., M. B. Baker, Jr., P. F. Ffolliott, F. R. Larson, and C. C. Avery. 1977. Wildfire effects on a ponderosa pine ecosystem: an Arizona case study. USDA Forest Service, Research Paper RM-191.

Copley, T. L., L. A. Forrest, A. G. McColl, and F. G. Bell. 1944. Investigations in erosion control and reclamation of eroded land in the Central Piedmont Conservation Experiment Station. *USDA Technical Bulletin 873.*

Crouse, R. P. 1961. First-year effects of land treatment on dryseason streamflow after a fire in southern California. USDA Forest Service, Pacific Southwest Forest and Range Experiment Station, Research Note 191.

Davis, E. A. 1987. Chaparral conversion and streamflow: nitrate increase is balanced mainly by a decrease in bicarbonate. *Water Resources Research* 23:215–224.

Davis, J. D. 1977. Southern California reservoir sedimentation. Preprint for the Annual Society of Civil Engineers Fall Convention and Exhibit.

DeBano, L. F. 1977. Influence of forest practices on water yield, channel stability, erosion, and sedimentation in the Southwest. In: *Forests for People: A Challenge in World Affairs. Proceedings of the Society of American Foresters 1977 National Convention.* Society of American Foresters, Washington, DC, pp. 74–78.

DeBano, L. F. 1981. *Water Repellent Soils: A State-of-the-Art.* USDA Forest Service, General Technical Report PWS-46.

DeBano, L. F., and C. E. Conrad. 1978. Effects of fire on nutrients in a chaparral ecosystem. *Ecology* 59:489–497.

Dunn, P. H., S. C. Barro, W. G. Wells III, M. A. Poth, P. M. Wohlgemuth, and C. G. Colver. 1988. *The San Dimas Experiment Forest: 50 years of Research.* USDA Forest Service, General Technical Report PSW-104.

Dunne, T., and L. B. Leopold. 1978. *Water in Environmental Planning.* W. H. Freeman and Company, San Francisco.

Ffolliott, P. F., and K. N. Brooks. 1996. Process studies in forest hydrology: a worldwide review. In: Singh, V. P., and B. Kumar, editors. *Surface-Water Hydrology.* Kluwer Academic Publishers, The Netherlands, pp. 1–18.

Ffolliott, P. F., and D. B. Thorud. 1977. Water yield improvement by vegetation management. *Water Resources Bulletin* 13:563–571.

Fisher, S. G., and W. L. Minckley. 1978. Chemical characteristics of a desert stream in flash flood. *Journal of Arid Environments* 1:25–33.

Fowler, W. B., and J. D. Helvey. 1978. Changes in the thermal regimes after prescribed burning and select tree removal (Grass Camp 1975). USDA Forest Service, Research Paper PNW-235.

Gifford, G. F., and F. E. Busby. 1973. Loss of particulate organic materials from semiarid watersheds as a result of extreme hydrologic events. *Water Resources Research* 9:1443–1449.

Gifford, G. F., J. C. Buckhouse, and F. E. Busby. 1976. Hydrologic impact of burning and grazing on a chained pinyon-juniper site in southeastern Utah. Utah Water Resources Laboratory, Publication PRJNR 012–1.

Glendening, G. E., C. P. Pase, and P. Ingebo. 1961. Preliminary hydrologic effects of wildfire in chaparral. In: *Proceedings of the 5th Annual Arizona Watershed Symposium,* September 21, 1961, Phoenix, Arizona, pp. 12–15.

Gottfried, G. J., and L. F. DeBano. 1990. Streamflow and water quality responses to preharvest prescribed burning in an undisturbed ponderosa pine watershed. In: Krammes, J. S., technical coordinator. *Effects of Fire Management of Southwestern Natural Resources.* USDA Forest Service, General Technical Report RM-191, pp. 222–228.

Gosz, J. R., C. S. White, and P. F. Ffolliott. 1980. Nutrient and heavy metal transport capacities of sediment in the southwestern United States. *Water Resources Bulletin* 16:927–933.

Grier, C. C. 1975. Wildfire effects on nutrient distribution and leaching in a coniferous ecosystem. *Canadian Journal of Forestry* 5:559–607.

Helvey, J. D. 1971. A summary of rainfall interception by certain conifers of North America. In: *Biological Effects in the Hydrological Cycle. Proceedings of the Third International Seminar for Hydrology Professors.* Purdue University, Lafayette, Indiana, pp. 103–112.

Helvey, J. D. 1973. Watershed behavior after forest fire in Washington. In: *Agriculture and Urban Considerations in Irrigation and Drainage. Proceedings of the Irrigation and Drainage Specialty Conference.* American Society of Civil Engineers, New York, pp. 403–422.

Helvey, J. D. 1980. Effects of a north central Washington wildfire on runoff and sedimentation. *Water Resources Bulletin* 16:627–634.

Helvey, J. D., A. R. Tiedemann, and W. B. Fowler. 1976. Some climatic and hydrologic effects of wildfire in Washington State. *Annual Proceedings of the Tall Timbers Fire Ecology Conference* 15:201–222.

Hibbert, A. R. 1984. Stormflows after fire and conversion of chaparral. In: *Proceedings of the Fourth International Conference on Mediterranean Ecosystems,* August 13–17, Perth, Australia, pp. 71–72.

Hibbert, A. R., E. A. Davis, and D. G. Scholl. 1974. Chaparral conversion potential in Arizona. Part I: Water response and effects on other resources. USDA Forest Service, Research Paper RM-126.

Hibbert, A. R., E. A. Davis, and O. D. Knipe. 1982. In: Conrad, C. E., and W. C.

Oechel, technical coordinators. *Proceedings of the Symposium on Dynamics and Management of Mediterranean-type Ecosystems.* USDA Forest Service, General Technical Report PSW-58, pp. 382–389.

Hoover, M. D., and C. F. Leaf. 1967. Process and significance of interception in Colorado subalpine forest. In: Sopper, W. E., and H. W. Lull, editors. *Proceedings of the International Symposium on Forest Hydrology.* Pergamon Press, New York, pp. 212–222.

Hudson, N. W. 1981. *Soil Conservation,* 2nd ed. Cornell University Press, New York.

Jones, M. E., P. F. Ffolliott, and W. O. Rasmussen. 1975. Precipitation and streamflow on 3 small Chilean watersheds. *Progressive Agriculture in Arizona* 27(1):13–16.

Klock, G. O., and J. D. Helvey. 1976. Soil-water trends following a wildfire on the Entiat Experimental Forest. *Annual Proceedings of the Tall Timber Fire Ecology Conference* 15:193–200.

Knighton, M. D. 1977. Hydrologic response and nutrient concentrations following spring burns in an oak-hickory forest. *Soil Science Society of America Journal* 41:627–632.

Krammes, J. S. 1960. Erosion from mountain side slopes after fire in southern California. USDA Forest Service, Pacific Southwestern Forest and Range Experiment Station, Research Note 171.

Lavabre, J., D. S. Torres, and F. Cernesson. 1993. Changes in the hydrological response of a small Mediterranean basin a year after a wildfire. *Journal of Hydrology* 142:273–299.

Lewis, W. M. Jr. 1974. Effects of fire on nutrient movement in a South Carolina pine forest. *Ecology* 55:1120–1127.

Longstreth, D. J., and D. T. Patten. 1975. Conversion of chaparral to grass in central Arizona: effects on selected ions in watershed runoff. *The American Midland Naturalist* 93:25–34.

Malmer, A. 1992. Water-yield changes after clear-felling tropical rainforest and establishment of forest plantation in Sabah, Malaysia. *Journal of Hydrology* 134:77–94.

McColl, J. G., and D. W. Cole. 1968. A mechanism of cation transport in a forest soil. *Northwest Science* 42:135–140.

McColl, J. G., and D. F. Grigal. 1975. Forest fire: effects of phosphorus movement to lakes. *Science* 188:1109–1111.

McNabb, D. H., F. Gaweda, and H. A. Froehlich. 1989. Infiltration, water repellency, and soil moisture content after broadcast burning a forest site in Southwest Oregon. *Journal of Soil and Water Conservation* 44:87–90.

Morton, F. I. 1990. Studies in evaporation and their lessons for the environmental sciences. *Canadian Water Resources Journal* 15:261–286.

Potter, L., D. Kidd, and D. Standiford. 1975. Mercury levels in Lake Powell: bioamplification of mercury in man-made desert reservoir. *Environmental Science and Technology* 9:41–46.

Reid, L. M. 1993. *Research and Cumulative Watershed Effects.* USDA Forest Service, General Technical Report PSW-GTR-141.

Rice, R. M. 1974. The hydrology of chaparral watersheds. In: Rosenthal, editor. *Proceedings of Symposium on Living with the Chaparral,* March 30–31, 1973, University of California, Riverside, California, pp. 27–34.

Rich, L. R. 1962. Erosion and sediment movement following a wildfire in a ponderosa pine forest in central Arizona. USDA Forest Service, Rocky Mountain Forest and Range Experiment Station, Research Note 76.

Robichaud, P. R., and T. A. Waldrop. 1994. A comparison of surface runoff and sediment yields from low- and high-severity site preparation burns. *Water Resources Bulletin* 30:27–34.

Rundel, P. W. 1977. Water balance in Mediterranean sclerophyll ecosystems. In: Mooney, H. A., and C. E. Conrad, technical coordinators. *Proceedings of the Symposium on the Environmental Consequences of Fire and Fuel Management in Mediterranean Ecosystems*. USDA Forest Service, General Technical Report WO-3, pp. 95–106.

Satterlund, D. R., and P. W. Adams. 1992. *Wildland Watershed Management*. John Wiley & Sons, New York.

Scott, D. F. 1993. The hydrological effect of fire in South African mountain catchments. *Journal of Hydrology* 150:409–432.

Scott, D. F., and D. B. Van Wyk. 1990. The effects of wildfire on soil wettability and hydrologic behaviour of an afforested catchment. *Journal of Hydrology* 121:239–256.

Sims, B. D., G. S. Lehman, and P. F. Ffolliott. 1981. Some effects of controlled burning on surface water quality. *Hydrology and Water Resources in Arizona and the Southwest* 11:87–90.

Sinclair, J. D., and E. L. Hamilton. 1955. Streamflow reactions to a fire-damaged watershed. In: *Proceedings of the Hydraulic Division*. American Society of Civil Engineering, New York.

Snyder, G. G., H. F. Haupt, and G. H. Belt, Jr. 1975. Clearcutting and burning slash alter quality of stream water in northern Idaho. USDA Forest Service, Research Paper INT-168.

Tennyson, L. C., P. F. Ffolliott, and D. B. Thorud. 1974. Use of time-lapse photography to assess potential interception in Arizona ponderosa pine. *Water Resources Bulletin* 10:1246–1254.

Tiedemann, A. R., and G. O. Klock. 1976. Development of vegetation after fire, reseeding, and fertilization on the Entiat Experimental Forest. *Annual Proceedings of the Tall Timbers Fire Ecology Conference* 15:171–192.

Tiedemann, A. R., J. D. Helvey, and T. D. Anderson. 1978. Stream chemistry and watershed nutrient economy following wildfire and fertilization in eastern Washington. *Journal of Environmental Quality* 7:580–588.

Tiedemann, A. R., C. E. Conrad, J. H. Dieterich, J. W. Hornbeck, W. F. Megahan, L. A. Viereck, and D. D. Wade. 1979. *Effects of Fire on Water: A State-of-Knowledge Review*. USDA Forest Service, General Technical Report WO-10.

Troendle, C. A. 1983. The potential for water yield augmentation from forest management in the Rocky Mountain region. *Water Resources Bulletin* 19:359–373.

Troendle, C. A., and J. R. Meiman. 1984. Options for harvesting timber to control snowpack accumulation. *Proceedings of the Annual Western Snow Conference* 52:86–97.

Wells, W. G. 1987. The effects of fire on the generation of debris flows in southern California. *Reviews in Engineering Geology* 7:105–114.

Wells, C. G., R. E. Campbell, L. F. DeBano, C. E. Lewis, R. L. Fredriksen, E. C. Franklin, R. C. Froelich, and P. H. Dunn. 1979. *Effects of fire on soil: A State-of-Knowledge Review*. USDA Forest Service, General Technical Report WO-7.

Whitehead, P. G., and M. Robinson. 1993. Experimental basin studies - an international and historical perspective of forest impacts. *Journal of Hydrology* 145:217–230.

Wolman, M. G. 1977. Changing needs and opportunities in the sediment field. *Water Resources Research* 13:50–54.

Wright, H. A., F. M. Churchill, and W. C. Stevens. 1976. Effect of prescribed burning on sediment, water yield, and water quality from dozed juniper lands in central Texas. *Journal of Range Management* 29:294–298.

Zwolinski, M. J. 1971. Effects of fire on water infiltration rates in a ponderosa pine stand. *Hydrology and Water Resources in Arizona and the Southwest* 1:107–112.

Chapter

9

Vegetation

INTRODUCTION

Effects of fire on vegetation are usually the most obvious impacts of burning. Fire, particularly fire of vegetation-modifying and vegetation-replacing severities, affects natural ecosystems by consuming plants, altering successional patterns, and changing vegetative resources such as timber, forage, and wildlife habitats. Understanding how fire affects vegetation is often a key to holistic, ecosystem-based natural resource management. A review of the more salient points of these effects on plants, succession, and vegetative resources is presented in this chapter.

PLANTS

Any plant can be killed directly by a fire of sufficient severity. With lower severity fires, where only part of the plant is likely to be killed, the ability of a plant to survive fire becomes important to managers trying to determine the effects of fire on vegetation. Inherent abilities of plants to survive a fire depend largely on their tolerance to heat and their resistance to fire. *Heat tolerance* refers to the ability of plant tissue to withstand high temperatures, while *fire resistance* is the ability of vegetation to survive the passage of fire.

Heat Tolerance

A plant is killed directly by fire when the temperature of internal living cells is raised to the lethal level. While precise information on the temperature necessary to kill living plant tissue is lacking (Brown and Davis 1973, Whelan 1995), the heat tolerance of plant protoplasm appears to be similar for most plant species. However, the temperature at which plant tissue is killed is largely dependent on the tissue moisture content (Wright 1970). Tissues with higher moisture content are generally killed at lower temperatures and in shorter time.

Not only is heat tolerance related to tissue moisture content, it is also different for different parts of a plant (Zwolinski 1990, Whelan 1995). In considering seeds, for example, some seeds are inherently tolerant of heat. Other seeds have hard, thick seed coats and, as a consequence, are able to tolerate high temperatures. Roots, bark, and foliage also vary in their relative heat tolerances. Variations in heat tolerance can occur with the presence of sugars, pectins, and other plant tissue substances.

Fire Resistance

Resistance of a plant to fire depends largely on its food reserves, the presence of fire adaptive traits, and the fire regimes to which the plant is exposed.

STATUS OF FOOD RESERVE

Plants are most susceptible to fire damage when they are flowering, forming seed, or actively growing, or when carbohydrate reserves are relatively low. Seasonal trends in food reserves vary greatly among plant species (Zwolinski 1990). Therefore, fire occurring in a particular part of a season can be more damaging to some plant species than others.

FIRE ADAPTIVE TRAITS OF WOODY PLANTS

Traits that confer fire resistance on woody plants include those that insulate or protect the buds and meristem tissues of the plant and those that add to, or conserve, food reserves of the plant (Zwolinski 1990, Whelan 1995, Pyne et al. 1996). These traits include:

- *Bark thickness.* A primary factor that determines whether a tree or shrub species is fire resistant is its bark thickness. Trees and shrubs suffer little heat damage if their bark thickness is 1.0 to 1.3 cm (Wright and Bailey 1982). Bark tends to be relatively thin in young, small trees and shrubs, becomes thicker as plants mature, and finally declines with senescence.
- *Protected buds.* Fire resistance of a plant is frequently attributed to the

location of meristems and protection that buds receive from lethal temperatures. Exposed locations of terminal and lateral apical buds of many small trees and shrub species make them highly susceptible to top-killing from fire, while plants with surface and subterranean buds are protected. Conversely, the basal meristems of many grass and forb species provide a fire-survival advantage.

- *Fire-stimulated sprouting.* Basal sprouting is common in some trees and shrubs after fire has destroyed their foliage and small twigs. An inhibiting factor of some kind apparently prevents bud activity while foliage is alive, but this inhibition disappears and dormant buds begin sprouting when the foliage is killed (Chandler et al. 1983). Resprouting after fire is generally related to age of the plant, stem size, season, fire frequency, and fire severity.
- *Seed retention and fire-stimulated seed dispersal.* Retention of seeds on plants and fire-stimulated seed dispersal are important for survival of many trees and shrubs in fire-prone environments. A number of pine species have cones that split open (dehiscence) under the effects of fire, a trait known as *serotiny.* Cone scales are held closed by a resinous material sensitive to high temperatures. Once a fire passes and melts the resins, the cone scales exfoliate and seeds are dispersed (Zwolinski 1990). When seeds are released, they often fall on a seedbed where ash and minerals provide nutrients, and reduction in overstory foliage allows increased sunlight to reach the soil surface to assist in germination and seedling growth.
- *Fire-stimulated germination of seeds.* Germination of seeds stored in the soil can be stimulated by the heat that is generated by a fire (Martin et al. 1975, Shea et al. 1979, Takahashi and Kikuchi 1986, Keeley 1987). The germination of seeds from some plant species on recently burned sites is also attributed to the release of seeds retained by the plants.

Other developmental patterns of trees and shrubs that can be fire adaptive traits include seed burial, plant longevity, chemical composition, time to initial flowering, and nature of shedding leaves (Gill 1981, Zwolinski 1990). However, while there are many adaptive mechanisms that facilitate reproduction and survival of certain plants in fire-prone environments, even a fire-tolerant plant species can lose vigor and, in time, die as a result of fire. Death is likely to happen if the pattern of fire frequency coincides with the general growth pattern of the plant, which tends to progressively reduce food reserves.

FIRE ADAPTIVE TRAITS OF HERBACEOUS PLANTS

Comparatively little is known of the factors that bestow fire tolerance to certain herbaceous plants. That herbage is generally less affected by fire than are trees and shrubs is often attributed to the protected position of their growing points. It is less clear, however, why certain herbaceous species increase after burning while others do not. It has been argued that seed of herbaceous species increasing in frequency after fire is less prone to fire, or that germination is improved by exposure to fire (Pratt and Gwynne 1977). But, it is difficult to distinguish between these two direct effects of fire on plant survival, and the possible indirect effects of selective prefire livestock grazing and drought mortality. Seeds of some grasses and other herbaceous plants are also more easily transported into surrounding sites than those of many shrubs and trees.

Another possible explanation of why certain herbaceous species are able to thrive after burning is that fire stimulates flowering of some species (Chandler et al., 1983, Gill 1981). While mechanisms responsible for the triggering of flowering by fire are largely unknown, this trait often results in prolific flowering followed by increased plant establishment. That this happens might be a consequence of greater seed set, lower on-plant seed predation, or less predation of seeds following dispersal.

Despite their general fire tolerance, perennial grasses can be susceptible to fire in early stages of their life and, therefore, experience high mortality at this time (McPherson 1995). However, many grasses and other herbaceous plants are relatively tolerant of fire that occurs at the end of the growing season, when they are dormant.

Example 9.1: Role of Lightning in Natural Fire Regimes (Lorimer 1990)

People have historically been a principal cause of fire in natural ecosystems, in both actual number of fires and area burned. Yet, lightning is the primary determinant of natural fire regimes in many ecosystems. Frequency of lightning fires is highly variable throughout the world, however. Lightning-caused fire is infrequent in tropical rainforests. Lightning is also relatively uncommon in much of Europe, eastern North America, and other moist temperate regions of the world. Lightning causes about 2 percent of all forest fires in the eastern United States. But, the incidence of lightning fires in the western United States is probably the highest in the world. In the Rocky Mountain region, for example, lightning causes nearly 60 percent of all forest fires. It is not uncommon for a single storm to ignite 30 or more fires. ■

NATURAL FIRE REGIMES

The effectiveness of fire adaptive traits in protecting plants should also be considered in terms of a sequence of natural fires, each with different characteristics and effects, rather than a single fire occurrence. Sequences of natural fire are reflected in the frequency of lightning-caused fire in many ecosystems (Example 9.1), especially ecosystems of dense vegetation and occasional dry weather. Life cycles of plant species and the fire regimes to which the species are subjected are equally important in assessing the effectiveness of these traits. Factors characterizing natural fire regimes and, therefore, affecting postfire plant survival include the following:

- *Fire frequency.* A plant species can be eliminated from a site if fire occurs too often, too early, or too late in the plant's life cycle. Therefore, fire frequency is important in determining species composition of a plant community on a site by selecting species that will continue to occupy the site (Heinselman 1981, Zwolinski 1990, Whelan 1995, Pyne et al. 1996). Species that sprout are generally able to withstand repeated fire, while those plants that produce seed are favored by infrequent fire (Keeley 1981).
- *Season of fire.* The season of fire occurrence also affects plant survival and flowering. Spring fire on rangelands in the western United States can damage annual grasses that emerge following winter rains but have not had time to produce seeds (Wright 1969, McPherson 1995). On the other hand, many perennial species are dormant during spring fire and could resprout later in the season.
- *Severity, duration, and extent of burning.* Plant responses to fire severity and duration of heating are variable. Low-severity fire can be damaging to fire-sensitive species, while moderate- to high-severity fire can stimulate reproduction in plant communities dependent on fire for renewal. The extent of the area burned can also affect plant survival and flowering when the species is unable to regenerate by sprouting. Wind, animals, or other carriers of seed might not provide adequate distribution when the burned area is extensive.

Fire Frequency

Fire frequency is a particularly important determinant of natural fire regimes (Kilgore 1981, Whelan 1995, Pyne et al. 1996). It has often been assumed, especially by laypeople, that the pristine vegetation of the United States first viewed by Euro-Americans was also a "climax vegetation" in which fire played little or no role (Riggs et al. 1996). On the contrary, fire has always been, and continues to be, a common element in the functioning of many ecosystems. Little vegetation is normally spared from the influence of naturally occurring fire in these ecosystems, although fire frequency varies widely in some vegetative types (Table 9.1).

FIRE-FREE INTERVALS

Effects of fire on composition of plant communities can be relatively minor when fire occurs at short intervals (<20 years), because the communities become dominated by fire-tolerant plant species. Under such fire regimes, tree seedlings and nonsprouting shrubs can be severely reduced in number or to-

TABLE 9.1 Fire-Free Intervals for Some Vegetative Types in Western North America

Vegetative Type	*Fire-Free Interval (yr)*	*Reference*
Ponderosa pine forests	7–10	Weaver (1951), Arno (1976), Kilgore and Taylor (1979)
Pinyon-juniper woodlands	10–30	Leopold (1924)
California chaparral	20–30	Byrne (1978), Radtke et al. (1982)
Mountain sagebrush	20–25	Houston (1973)
Tallgrass prairie	5–10	Wright and Bailey (1982)
Desert grassland	7–10	Leopold (1924), Schmutz et al. (1982), Wright and Bailey (1982)

tally eliminated on a site. Sprouting trees and shrubs can be top-killed (with roots still alive) or totally killed when they are burned at a young age (Wright et al. 1976). Fire in shrub-dominated communities and grasslands generally kills aerial parts of small trees and shrubs, encouraging dominance by fire-tolerant herbaceous species. Sustained short fire intervals serve to maintain the dominant herbaceous character of the vegetation and, in doing so, slow succession to advanced plant communities.

Short fire-free intervals in tree and shrub communities tend to result in low-severity surface fires (Riggs et al. 1996). While woody overstories in these plant communities can be affected at specific locations where fire severity is high enough to cause some mortality, much of the stand remains intact. Fuel loads continue to accumulate as fire intervals increase in length, increasing the probability that eventual fire will be high in severity. Vegetation replacement is often initiated with this fire when it occurs.

Seral tree species tend to be maintained in communities that are characterized by intermediate fire-free intervals (20–75 years). When fire does occurs at these intervals, it tends to be moderate in severity. Fire-sensitive plant species often become established and, in some situations, eventually dominate the site. Fuel buildups continue and become more continuous in distribution. As a consequence, subsequent occurrence of high-severity fire results in generally greater changes in plant compositions and structures than would occur if the communities had been subjected to more frequent low-severity fires.

Long intervals (>100 years) between fire are typically associated with high-severity, vegetation-replacing fire. Initial postfire vegetation is likely to differ markedly in composition and structure from prefire vegetation. Because of the high mortality of species found originally in prefire plant communities, intensely burned sites become dominated by annuals and short-lived perennials that rapidly establish themselves on disturbed sites.

Long fire intervals, however, do not necessarily reduce the importance of fire to long-term community persistence on a site (Agee 1991). Fire-dependent forests of coastal Douglas-fir, jack pine *(P. banksiana)*, and lodgepole pine are evidence of this point. Recurring fire that maintains open canopies is necessary for the inclusion of plant species that are intolerant to shade and, therefore, require full sunlight to compete and survive as components in uneven-aged, mixed-species communities.

EFFECTS OF MANAGEMENT PRACTICES

Management practices increase fire-free intervals in some instances. For example, fire suppression activities, particularly in areas of high human population and cultural developments, have often lengthened intervals between fire (Riggs et al. 1996). Livestock grazing reduces accumulations of fine fuels and, as a consequence, the occurrence of fire on many rangelands. Livestock grazing can substantially reduce fire potentials on arid and semiarid rangelands, where natural accumulations of fine fuels are usually near or below levels necessary to sustain fire anyway. Clearing of land for agricultural cropping, road systems, and transmission lines can decrease the continuity of fuels across many landscapes, also increasing fire-free intervals in many instances.

Increases in fire-free intervals because of management practices have resulted in major changes in compositions of many plant communities. Woody plant species frequently advance onto sites occupied by herbaceous plants, especially on sites subjected to intensive livestock grazing pressure. Some seral forest types have also been affected by establishment of late-seral tree species that normally would not dominate. Succession in some plant communities has also been changed by increases in fire-free intervals. Long-term protection from fire can change canopy cover and vertical structure of plant communities, resulting in increasing densities and reproduction of "climax" plant species, adding diversity to the communities.

SUCCESSION

There are two schools of thought on the perspective to take in analyzing the effects of fire on succession in natural ecosystems. One school fosters a community-replacement concept of succession originally proposed by Clements (1916, 1936), who postulated that continuing successions of plant communities ultimately reach biologically stable stages. Ability of individual plant species to persist in postfire environments is the basis for a second school of thought in interpreting effects of fire on succession. Here, the replacement of individual species, as opposed to community replacement, is the focus. It is suggested that both approaches to studying succession have merit and, depending on the ecosystem, stage of succession, and nature of the fire, should be considered in analyzing fire effects on successional patterns.

Community Replacement

The Clementsian concept of succession simply states that one plant community is replaced by another plant community through a largely orderly and predictable process resulting from naturally occurring modifications in conditions that create a more favorable environment for the establishment of the other community. This progression eventually leads to a biologically stable ecosystem called the *climax plant community.* Following a fire disturbance, the prefire plant community is likely to disappear, being replaced by a plant community earlier in the successional progression (Figure 9.1). Once this retrogression has occurred, succession then moves forward once again through seral stages toward the climax plant community.

Assumptions on which the community-replacement concept of succession is based cannot always be met in considering how fire disturbances affect successional patterns, however. In fact, this concept of succession is probably not appropriate in some situations. Not all fire-induced successions necessarily move in an orderly progression to a climax plant community (Zwolinski 1990). When fire is an integral part of the natural environment for a plant community, the term *climax* has little meaning, since plants that persist are the climax species.

Furthermore, climax implies stability, and plant communities are rarely stable. With a range of ages and life spans, weak and overmature plants often disappear and are replaced by new plants and other species. Openings that are created by fire, or other damaging agents, also allow establishment of other plants. Therefore, plant communities are largely characterized by continual change rather than stability. Interestingly enough, Patterson (1986) recommended that the word *climax* be stricken from forestry terminology largely for this reason.

Even though the community-replacement concept of succession does not always deal effectively with plant changes following fire (Zwolinski 1990), the following generalizations can be made on effects of fire on successional patterns of individual plant species (Noble and Slatyer 1980, 1981):

- Immediate postfire species composition is largely dependent on propagules that have arrived from adjacent unburned areas, persisted through the fire, or resprouted from organs surviving the fire. Communities of postfire plant species, therefore, are often similar to prefire communities or communities existing on adjacent unburned areas.
- There is a surge of plant recruitment into a community under conditions of low competition shortly after fire. However, a slowing of recruitment

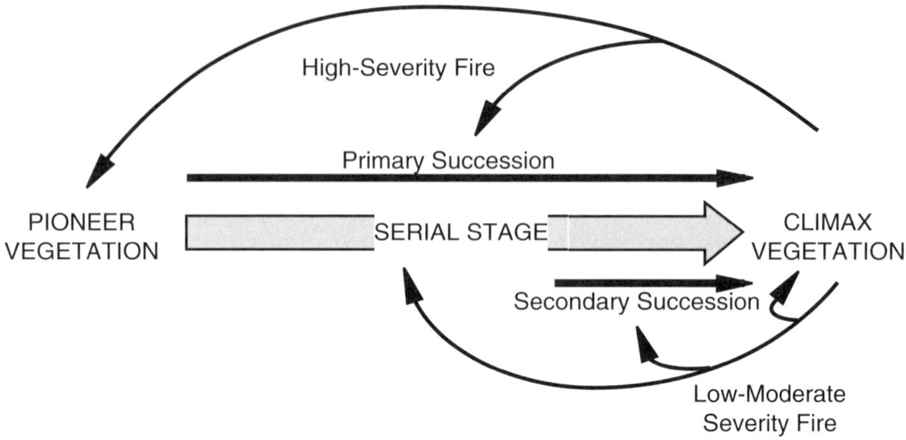

FIGURE 9.1 Community-replacement progresses to a climax plant community in the classical Clementsian concept of succession. *Primary succession* is initiated on lava flow, sand dunes, alluvial deposits, and other newly exposed sites. *Secondary succession* follows disturbance such as fire.

often follows this initial surge as plants become established and, as a consequence, more difficult to displace.

- Plant species that are long-lived and those species that can regenerate and grow under canopy cover are likely to dominate in the absence of further fire.

These generalizations can be helpful to understanding successional pathways following fire. However, evaluation of responses of individual plant species leading to community development is often more useful in this regard than the community-replacement concept of succession. Fire-induced succession is largely related to abilities of individual plant species to colonize, grow, and ultimately survive in postfire environments (Zwolinski 1990, Pyne et al. 1996). Therefore, life-history characteristics of key plant species in a community can often be used to describe successional patterns following fire of varying frequencies and severities, as discussed next.

Individual Plant Species

Fire adaptive traits of individual plant species can help to predict major shifts in postfire species composition and dominance in fire-prone ecosystems. Therefore, one should become familiar with individual plant species in their area, and identify the attributes that may permit these species to persist following fire. Noble and Slatyer (1977, 1980, 1981), Cattelino et al. (1979), Noble (1981), and Pyne et al. (1996) identified a number of attributes that are vital to the status of plant species on sites subjected to recurrent fire. These authors also proposed a *multiple-pathways model* based on these attributes for predicting major shifts in species composition and dominance on fire-prone sites and, thereby, providing information on postfire successional patterns. The model reflects species attributes that are vital to reproduction and survival of the plant species within the postfire community. These attributes and the authors' nomenclature for the attributes include the following:

- *Method of persistence.* This attribute describes the method of persistence through fire (or other) disturbance. Four mechanisms are recognized:

  Persistence by arrival of widely dispersed seeds *(D species).* Seeds are dispersed from surrounding unburned areas and always available at the site.

  Persistence by seeds with long viability that are stored in soil *(S species).* Not all seeds germinate after fire and the seed pool can persist through a series of fires without replenishment.

  Persistence of seeds with short viability that survive fire disturbance with protective fruits or cones that are stored in canopies *(C species).* Seeds are available only if adults were present before the fire.

  Persistence by part of the individual plant surviving fire and recovery by vegetative regrowth *(V species).*

- *Conditions for establishment.* This attribute describes the conditions necessary for establishment. Three mechanisms are recognized:

  Able to become established at any time, with adults of same and other species occurring on the site. These species can tolerate competition *(T species). Tolerance* applies to light, soil nutrients, or other physiological requirements.

  Able to become established only immediately after a fire when competition is usually reduced. These species are intolerant of competition in established plant communities *(I species).*

  Unable to become established immediately after fire, but are able to do so once mature individuals of either same or other species are present. These species have some requirement that is only provided by established plant communities *(R species).*

- *Life history.* This attribute plays an important part in the replacement sequence and in describing the role of a plant species after fire. There are four critical events in the life history of the species that can be arranged in order of their occurrence following fire occurrence (time *0*).

  Replenishment of sufficient propagules to survive another fire *(p).*

  Maturity *(m)* is the time at which individual plants will have recovered or grown sufficiently to be established in the community. *Maturity,* therefore, is the time when the individual is able to contribute propagules

	Critical Events
	0—p—m—l—e—∞
DT	0p——m——le∞
DI	0p——m—l——e∞
DR	0p——(m)——le∞
ST	0p——m——le∞
SI	0p——m—l—e—∞
SR	0p——(m)——le∞
CT	0——pm——le∞
CI	0——pm—le——∞
VT	0pm————le∞
VI	0pm——le——∞
(VD)I	0pm——l——e∞
(VS)I	0pm——l—e—∞

FIGURE 9.2 Life-history characteristics for various combinations of mechanisms for method of persistence and conditions for establishment of plant species. The critical events are described in the text. The *(m)* for the R species indicates that their reaching maturity is dependent on the presence of other mature individuals. (VD) and (VS) species are those that can reproduce either vegetatively or from seeds. (Adapted from Noble and Slatyer 1977.)

into the pool that will enable the species to persist through another fire disturbance.

Senescence and loss of the species from the community *(l)*.

Loss of propagules from the site so that the species is extinct *(e)*.

The first two attributes—that is, method of persistence and conditions for establishment—can be combined with their critical events to represent the life history characteristics shown in Figure 9.2. The replacement sequence of a plant species on a site of recurring fire occurrence can be predicted with knowledge of the life history characteristics presented in this figure.

Example 9.2: Application of Multiple Pathway Model to Predict Successional Patterns in a Mallee **(Eucalyptus)-Callitris** *Community in Australia (Noble and Slatyer 1977, Cattelino et al. 1979)*

The mallee species are representative of the shrub-like, multiple-stemmed eucalypti in western South Wales and, therefore, are described as a VI species (see Figure 9.2). The only information required for the life history of mallee species is longevity *(l)*, which is probably of the order of several centuries. All other critical events either occur at time 0 or coincide with time 1. *Callitris* is a conifer tree species that regenerates from seed and reaches maturity in 6 to 8 years. *Callitris* is classified as a CT species. No further information is needed on its life history, since *p* and *m* coincide, and events *l* and *e* occur at infinity.

Now, let us consider a site that is currently supporting both mature mallee and *Callitris,* that is, both a VI and CT species, respectively. It is assumed that aboveground tissues of both species will be killed if a hot fire burns the site. The VI mallee will regenerate immediately after the fire and, as a consequence, will become the first stage of the replacement sequence. Seeds of the CT *Callitris* will germinate when suitable conditions occur, but mallee will remain as the only mature species in the plant community until the *Callitris* seedlings attain maturity, 6 to 8 years after this fire, and the site is returned to a VI and CT community. The mallee will die off if the site remains unburned for a long time period, leaving a *Callitris* (CT) community. The successional pattern for this sequence is shown in Figure A below.

Fire periodicity does not always follow this sequence, however. If a fire occurs during the VI stage, within 6 to 8 years following an earlier burn, the immature *Callitris* will be destroyed and, therefore, lost to the community or (as a minimum) substantially reduced in its importance. Similarly, if a CT community burns after a long time period without fire, the *Callitris* will regenerate from its seed source, but the VI mallee will be lost because of its V mechanism and its relatively short life span. These critical events for both species are illustrated in Figure B. The overall successional possibilities for this example are shown in Figure C.

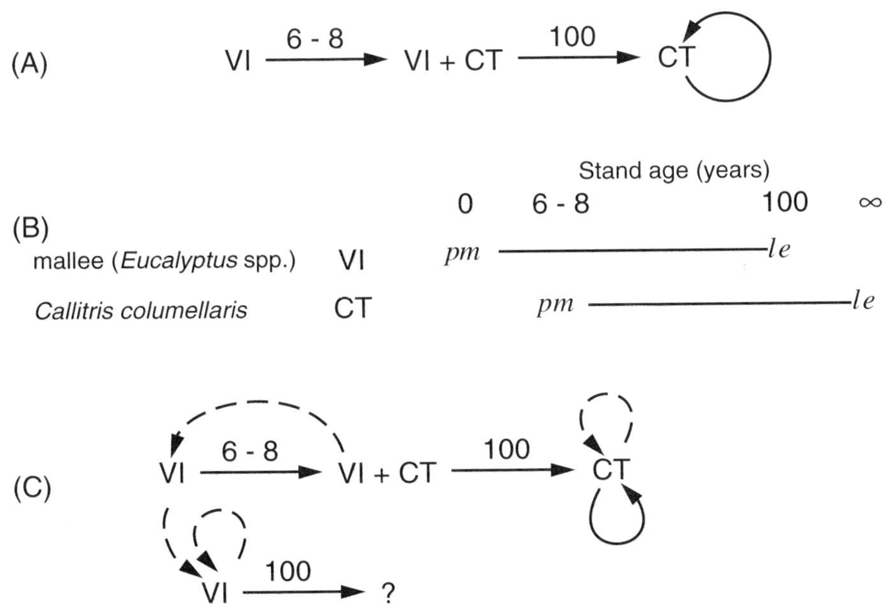

Successional patterns in a mallee–*Callitris* community: (A) mallee (VI)–*Callitris* (CT) succession without fire; (B) life-history characteristics for mallee–*Callitris;* (C) species replacement sequences in a mallee–*Callitris* community. The approximate time (in years) for transitions in undisturbed plant communities is shown by solid line arrows. The broken line arrows show the transitions caused by fire. (Adapted from Noble and Slatyer 1977.) ■

Application of the multiple-pathway model to predict a postfire successional pattern was originally presented by Noble and Slatyer (1977) and Cattelino et al. (1979) for a mallee *(Eucalyptus)–Callitris* community in western New South Wales, Australia (Example 9.2). Subsequently, increasing numbers of investigators have used the vital attributes approach to describing successional patterns following fire (Rowe 1983, Noste and Bushey 1987, Crane and Fischer 1986). For example, fire ecology information and fire management guidelines based largely on tree species attributes vital to replacement sequences and fire characteristics in forest communities of the northern Rocky Mountains are available (Cattelino et al. 1979, Kessell and Fischer 1981, Crane and Fischer 1986).

Fire Frequency

Fire frequency is important to successional patterns in natural ecosystems, regardless of whether a community-replacement or individual plant species perspective is taken on postfire succession. Compositions of plant species in the varying stages of community succession are often affected by fire frequency. Plant compositions that have been studied the most are those represented by climax species.

Prolonged protection from fire of climax communities comprised of trees and shrubs often changes the canopy cover and vertical structures of these communities. Increases in fire-free intervals also augment the densities of climax tree and shrub species until they fully occupy the site (Riggs et al. 1996). These species can continue to dominate in the absence of fire, especially in arid and semiarid environments, and additional community change is generally minimal once they achieve dominance. Reproduction of climax tree and shrub species can add the dimension of an uneven-aged structure to a fire-prone community.

Diversities and productivities of herbaceous understories of grasses and forbs often decline as trees and shrubs attain dominance when successional patterns progress toward a climax status. This decline is attributed largely to changes in light quality, increases in competition for water and nutrients, interruptions in establishment of herbaceous plants, and allelopathy (Gruell et al. 1986, Bunting et al. 1987). Reintroduction of recurring fire can reverse this decline, generally at the expense of the dominant tree and shrub species. Effectiveness of fire adaptive traits in some plant species declines in prolonged periods of fire exclusion, a phenomenon also affecting succession. An example of this decline in effectiveness is older bitterbrush *(Purshia tridentata)* growing under the full ponderosa pine canopies of later successional stages in the western United States (Bunting et al. 1985). These less vigorous plants are often less able to resprout following fire than younger and often more vigorous bitterbrush plants growing in the more open conditions of earlier successional stages.

VEGETATIVE RESOURCES

Vegetative resources are contained in plant communities of value to people. When demanded, they are available through the implementation of prescribed management practices. Fire impacts vegetative resources such as timber, forage, and wildlife habitats in numerous ways, some of which are beneficial while others are detrimental to the sustainability of these resources.

Timber

Impacts of fire on timber resources are largely reflected by immediate and delayed tree mortality, changes in postfire growth of residual trees, and changes in the potential for tree regeneration.

MORTALITY

One of the more obvious and often immediate effects of fire is frequently a loss of timber resources. There can be immediate and complete consumption of trees by fire, or a delayed killing of trees that remain standing for some time after the fire occurrence (Harrington 1996). Fire-injured trees that remain standing can often be salvaged for utilization before falling to the ground and decomposing.

The season of the fire occurrence can affect resistance of a tree to mortality. Some tree species appear more susceptible to mortality following early-growing season burns than to mortality from late-growing-season or dormant-season burns (Harrington 1987, Regelbrugge and Conard 1993). Bark moisture content is often higher in the early growing season, increasing thermal conductivity and exposing vulnerable tissues such as elongating meristems (Wagener 1961, Dieterich 1979). Carbohydrate reserves are also lower early in the growing season. Insulation of meristematic tissue by well-developed buds in tree species with large buds is believed to facilitate survival of trees with extensive crown scorch.

Delayed mortality can occur in fire-injured trees standing after fire. The elapsed time before death generally varies with the species, prefire vigor of the tree, and fire severity. Severely damaged trees often die within 2 or 3 years of fire, however. Because of their weakened physiological condition, fire-injured trees can also be subsequently infested by insects or infected by disease (Dixon et al. 1984, Harrington and Hawksworth 1990, Savage 1994), increasing mortality (Example 9.3). This problem is compounded when the pests subsequently attack uninjured, healthy trees.

Example 9.3: Effects of Postfire Insect Infestations on Tree Mortality: A Case Study (Rasmussen et al. 1996)

Surveys of bark beetle and wood borer infestations in the Greater Yellowstone Area were conducted from 1991 through 1993 to determine the effect of delayed tree mortality in mosaics of fire-killed and living stands after the devastating 1988 fires; the relationships between fire injury and infestation; and the effects of insect buildup in fire-injured trees on infestation rates for uninjured trees. It was found that fire injury accounted for more delayed mortality than insect infestation, but both types of mortality greatly altered the mosaics of trees immediately apparent after the 1988 Yellowstone fires. The high level of infestation suggests that insects built up in fire-injured trees and then caused increased infestation of uninjured, healthy trees. ■

The magnitude of crown scorch, of bole injury, or of a combination of the two types of injuries usually determines postfire mortality of standing trees injured by fire. Mortality of some tree species is more closely related to crown scorch than bole injury (Ryan et al. 1988, Finney and Martin 1993), while the reverse is true for other species (Peterson et al. 1991). In general, trees experiencing 60 to 70 percent or more of crown scorch are likely to die within a few years of fire occurrence as either a direct result of fire or from the compounding effect of insect infestation or disease incidence. The importance of bole injury to postfire tree survival is related mostly to bark thickness of the species. For example, the greater relative importance of bole injury to postfire survival of lodgepole pine is likely to be due to its thin bark.

Char-height, the height of stem charring as a proportion of total tree height, is also an indicator of postfire tree mortality in some species (Figure 9.3). Other predictors of mortality can include tree diameter, flame length, and surface fuel consumption (Ryan et al. 1988, Finney and Martin 1993). Regardless of the form of injury, however, the process by which trees die as a result of fire-induced injury is poorly understood (Regelbrugge and Conard 1993). A better understanding of relationships among season of burning, fire behavior and severity, fire injury to individual trees, and physiological responses to these injuries will improve the ability to predict tree mortality and, as a converse, survival.

Trees of almost all size classes can be killed, and much of the merchantable volume can be lost in a high-severity crown fire. However, mortality after low-severity surface fire and prescribed burning is often confined to trees in smaller diameter classes and those intermediate or suppressed in their relative stand position (Loomis 1973, Abbott and Loneragan 1983, Ffolliott and Guertin 1990). There can also be a minimum tree density level below which a low-severity surface fire will have little effect on tree mortality in some forest types (Pearson et al. 1972).

Mortality is frequently increased due to basal injuries caused by the burning of accumulations of litter at the base of trees. The extended heating that is associated with combustion of these deep mounds of fuel often results in greater cambium mortality in larger, old-growth trees than in smaller, young-growth trees (Harrington 1981, Ryan and Frandsen 1991). Greater mortality of larger trees on sites of infrequent fire occurrence, where buildups of litter

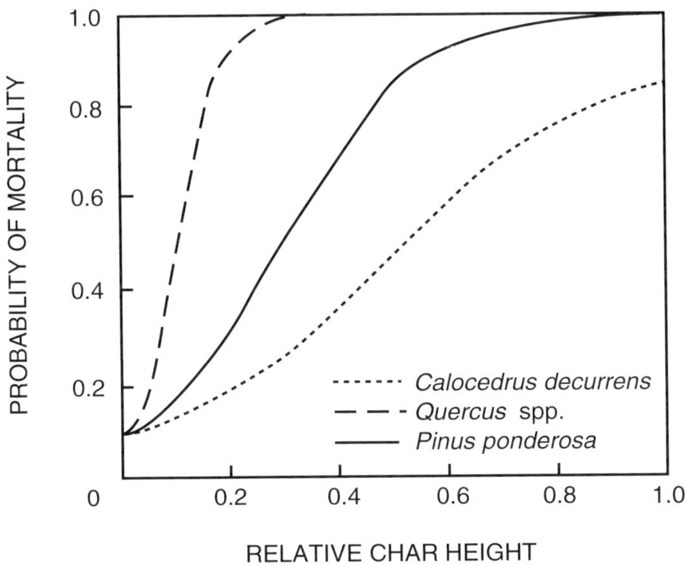

FIGURE 9.3 Probability of postfire mortality as a function of relative char-height for tree species common in ponderosa pine forests in the Sierra Nevada of California. (Adapted from Regelbrugge and Conard 1993.)

at the bases of trees are greatest, seems contrary to the general belief that fire resistance generally increases with tree size. Therefore, evaluation of the trade-off between greater fire resistance due to tree size and a greater heat potential because of increased fuel accumulations at the bases of these trees is necessary before fire is reintroduced into ecosystems characterized by long fire-free intervals.

POSTFIRE GROWTH

Information on postfire growth of trees surviving fire is conflicting. Some investigators have reported increases in growth (Cooper 1960, Wyant et al. 1983, Reinhardt and Ryan 1988), others decreases (Wooldridge and Weaver 1965, Johansen and Wade 1987, Peterson et al. 1991), and still others no change (Waldrop and van Lear 1984, Hunt and Simpson 1985). Unfortunately, little information is available on long-term tree growth rates after fire, and, as a consequence, the effects of fire on long-term postfire timber production are largely unknown.

The factor likely to most affect postfire growth is the physiological status of the surviving trees after burning. Crown scorch and bole injury can both weaken trees, with some weakened trees dying and others surviving, but in a less vigorous condition. Reduced tree density after fire can enhance growth of

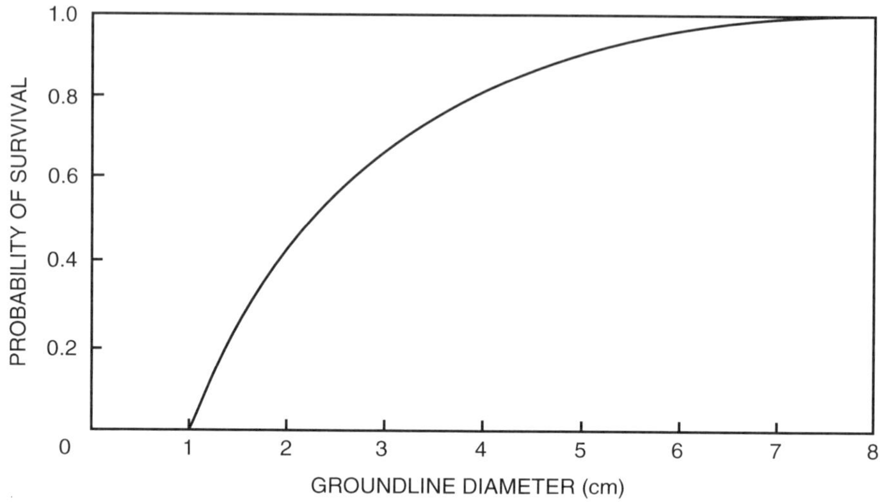

FIGURE 9.4 Survival of loblolly pine following prescribed burning is related to groundline diameter in the southeastern United States. (Adapted from Wade 1993.)

surviving trees, although the overall stand growth often remains unchanged. Increased postfire tree growth, when it occurs, is offset by delayed mortality in many situations, resulting in little or no change.

Stands can become overstocked with smaller trees and, as a consequence, stagnate in their growth. This condition is often remedied by manual thinning treatments employing hand-labor, machines, or chemicals, although costs can be high and the problem of slash disposal must be resolved. As an alternative to such thinning, low-severity prescribed fire has been used in some forest types to thin overstocked stands from below (Ffolliott and Guertin 1990, Wade 1993, Peterson et al. 1994). Smaller suppressed trees are eliminated from the stands by burning, releasing larger trees to accelerated growth rates (Figure 9.4). Average tree diameter in the affected stands is also increased due to removal of smaller trees by burning.

Effectiveness of fire as a thinning tool is related to the concentrations and spatial distributions of fuels on the soil surface before burning. While satisfactory thinning can be achieved on some burned sites, density of trees on nearby sites in the burned area might be little affected. This unevenness of thinning is attributed to patterns of fuels on the soil surface in most instances.

REGENERATION

Fire-stimulated germination of seeds that have been stored in the soil can contribute to the regeneration of many tree species. Factors affecting the response of seeds to heating are the tree species itself and the moisture contents of the seedbed and seeds (Martin et al. 1975, Shea et al. 1979, Takahashi and Kikuchi 1986, Keeley 1987). However, the seeds buried in the soil can also be insulated from the effects of fire and, therefore, not affected by burning.

Burning has a potential to prepare seedbeds for natural tree regeneration in openings that are too small for planting of seedlings. Environmental conditions affected by burning that favor germination of seed include the creation of a more receptive seedbed because:

- Competing vegetation is often eliminated.
- Decomposed organic matter on the soil surface is reduced or eliminated.
- Nutrient availability is often increased (see Chapter 6).
- Soil moisture and temperature conditions become more favorable.

A good seed crop is needed in addition to a receptive seedbed for natural regeneration to be successful. However, good seed crops generally do not occur every year. Repeated burning can be necessary for seedbeds to remain receptive of seed from surviving trees (Sackett 1984, Cain 1993, Sackett et al. 1996). It is also important that potential seed trees are not severely damaged in prescribing fire to prepare a seedbed.

Burning can also be useful in preparing a site for the planting of tree seedlings by eliminating or effectively reducing competing vegetation. Such site preparation conserves soil moisture and nutrients for establishment and rapid early growth of the seedlings. Tree seedlings planted immediately following removal of competing vegetation generally attain the highest survival rates.

While successful establishment and high initial survival of planted tree seedlings can be possible, many postfire tree plantlets are short-lived as a result of drought, frost heaving, animal damage, or postfire litter accumulations (Example 9.4). Therefore, long-term survival of tree seedlings established on burned sites is often questionable.

Example 9.4: Survival and Early Growth of Tree Seedlings Following Prescribed Fire: A Case Study (Ffolliott and Guertin 1990)

More postfire ponderosa pine seedlings initially survived on a burned site than on an adjacent unburned area in northern Arizona. Newly started seedlings occupied 85 percent of the mil-acre plots on the burned site, compared to 20 percent of the same number of unburned plots in the first year after the prescribed fire. However, many of the new seedlings died a year later, due largely to insufficient soil moisture. Twenty-four years after the prescribed fire, postfire seedlings still occupied 25 percent of the plots on the burned site. No seedlings stocked the plots on the unburned area. The failure of more tree seedlings to survive on the burned site was attributed to litter accumulations, which were returning to prefire conditions. ■

Forage

Fire can alter the production, species composition, and quality of *forage* on a site, that is, grass and grass-like plants, forbs, and half-shrubs available to, and eaten by, livestock or other herbivores. Forage resources found in forests, woodlands, shrub-dominated communities, and grassland ecosystems are potentially affected by fire occurring in these communities. Changes in forage resources following fire, in turn, affect capabilities of burned areas to support livestock and indigenous wildlife species.

PRODUCTION

Forage production generally increases in forests, woodlands, and shrub-dominated communities as densities of trees and shrubs decrease. These increases

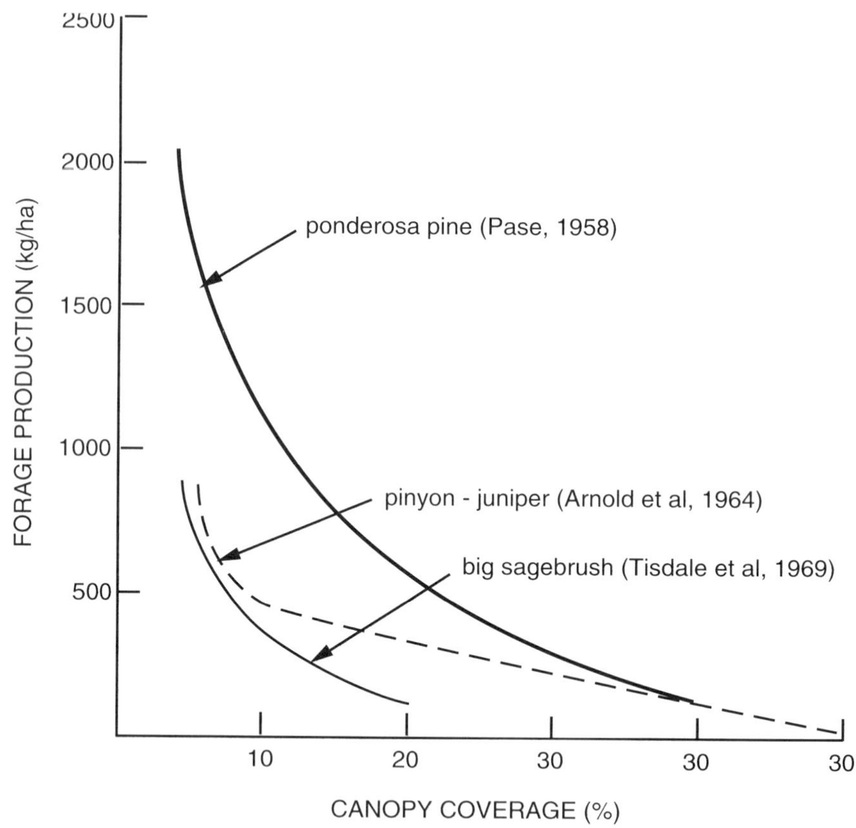

FIGURE 9.5 Relationships between forage production and densities of trees and shrubs are generally curvilinear in form, although mathematical expressions of the relationships differ. Curves are adapted from cited references. (Adapted from Riggs et al. 1996. Copyright © 1996 Society for Range Management.)

are attributed to decreases in competition for light, water, and nutrients between overstory vegetation of trees and shrubs and understory forage vegetation (Figure 9.5). Therefore, forage production often increases on those sites where fire has eliminated or reduced the densities of overstory trees and shrubs.

Increases in forage production of varying magnitudes have widely been reported in forests, woodlands, and shrub-dominated communities after occurrences of wildfire (Lemon 1949, Campbell et al. 1977, Ohmann and Grigal 1979, Larson and Duncan 1982); following prescribed fire (Reynolds and Bohning 1956, Halls and Alcaniz 1971, Perala 1974, Ralphs and Busby 1979, Ffolliott and Guertin 1990); and where burning has been used in conjunction with other management activities to eliminate competing vegetation, reduce fuel buildups, and improve wildlife habitats (Lay 1967, Pratt and Knight 1971, Brender et al. 1976, Springfield 1976, Bailey and Anderson 1979). However, regardless of the type of fire, forage production often returns to prefire levels in a few years in the absence of recurring burning.

Factors affecting postfire forage production, in general, include the severity of burning, whether the fire occurred in the growing or dormant season, and prefire species composition and production of forage plants. Forage production in almost any natural ecosystem is also stimulated by fire because:

- Soils are generally warmer on burned areas and, therefore, plant growth starts earlier in the season.
- Burning of litter on the soil surface reduces interception of light and water, also favoring plant growth.
- Release of nutrients by burning increases soil fertility, once again encouraging growth of forage plants.
- Plant growth and vigor are often promoted by removal of senescent shoots and foliage.

Increases in herbaceous biomass following fire on some sites are often attributed to the production of unpalatable nonforage plants, which invade burned sites in early stages of postfire succession (Ffolliott and Guertin 1990). Therefore, while total production of all herbaceous vegetation (nonforage and forage plants) might increase, forage production can remain unchanged or even decline immediately after fire. Seeding of forage species immediately after fire is often prescribed to control postfire soil loss and the invasion of nonforage plants.

SPECIES COMPOSITION

Fire can alter species composition of plant communities. Plant species are frequently eliminated by burning, relative frequencies of species occurrence are changed, and plant species representing earlier successional stages are introduced into burned ecosystems (Neiland 1958, Cushwa and Redd 1966, Taylor 1973, Krefting and Ahlgren 1974, Stransky and Harlow 1981), depending largely on the fire severity, successional stage at the time of burning, and season of burning. Changes in species compositions of plant communities following fire are site-specific for the most part.

Impacts of changes in species composition on the character of postfire forage resources are largely reflected by the relative changes in occurrences of forage species within burned plant communities. The magnitude of these impacts is determined by the combined effects of eliminations, changes in frequencies, and introductions of forage species following fire in relation to the composition of all herbaceous plant species before burning. Changes in forage components of the herbaceous biomass by fire are generally less than the compositional changes in all herbaceous vegetation.

QUALITY

Burning often eliminates old and poorly digestible plant parts containing low concentrations of digestible fiber, proteins, and minerals, depositing some of the residues on soil surfaces as ash (Woodmansee and Wallach 1981, Riggs et al. 1996). As a consequence, increased nutrient availability in the soil (with increased postfire light and temperature) can facilitate increased nutrient uptake by surviving forage plants. Grazing livestock and other herbivores often benefit from the resulting postfire changes in forage quality.

Crude protein, a number of nutrients including P, N, Ca, Mg, and K, and *in vitro* digestibility contents are often higher in forage plants growing on burned sites than in plants growing on unburned sites, at least in the first postfire growing season (Pearson et al. 1972, Christensen 1977, Hobbs and Spowart 1984, Cox 1988, DeByle et al. 1989, Carreira and Niell 1992). Magnitudes of the higher contents are also species-specific and vary generally with the type of ecosystem burned, its successional status, the plant-part burned, and the fire severity and season of burning. Durations of these increases are usually short in the absence of repeated fire, often lasting only through the second growing season.

Burning can also affect forage quality by altering the phenological profiles of plant communities. As burning changes species composition of plant communities, it is also likely to alter the phenological development of the communities through diversification of postfire phenologies (Riggs et al. 1996). Therefore, springtime growth of the forage resources might occur earlier, with forage of higher concentrations of digestible fiber, proteins, and minerals available for herbivore consumption earlier in the season. Availability of other forage components of critical importance to herbivores can be delayed, however, with a negative consequence. The importance of fire-induced phenological shifts to nutritional limitations depends largely on their severity, scale of burning, and the seasonal pattern of nutritional constraints on the herbivores involved.

Despite the somewhat inconsistent and often transient improvements in forage resources, fire can be a useful tool for altering species composition and quality in many plant communities. Availability of more palatable and nutritious forage is often obtained in the process. Duration of these enhancements is generally a function of the time required for successional patterns to restore prefire community structure and composition.

Wildlife Habitats

Fire occasionally injures and kills wildlife directly by burning and suffocation (Chew et al. 1959, Bock and Lynch 1970, Bulan and Barrett 1971, Harrison and Murad 1972, Buech et al. 1977). However, these losses are negligible in most instances. More important influences of fire on wildlife populations are largely mediated indirectly through changes in food, cover, and structural diversity (Lyon et al. 1978, Patton 1992).

FOOD

Burning can substantially alter the production, species composition, and quality of food resources that are available to wildlife by creating openings in otherwise dense overstory vegetation or reducing overstory densities, as discussed earlier in the context of fire effects on forage resources. Much of the impetus for prescribed burning on behalf of wildlife is rooted in knowledge that fire often enhances food resources, increasing growth of postfire wildlife populations (Roppe and Hein 1978, Niering 1981, Wood 1981, Huntley and McGee 1982, Klein 1982). Successional progressions slowed by fire occurrence can also benefit wildlife populations by maintaining preferred successional stages (Carlson et al. 1993). On the other hand, however, fire-induced reversals of successional patterns can exert detrimental influences on wildlife populations when preferred foods are eliminated from the burned site.

The extent of the area that must burn to significantly affect food and other habitat components of a wildlife population increases with increasing size of individuals in the population, and the population size itself, and with its spatial density. Therefore, it is common that large herbivores (deer, elk, moose) are influenced to a greater extent by larger burned areas than smaller burns. Smaller burns can have immediate impacts on the well-being of small mammal populations (rabbits, voles, mice), however.

It has been difficult for biologists to demonstrate the extent to which fire-induced changes in quantities and qualities of food resources actually affect growth and productivity of wildlife populations (Hobbs and Spowart 1984, Peek 1989, Riggs et al. 1996). While results of experiments with cattle show the effects of fire to be striking, with liveweight gains on burned pastures often greater than those on unburned pastures (Angell et al. 1986, Svejcar

1989), comparable information on indigenous wildlife species is sparse. Nevertheless, benefits to wildlife are likely to include higher birthweights, higher survival of adults, increased recruitments into the postfire populations, and, for hunted wildlife species, improved harvestable yields (Riggs et al. 1996).

COVER

Immediate postfire environments frequently present wildlife with a sudden and often drastic change in *protective cover,* important to wildlife when escape from predators becomes necessary, and *thermal cover,* critical to the maintenance of body heat or to cooling. Depending on the severity of the fire, protective or thermal cover in different habitat strata can be destroyed or severely modified in its structure (Walter 1977, Lyon et al. 1978, Roppe and Hein 1978, Wood 1981). While changes of cover strata important for a particular wildlife species can cause that species to become displaced from the burned area, removal of the same cover can also encourage movement of other wildlife species into the burned area.

Duration of changes in habitat structure and microclimate is related to the speed at which succession returns to prefire compositions (Lyon et al. 1978, Patton 1992, Carlson et al. 1993). Postfire succession is variable in time and largely site-specific. Fire severity, botanical complexity of burned plant community, and postfire management practices are among the factors affecting successional patterns and resulting wildlife habitats following burning.

STRUCTURAL DIVERSITY

Mosaics of food resources and cover create structural diversity, an important element in wildlife habitat quality. A large fire of high severity often reduces interspersion of food and cover, producing a uniformity in habitat condition (Lyon et al. 1978). However, as mentioned earlier, fire rarely burns evenly and, as a consequence, postfire clumps of vegetation, burned logs, and open spaces usually result. These patterns of structural diversity within a burn help to determine where wildlife species will live and the densities that they might attain when their habitat requirements are satisfied (Example 9.5). Clumped vegetation in a postfire ecosystem can also influence predator–prey relationships, however, by providing cover for predators.

Example 9.5: Effects of Fire on Structural Diversity (Lyon et al. 1978, Severson and Rinne 1990, Patton 1992)

ECOTONES

Fire often creates abrupt edges between adjacent vegetative types, called *ecotones,* and a variety of postfire successional plant communities. Edge and vegetative interspersion where two vegetative types come together is a valuable habitat component for large herbivores (deer), upland game (grouse), and other wildlife populations in many situations. However, wildlife species needing large blocks of land with interior habitats might not benefit from the creation of edge.

SNAGS

Some 85 species of North American birds excavate nesting holes, use cavities resulting from wood decay, or use holes created by other species in deteriorating or dead trees (Davis et al. 1983). Such trees, called *snags,* are also important to numerous species of mammals, reptiles, amphibians, and invertebrates. Fire can create snags favorable to a wide variety of avifauna and other wildlife species. Removal of snags through their consumption by severe fire can also be a factor in the elimination of these species.

ENVIRONMENTAL HETEROGENEITY

Of particular importance to maintaining environmental heterogeneity in natural ecosystems are vegetative interspersions created by fire, each of which is capable of supporting a specific faunal community. Individual wildlife species are usually associated with a particular stage of vegetative growth. Loss of a specific postfire successional stage is often correlated with a decline of wildlife species dependent on the particular vegetation represented. Retention of all successional stages can ensure minimal levels of all species in an area (Thomas et al. 1976). ■

COMPOSITE IMPACTS

It is difficult to isolate effects of fire on structural diversity from the effects of fire on food and cover. In terms of prescribing management activities to modify wildlife habitats, therefore, it is often helpful to consider the effects of fire on wildlife habitats in terms of the composite impacts of fire on all habitat components and, in turn, on affected wildlife populations. The following generalizations can be made in this regard:

- Large herbivores (deer, moose, elk) often increase in numbers after fire if the burned area is large, minimal cover is retained, and appropriate structural diversity is maintained (Keay and Peek 1980, Pierce and Peek 1984, Canon et al. 1987, Bentz and Woodward 1988). Large herbivores are likely to leave the area, at least temporarily, otherwise.
- Avifauna (nongame birds) that forage on herbaceous vegetation or the

ground frequently increase in numbers following fire, while birds dependent on boles and canopies of trees for foraging are likely to decline (Lawrence 1966, Bock and Webb 1984, Peterson and Best 1987). A richer avifauna in reference to species numbers is often found in postfire habitats.

- There is generally little change in the total numbers of small mammals after burning, although shifts in species compositions can often occur. Some small mammals survive fire within the confines of the burn, others escape to unburned areas, and still others will die (Lawrence 1966, Vogel 1973, Buech et al. 1977, Lyon et al. 1978). Survival of residents within a burned area is usually associated with the existence of small inclusions of unburned habitat.

Furthermore:

- Assessing impacts of fire on wildlife habitats immediately after suppression can be misleading. While there are damages, sometimes serious ones, associated with most burning, benefits can also be apparent when fire effects are evaluated on both a short- and long-term basis (Lowe et al. 1978). Such benefits and damages should be considered to arrive at a more realistic picture of postfire habitats.
- Many wildlife populations remain mostly unchanged by fire in terms of density, and in terms of whether they are expanding, stable, or declining in number. The major difference at the species level appears to be that burn areas and fire-dependent ecosystems tend to support larger-bodied avifauna and small mammals than unburned or relatively fire-free ecosystems.

Whether fire significantly affects food, cover, and diversity components of wildlife habitats and, in doing so, wildlife populations depends largely on long-term, landscape-scale interactions among the affected plant communities and wildlife populations (Riggs et al. 1996). Interactions of morphological, physiological, and behavioral patterns of wildlife species with the plants and climate of postfire environments define capacities of the wildlife species to adapt to the habitat changes brought about by fire.

SUMMARY

How a fire affects plants, successional cycles, and timber, forage, and wildlife habitats is determined largely by the severity of the fire. In this regard, the purpose of this chapter has been to provide the general information necessary so that you are able to:

1. Discuss different ways that plants can tolerate fire and, when fire occurs, how these adaptations can enhance their role in postfire succession.

2. Describe the community-replacement concept of succession in comparison to the abilities of individual plant species to persist in postfire environments, a second approach to studying succession after fire.
3. Describe the possible beneficial and detrimental impacts of fire on those plant communities with value to people, for example:
   a. Explain how a fire can cause immediate or, some cases, delayed tree mortality, affect the postfire growth of residual trees, and enhance the chances for successful tree regeneration.
   b. Identify the situations in which fire can alter the production, species composition, and quality of grasses and grass-like plants, forbs, and half-shrubs available to, or eaten by, herbivores.
   c. Appreciate the impacts of fire on the components (food and cover) and structural diversity of wildlife habitats.

REFERENCES

Abbott, L., and O. Loneragan. 1983. Influence of fire on growth rate, mortality, and butt damage in Mediterranean forest of western Australia. *Forest Ecology and Management* 6:139–153.

Agee, J. K. 1991. Fire history along an elevational gradient in the Siskiyou Mountains, Oregon. *Northwest Science* 65:188–199.

Angell, R. F., J. W. Stuth, and D. L. Drawe. 1986. Diets and liveweight changes of cattle grazing fall burned gulf cordgrass. *Journal of Range Management* 39:233–236.

Arno, S. F. 1976. *The Historical Role of Fire on the Bitteroot National Forest.* USDA Forest Service, General Technical Report INT-187.

Arnold, J. F., D. A. Jameson, and E. H. Reid. 1964. The pinyon-juniper type of Arizona: Effects of grazing, fire, and tree control. USDA Forest Service, Production Research Paper 84.

Bailey, A. W., and H. G. Anderson. 1979. Brush control on sandy rangelands in central Alberta. *Journal of Range Management* 32:29–32.

Bentz, J. A., and P. M. Woodward. 1988. Vegetation characteristics and bighorn sheep use on burned and unburned areas in Alberta. *Wildlife Society Bulletin* 16:186–193.

Bock, C. E., and J. F. Lynch. 1970. Breeding bird populations of burned and unburned conifer forest in the Sierra Nevada. *Condor* 72:182–189.

Bock, C. E., and B. Webb. 1984. Birds as grazing indicator species in southeastern Arizona. *Journal of Wildlife Management* 48:1045–1049.

Brender, E. V., W. H. McNab, and S. Williams. 1976. Fuel accumulations in Piedmont loblolly pine plantations. USDA Forest Service, Research Note SE-233.

Brown, A. A., and K. P. Davis. 1973. *Forest Fire: Control and Use.* McGraw-Hill Book Company, New York.

Buech, R. R., K. Siderits, R. E. Radtke, H. L. Sheldon, and D. Elsing. 1977. Small mammal populations after a wildfire in Northeast Minnesota. USDA Forest Service, Research Paper NC-151.

Bulan, C. A., and G. W. Barrett. 1971. The effects of two acute stresses on the arthropod component of an experimental grassland ecosystem. *Ecology* 52:597–605.

Bunting, S. C., B. M. Kilgore, and C. L. Bushey. 1987. *Guidelines for Prescribed Burning Sagebrush-Grass Rangelands in the Northern Great Basin.* USDA Forest Service, General Technical Report INT-231.

Bunting, S. C., L. F. Neuenschwander, and G. E. Gruell. 1985. Fire ecology in antelope bitterbrush in the northern Rocky Mountains. In: Lotan, J. E., and J. K. Brown, compilers. *Fire's Effect on Wildlife Habitat: Proceedings of a Symposium.* USDA Forest Service, General Technical Report INT-186, pp. 48–57.

Byrne, R. 1978. Fossil record discloses wildfire history. *California Agriculture* 10:13–14.

Cain, M. D. 1993. A 10–year evaluation of prescribed winter burns in uneven-aged stands of *Pinus taeda* L., and *P. echinata* Mill.: woody understory vegetation response. *International Journal of Wildland Fire* 3:13–20.

Campbell, R. E., M. B. Baker, Jr., P. F. Ffolliott, F. R. Larson, and C. C. Avery. 1977. Wildfire effects on a ponderosa pine ecosystem: an Arizona case study. USDA Forest Service, Research Paper RM-191.

Canon, S. K., P. J. Urness, and N. V. DeByle. 1987. Habitat selection, foraging behavior, and dietary nutrition of elk in burned aspen forest. *Journal of Range Management* 49:433–437.

Carlson, P. C., G. W. Tanner, J. M. Wood, and S. R. Humphrey. 1993. Fire in key deer habitat improves browse, prevents succession, and preserves endemic herbs. *Journal of Wildlife Management* 57:914–928.

Carreira, J. A., and F. X. Niell. 1992. Plant nutrient changes in a semi-arid Mediterranean shrubland after fire. *Journal of Vegetation Science* 3:457–466.

Cattelino, P. J., I. R. Noble, R. O. Slatyer, and S. R. Kessell. 1979. Predicting the multiple pathways of plant succession. *Environmental Management* 3:41–50.

Chandler, C., P. Cheney, P. Thomas, L. Trabaud, and D. Williams. 1983. *Fire in Forests. Volume I. Forest Fire Behavior and Effects.* John Wiley & Sons, New York.

Chew, R. M., B. B. Butterworth, and R. Grechman. 1959. The effects of fire on the small mammal population of chaparral. *Journal of Mammalogy* 40:253.

Christensen, N. L. 1977. Fire and soil-plant nutrient relations in a pine-wiregrass savanna on the coastal plain of North Carolina. *Oecologia* 31:31–27.

Clements, F. E. 1916. *Plant Succession: An Analysis of the Development of Vegetation.* Carnegie Institute, Publication 242, Washington, DC.

Clements, F. E. 1936. Nature and structure of the climax. *Journal of Ecology* 24:252–284.

Cooper, C. F. 1960. Changes in vegetation, structure and growth of southwestern ponderosa pine forests since white settlement. *Ecological Monographs* 30:129–164.

Cox, J. R. 1988. Season burning and mowing impacts on *Sporobolus wrightii* grasslands. *Journal of Range Management* 41:12–15.

Crane, M. F., and W. C. Fischer. 1986. *Fire Ecology of the Forest Habitat Types in Central Idaho.* USDA Forest Service, General Technical Bulletin INT-218.

Cushwa, C. T., and J. B. Redd. 1966. One prescribed burn and its effect on habitat on the Powhatan game management area. USDA Forest Service, Research Note SE-61.

Davis, J. W., G. A. Goodwin, and R. A. Ockenfels, technical coordinators. 1983. *Snag Habitat Management: Proceedings of the Symposium.* USDA Forest Service, General Technical Report RM-99.

DeByle, N. V., P. J. Urness, and D. L. Blank. 1989. Forage quality in burned and unburned aspen communities. USDA Forest Service, Research Paper INT-404.

Dieterich, J. H. 1979. Recovery potential of fire-damaged ponderosa pine. USDA Forest Service, Research Note RM-379.

Dixon, W. N., J. A. Cornell, R. C. Wilinson, and J. L. Foltz. 1984. Using stem char to predict mortality and insect infestation of fire-damaged slash pines. *Southern Journal of Applied Forestry* 8:85–88.

Ffolliott, P. F., and D. P. Guertin. 1990. Prescribed fire in Arizona ponderosa pine forests: a 24-year case study. In: Krammes, J. S., technical coordinator. *Effects of Fire Management of Southwestern Natural Resources.* USDA Forest Service, General Technical Report RM-191, pp. 250–254.

Finney, M. A., and R. E. Martin. 1993. Modeling effects of prescribed fire on young-growth coast redwood trees. *Canadian Journal of Forest Research* 23:1125–1135.

Gill, A. M. 1981. Fire adaptive traits of vascular plants. In: Mooney, H. A., T. M. Bonnicksen, N. L. Christensen, J. E. Lotan, and W. A. Reiners, technical coordinators. *Fire Regimes and Ecosystem Properties.* USDA Forest Service, General Technical Report WO-26, pp. 208–230.

Gruell, G. E., J. K. Brown, and C. L. Bushey. 1986. *Prescribed Fire Opportunities in Grassland Invaded by Douglas-Fir: State-of-the-Art Guidelines.* USDA Forest Service, General Technical Report INT-198.

Halls, L. K., and R. Alcaniz. 1971. Forage yields in an east Texas pine-hardwood forest. *Journal of Forestry* 69:25–26.

Harrington, M. G. 1981. Preliminary burning prescriptions for ponderosa pine fuel reductions in southeastern Arizona. USDA Forest Service, Research Note RM-472.

Harrington, M. G. 1987. Ponderosa pine mortality from spring, summer, and fall crown scorching. *Western Journal of Applied Forestry* 2:14–16.

Harrington, M. G. 1996. Fall rates of prescribed fire-killed ponderosa pine. USDA Forest Service, Research Paper INT-469.

Harrington, M. G., and F. G. Hawksworth. 1990. Interactions of fire and dwarf mistletoe on mortality of southwestern ponderosa pine. In: Krammes, J. S., technical coordinator. *Effects of Fire Management of Southwestern Natural Resources.* USDA Forest Service, General Technical Report RM-191, pp. 234–240.

Harrison, R. E., and J. L. Murad. 1972. Effects of annual prescribed burning on nematode populations from a Louisiana pine forest. *Journal of Nematology* 4:225–226.

Heinselman, M. L. 1981. Fire intensity and frequency as factors in the distribution and structure of northern ecosystems. In: Mooney, H. A., T. M. Bonnicksen, N. L. Christensen, J. E. Lotan, and W. A. Reiners, technical coordinators. *Fire Regimes and Ecosystem Properties.* USDA Forest Service, General Technical Report WO-26, pp. 7–57.

Hobbs, N. T., and R. A. Spowart. 1984. Effects of prescribed fire on nutrition of mountain sheep and mule deer during winter and spring. *Journal of Wildlife Management* 48:531–560.

Houston, D. B. 1973. Wildfires in northern Yellowstone National Park. *Ecology* 54:1111–1117.

Hunt, S. M., and J. A. Simpson. 1985. Effects of low intensity prescribed fire on the growth and nutrition of slash pine plantation. *Australian Forest Research* 15:67–77.

Huntley, J. C., and C. E. McGee. 1982. Impact of fire on regeneration and wildlife habitat in upland hardwood forests. In: *Proceedings of the 1982 Annual Meeting of the Society of American Foresters.* Society of American Foresters, Washington DC, pp. 158–162.

Johansen, R. W., and D. D. Wade. 1987. Effects of crown scorch on survival and diameter growth of slash pines. *Southern Journal of Applied Forestry* 11:180–184.

Keay, J. A., and J. M. Peek. 1980. Relationships between fires and winter habitat use of deer in Idaho. *Journal of Wildlife Management* 44:372–380.

Keeley, J. E. 1981. Reproductive cycles and fire regimes. In: Mooney, H. A., T. M. Bonnicksen, N. L. Christensen, J. E. Lotan, and W. A. Reiners, technical coordinators. *Fire Regimes and Ecosystem Properties.* USDA Forest Service, General Technical Report WO-26, pp. 231–277.

Keeley, J. E. 1987. Role of fire in seed germination of woody taxa in California chaparral. *Ecology* 68:434–443.

Kessell, S. R., and W. C. Fischer. 1981. *Predicting Postfire Plant Succession for Fire Management Planning.* USDA Forest Service, General Technical Report INT-94.

Kilgore, B. M. 1981. Fire in ecosystem structure: western forests and scrublands. In: Mooney, H. A., T. M. Bonnicksen, N. L. Christensen, J. E. Lotan, and W. A. Reiners, technical coordinators. *Fire Regimes and Ecosystem Properties.* USDA Forest Service, General Technical Report WO-26, pp. 58–89.

Kilgore, B. M., and D. Taylor. 1979. Fire history of a sequoia–mixed conifer forest. *Ecology* 60:129–142.

Klein, D. R. 1982. Fire, lichens, and caribou. *Journal of Range Management* 35:390–395.

Krefting, L. W., and C. E. Ahlgren. 1974. Small mammals and vegetation changes after fire in a mixed conifer–hardwood forest. *Ecology* 55:1391–1398.

Larson, J. R., and D. A. Duncan. 1982. Annual grassland response to fire retardant and wildfire. *Journal of Range Management* 35:700–703.

Lawrence, G. E. 1966. Ecology of vertebrate animals in relation to chaparral fire in the Sierra Nevada foothills. *Ecology* 47:278–291.

Lay, D. W. 1967. Browse palatability and the effects of prescribed burning in southern pine forests. *Journal of Forestry* 54:342–347.

Lemon, P. C. 1949. Successional responses of herbs in the longleaf-slash pine forest after fire. *Ecology* 30:135–145.

Leopold, A. 1924. Grass, brush, and fire in southern Arizona. *Journal of Forestry* 22:1–10.

Loomis, R. M. 1973. Estimating fire-caused mortality and injury in oak–hickory forests. USDA Forest Service, Research Paper NC-94.

Lorimer, C. G. 1990. Behavior and management of forest fires. In: Young, R. A., and R. L. Giese, editors. *Introduction to Forest Science.* John Wiley & Sons, New York, pp. 427–448.

Lowe, P. O., P. F. Ffolliott, J. H. Dieterich, and D. R. Patton. 1978. *Determining Potential Benefits from Wildfire in Arizona Ponderosa Pine Forests.* USDA Forest Service, General Technical Report RM-52.

Lyon, L. J., H. S. Crawford, E. Czuhai, R. L. Fredriksen, R. F. Harlow, L. J. Metz, and H. A. Pearson. 1978. *Effects of Fire on Fauna: A State-of-Knowledge Review.* USDA Forest Service, General Technical Report WO-6.

Martin, R. E., R. L. Miller, and C. T. Cushwa. 1975. Germination response of legume seeds subject to moist and dry heat. *Ecology* 56:1441–1445.

McPherson, G. R. 1995. The role of fire in the desert grasslands. In: McClaran, M. P., and T. R. Van Devender, editors. *The Desert Grassland*. University of Arizona Press, Tucson, Arizona, pp. 130–151.

Neiland, B. J. 1958. Forest and adjacent burn in the Tillamook burn area of northwestern Oregon. *Ecology* 39:660–671.

Niering, W. A. 1981. The role of fire management in altering ecosystems. In: Mooney, H. A., T. M. Bonnicksen, N. L. Christensen, J. E. Lotan, and W. A. Reiners, technical coordinators. *Fire Regimes and Ecosystem Properties*. USDA Forest Service, General Technical Report WO-26, pp. 489–510.

Noble, I. R. 1981. Predicting successional change. In: Mooney, H. A., T. M. Bonnicksen, N. L. Christensen, J. E. Lotan, and W. A. Reiners, technical coordinators. *Fire Regimes and Ecosystem Properties*. USDA Forest Service, General Technical Report WO-26, pp. 278–300.

Noble, I. R., and R. O. Slatyer. 1977. Post-fire succession of plants in Mediterranean ecosystems. In: Mooney, H. A., and C. E. Conrad, technical coordinators. *Environmental Consequences of Fire and Fuel Management in Mediterranean Ecosystems*. USDA Forest Service, General Technical Report WO-3, pp. 27–36.

Noble, I. R., and R. O. Slatyer. 1980. The use of vital attributes to predict successional changes in plant communities subject to recurrent disturbances. *Vegetatio* 43:5–21.

Noble, I. R., and R. O. Slatyer. 1981. Concepts and models of succession in vascular plant communities subject to recurrent fire. In: Gill, A. M., R. H. Groves, and I. R. Noble, editors. *Fire and the Australian Biota*. Australian Academy of Science, Canberra City, Australia, pp. 311–335.

Noste, N. V., and C. L. Bushey. 1987. *Fire Response of Shrubs of Dry Forest Habitats in Montana and Idaho*. USDA Forest Service, General Technical Report INT-239.

Ohmann, L. F., and D. F. Grigal. 1979. Early revegetation and nutrient dynamics following the 1971 Little Sioux forest fire in northeastern Minnesota. *Forest Science Monograph* 21.

Pase, C. P. 1958. Herbage production and composition under immature ponderosa pine stands in the Black Hills. *Journal of Range Management* 11:238–243.

Patterson, W. A. III. 1986. A "new ecology"—implications for modern forest management. *Journal of Forestry* 84(2):93.

Patton, D. R. 1992. *Wildlife Habitat Relationships in Forested Ecosystems*. Timber Press, Portland, Oregon.

Pearson, H. A., J. R. Davis, and G. H. Schubert. 1972. Effects of wildfire on timber and forage production in Arizona. *Journal of Range Management* 25:250–253.

Peek, J. M. 1989. Another look at burning shrubs in northern Idaho. In: Baumgartner, D. W., D. W. Breuer, B. A. Zamora, L. F. Neuenschwander, and R. H. Wakimoto, editors. *Proceedings of Symposium on Prescribed Fire in the Intermountain Region*. Washington State University, Pullman, Washington, pp. 157–159.

Perala, D. A. 1974. Repeated prescribed burning in aspen. USDA Forest Service, Research Note NC-171.

Peterson, D. L., M. J. Arbaugh, G. H. Pollock, and L. J. Robinson. 1991. Postfire growth of *Pseudotsuga menziesii* and *Pinus contorta* in the northern Rocky Mountains, USA. *International Journal of Wildland Fire* 1:63–71.

Peterson, D. L., S. S. Sackett, L. J. Robinson, and S. M. Haase. 1994. The effects of repeated prescribed burning on *Pinus ponderosa* growth. *International Journal of Wildland Fire* 4:239–247.

Peterson, K. L., and L. B. Best. 1987. Effects of prescribed burning on nongame birds in a sagebrush community. *Wildlife Society Bulletin* 15:317–329.

Pierce, J. D., and J. M. Peek. 1984. Moose habitat use and selection patterns in north-central Idaho. *Journal of Wildlife Management* 48:1335–1343.

Pratt, D. J., and M. D. Gwynne. 1977. *Rangeland Management and Ecology in East Africa.* Robert E. Krieger Publishing Company, Huntington, New York.

Pratt, D. J., and J. Knight. 1971. Brush-control studies in the drier areas of Kenya vs. effects of controlled burning and grazing management of *Tarchonanthus/Acacia* thicket. *Journal of Applied Ecology* 8:217–237.

Pyne, S. J., P. L. Andrews, and R. D. Laven. 1996. *Introduction to Wildland Fire.* John Wiley & Sons, New York.

Radtke, K. W. H., A. M. Arndt, and R. H. Wakimoto. 1982. Fire history of the Santa Monica Mountains. In: Conrad, C. E., and W. C. Oechel, technical coordinators. *Dynamics and Management of Mediterranean-Type Ecosystems.* USDA Forest Service, General Technical Report PSW-58, pp. 438–443.

Ralphs, M. H., and F. E. Busby. 1979. Prescribed burning: vegetation change, forage production, cost, and returns on six demonstration burns in Utah. *Journal of Range Management* 32:267–270.

Rasmussen, L. A., G. D. Amman, J. C. Vandygriff, R. D. Oakes, A. S. Munson, and K. E. Gibson. 1996. Bark beetle and wood borer infestation in the Greater Yellowstone Area during four postfire years. USDA Forest Service, Research Paper INT-487.

Regelbrugge, J. C., and S. G. Conard. 1993. Modeling tree mortality following wildfire in *Pinus ponderosa* forests in the central Sierra Nevada of California. *International Journal of Wildland Fire* 3:139–148.

Reinhardt, E. D., and K. C. Ryan. 1988. Eight-year tree growth following prescribed underburning in a western Montana Douglas-fir/western larch stand. USDA Forest Service, Research Note INT-387.

Reynolds, H. G., and J. W. Bohning. 1956. Effects of burning on a desert grass–shrub range in southern Arizona. *Ecology* 37:769–777.

Riggs, R. A., S. C. Bunting, and S. E. Daniels. 1996. Prescribed fire. In: Krausman, P. R., editor. *Rangeland Wildlife.* Society for Range Management, Denver, Colorado, pp. 295–319.

Roppe, J. A., and D. Hein. 1978. Effects of fire on wildlife in lodgepole pine forests. *The Southwestern Naturalist* 23:279–287.

Rowe, J. S. 1983. Concepts of fire effects on plant individuals and species. In: Wein, R. W., and D. A. MacLean, editors. *The Role of Fire in Northern Circumpolar Ecosystems.* John Wiley & Sons, New York, pp. 135–154.

Ryan, K. C., and W. H. Frandsen. 1991. Basal injury from smoldering fires in mature *Pinus ponderosa* Laws. *International Journal of Wildland Fire* 2:107–118.

Ryan, K. C., D. L. Peterson, and E. D. Reinhardt. 1988. Modeling long-term fire-caused mortality of Douglas-fir. *Forest Science* 34:190–199.

Sackett, S. S. 1984. Observations on natural regeneration in ponderosa pine following a prescribed fire in Arizona. USDA Forest Service, Research Note RM-435.

Sackett, S. S., S. M. Haase, and M. G. Harrington. 1996. Lessons learned from fire use for restoring southwestern ponderosa pine ecosystems. In: Covington, W., and P. K. Wagner, technical coordinators. *Conference on Adaptive Ecosystem Restoration and Management: Restoration of Cordilleran Conifer Landscapes of North*

America. USDA Forest Service, General Technical Report RM-GTR-278, pp. 54–61.

Savage, M. 1994. Anthropogenic and natural disturbance and patterns of mortality in a mixed conifer forest in California. *Canadian Journal of Forest Research* 24:1149–1159.

Schmutz, E. M., M. K. Sourabie, and D. A. Smith. 1982. The Page Ranch story: its vegetative history and management implications. *Desert Plants* 7:13–21.

Severson, K. E., and J. N. Rinne. 1990. Increasing habitat diversity in southwestern forests and woodlands via prescribed fire. In: Krammes, J. S., technical coordinator. *Effects of Fire Management of Southwestern Natural Resources*. USDA Forest Service, General Technical Report RM-191, pp. 94–104.

Shea, S. R., J. McCormick, and C. C. Portlock. 1979. The effect of fires on regeneration of leguminous species in the northern jarrah (*Eucalyptus marginata* Sm) forest of western Australia. *Australian Journal of Ecology* 4:195–205.

Springfield, H. W. 1976. Characteristics and management of southwestern pinyon-juniper ranges: the status of our knowledge. USDA Forest Service, Research Paper RM-160.

Stransky, J. J., and R. F. Harlow. 1981. Effects of fire on deer habitat in the Southeast prescribed burning. In: Wood, G. W., editor. *Prescribed Fire and Wildlife in Southern Forests*. Belle W. Baruch Forest Science Institute, Clemson University, Georgetown, South Carolina, pp. 135–142.

Svejcar, T. J. 1989. Animal performance and diet quality as influenced by burning tallgrass prairie. *Journal of Range Management* 42:11–15.

Takahashi, M., and T. Kikuchi. 1986. The heat effect on seed germination of some species in the initial stage of post-fire vegetation. *Ecological Review* 21:11–14.

Taylor, D. L. 1973. Some ecological implications of forest fire control in Yellowstone National Park, Wyoming. *Ecology* 54:1394–1396.

Thomas, J. W., R. J. Miller, H. Black, J. E. Rodiek, and C. Maser. 1976. Guidelines for maintaining and enhancing wildlife habitat in forest management in the Blue Mountains of Oregon and Washington. *Transactions of the North American Wildlife and Natural Resources Conference* 41:452–476.

Tisdale, E. W., M. Hironaka, and M. A. Fosberg. 1969. *The Sagebrush Regions of Idaho: A Problem in Range Resources Management*. Utah Agricultural Experiment Station, Logan, Utah, Bulletin 512.

Vogel, R. J. 1973. Effects of fire on plants and animals of a Florida wetland. *The American Midland Naturalist* 89:334–347.

Wade, D. D. 1993. Thinning young loblolly pine stands with fire. *International Journal of Wildland Fire* 3:169–178.

Wagener, W. W. 1961. Guidelines for estimating the survival of fire-damaged trees in California. USDA Forest Service, California Forest and Range Experiment Station, Miscellaneous Paper 60.

Waldrop, T. A., and D. H. van Lear. 1984. Effect of crown scorch on survival and growth of young loblolly pine. *Southern Journal of Applied Forestry* 8:35–40.

Walter, H. 1977. Effects of fire on wildlife communities. In: Mooney, H. A., and C. E. Conrad, technical coordinators. *Environmental Consequences of Fire and Fuel Management in Mediterranean Ecosystems*. USDA Forest Service, General Technical Report WO-3, pp. 183–192.

Weaver, H. 1951. Fire as an ecological factor in the southwestern ponderosa pine forests. *Journal of Forestry* 49:93–98.
Whelan, R. J. 1995. *The Ecology of Fire.* Cambridge University Press, Cambridge, England.
Wood, G. W., editor. 1981. *Prescribed Fire and Wildlife in Southern Forests.* The Belle W. Baruch Forest Science Institute, Clemson University, Georgetown, South Carolina.
Woodmansee, R. G., and L. S. Wallach. 1981. Effects of fire regimes on biogeochemical cycles. In: Clark, F. E., and T. Rosswell, editors. *Terrestrial Nitrogen Cycles.* Ecological Bulletin 33, pp. 649–669.
Wooldridge, D. D., and H. Weaver. 1965. Some effects of thinning a ponderosa pine thicket with a prescribed fire, II. *Journal of Forestry* 63:92–95.
Wright, H. A. 1969. Effects of spring burning on tobosa grass. *Journal of Range Management* 22:425–427.
Wright, H. A. 1970. A method to determine heat-caused mortality in bunchgrass. *Ecology* 51:582–587.
Wright, H. A., and A. W. Bailey, 1982. *Fire Ecology—United States and Southern Canada.* John Wiley & Sons, New York.
Wright, H. A., S. C. Bunting, and L. F. Neuenschwander. 1976. Effect of fire on honey mesquite. *Journal of Range Management* 29:467–471.
Wyant, J. G., R. D. Laven, and P. N. Omi. 1983. Fire effects on shoot growth characteristics of ponderosa pine in Colorado. *Canadian Journal of Forestry* 63:92–95.
Zwolinski, M. J. 1990. Fire effects on vegetation and succession. In: Krammes, J. S., technical coordinator. *Effects of Fire Management of Southwestern Natural Resources.* USDA Forest Service, General Technical Report RM-191, pp. 18–24.

Chapter

10

Wetlands and Riparian Ecosystems

INTRODUCTION

Wetlands and riparian ecosystems are biotic communities that develop as a result of the regular presence of water. Undisturbed or well-managed wetlands and riparian ecosystems provide benefits far in excess of the relatively small portion of the total land area that they occupy by supporting a diversity of plants and animals that are not found elsewhere. Wetlands and riparian ecosystems also play important, but often not obvious, roles relating to enhanced water quality, flood peak attenuation, and reduced erosion and sediment transport.

Fire affects soil, water, and vegetation in wetlands and riparian ecosystems in much the same way as discussed in earlier chapters of this book. Because of their unique ecological settings and often fragile environments, however, these systems warrant additional consideration. Although the terms *wetland* and *riparian* are sometimes used interchangeably, it is important to distinguish between the two systems from an ecological perspective and in terms of their response to fire.

WETLANDS

Wetlands are saturated by surface water, groundwater, or combinations of both at a frequency and duration sufficient to support a prevalence of vegetation adapted to saturated soil conditions. Wetlands frequently form in low-lying areas where the groundwater table intersects the surface either permanently or frequently enough that saturated and oxygen-deficient soil and vegetation that can subsist on these soils are common (Hendrickson and Minkley 1984, Mitsch and Gosselink 1993). Standing water, therefore, exerts a major influence on the formation of wetland soils. Wetland ecosystems generally occur where annual precipitation exceeds potential evapotranspiration over large areas with little topographic relief. These ecosystems are either inland or coastal and include marshes, shallows, swamps, lake shores, peatlands, wet meadows, estuaries, and some riparian areas.

Ecological Significance

Until their ecological functions were recognized, wetlands were often considered lands with little value and, as a consequence, were converted to other land uses. However, wetlands can modify the effects of flooding, aquifer recharge, and water quality. Wetlands are also effective in removing excess sediments and nutrients from water by detention and settling of stormflow, denitrification, burial of chemical precipitation, plant uptake and transformation, biomass decomposition, and organic matter accumulation in peat, muck, and living biomass.

Wetlands play important roles in N, S, and C cycles. Because of their frequently anaerobic conditions, wetlands are effective in supporting denitrification involving the conversion of NO_3-N and NH_4-N to N_2. Sulfur gases released by burning fossil fuels can be transformed and fixed as phosphate and metal complexes in wetland soils. While C fixation is only a small part of the net annual primary productivity in wetlands, these ecosystems can be important sinks for C. In this regard, there is concern among scientists studying global climate change that a trend in wetland loss has shifted wetlands from being net sinks of important elements such as C and N to being net sources of these elements. Wetlands (peatlands) are also a source of methane, an important greenhouse gas.

Wetlands are important to many wildlife and fish populations. Some of the animals that rely on, or are associated with, wetlands include beavers, alligators, and a diversity of waterfowl species. Over one-half of the 800 species of the migratory birds in the United States rely on wetland habitats and breeding territory (Wharton et al. 1982). Depending on the wetland system involved, between 20 and 75 percent of the threatened and endangered species (TES) in the United States rely on wetlands for maintaining viable populations (Niering 1988). TES amphibians are the animals most dependent on wetlands; mammals are the least dependent. Wetland areas are attractive to tourists who visit wetland wildlife refuges, riparian corridors, and national seashores.

Prairie pothole wetlands in midwestern North America and other northerly latitudes of the world are only a small percentage of their original area, largely because of their drainage for agricultural crop cultivation. Some of the world's most productive habitats for waterfowl have been reduced substantially as a consequence of this drainage. What is not as well recognized, however, are the hydrologic implications associated with the loss of these wetlands. Other wetlands, such as peatlands, have likewise been reduced in area in many parts of the world. Drainage and the subsequent mining of peatlands have occurred in the upper Midwest of the United States, Canada, Finland, and Russia.

Wetlands that occur in coastal areas and, therefore, are influenced by salt water can also be influenced by freshwater runoff and river discharges. Coastal wetlands furnish food, spawning areas, and habitats for fish species that spend part or all of their life in salt water. These wetlands can be impacted by changes in discharge, sediment, and pollutants from rivers and coastal areas.

Responses to Fire

That wetlands frequently burn almost seems contradictory. However, wetlands do dry during drought periods and when the water table recedes, and, as a consequence, they become subject to fire occurrences (Example 10.1). Small wetland areas that are inclusions in larger upland watersheds can burn at frequencies that are representative of adjacent terrestrial vegetative types. Permafrost peatlands in the boreal zone of North America burn naturally at an interval averaging 200 years (Caljouw 1995). The extensive bogs in the continental interior or maritime fringes can burn at frequencies of 200 to 800 years, while their smaller isolated remnants burn more frequently. Shrub bogs (pocosins) of the Atlantic Coastal Plain burn at shorter intervals (5 to 25 years) than their boreal counterparts. The fire cycle in Atlantic Coastal Plains white-cedar swamps (25 to 200 years) is not too different from that of the interior Alaskan black spruce *(Picea mariana)* swamps (50 to 200 years). Wetlands in the wet pine flatwoods of the southeastern coastal plain burn at intervals of 2 to 8 years. Occurrences of fire usually follow drought cycles.

Example 10.1: Wetlands Affected by Fire

PEATLANDS

Peatlands cover vast areas across northern North America, Asia, and Europe that are characterized by deep organic soils, flat topography, and shallow water tables. Peatlands develop on sites where water moves slowly, causing anaerobic conditions. Organic soils accumulate to several meters in some areas as a result. Compositions of vegetative communities reflect the regional groundwater or perched water tables that sustain these systems. Fire has been a significant process in the evolution and subsequent sustainability of peatlands, affecting nutrient cycling, vegetative development, and wildlife habitats.

Shrub bogs of the Middle Atlantic Coastal Plain developed under the influence of prolonged periods of saturated conditions and accumulations of organic soils. Organic matter in these shrub bogs can range in depth from a few centimeters to over 9 m (Little 1950). Vegetative successional patterns in both the shrub bog interior and surrounding pine forests are influenced by the nature of the occurrence of fire, periods of saturated conditions, soil type, and peat depth (Sharitz and Gibbons 1982). Elsewhere, savannas of the southeastern United States, particularly those on mineral soils, are largely maintained by frequent low-intensity fire (Christensen 1981). Atlantic white-cedar *(Chamaecyparis thyoides)* forests occurring on deep organic soils are the product of a low frequency, relatively high-intensity fire regime. Frequent fire converts these areas to shrub bogs, while infrequent fire results in a succession to hardwoods.

FRESHWATER MARSHES

Freshwater marshes are a diverse group of wetlands that have relatively high pH substrates, high productivity and biological activity, and rapid cycling of nutrients and organic matter. Examples of freshwater marshes in North America include the prairie potholes of the Great Plains, coastal shallow freshwater marshes, the central and vernal marshes of California, the extensive marshes that occur in the Great Lakes region, playas in the high plains of northwestern Texas and eastern New Mexico, and the largest marsh system in the United States—the Everglades of southern Florida.

Freshwater marshes can burn for extended periods of time in drought. Therefore, the plants and animals that inhabit these marshes are attuned to fire. Burning has been used as a tool in the management of freshwater marshes to remove annuals and dead herbaceous cover, to prevent the buildup of debris on a marsh floor, to reduce the surface of the marsh floor by burning into the organic soils, and to reduce or eliminate the invasion of woody vegetation into impoundments (Linde 1985, Kirby et al. 1988). Other purposes of prescribed burning include destroying sphagnum moss to bring about succession to sedges and grasses to create nesting sites for waterfowl, cleaning impoundment basins prior to possible flooding, producing open areas that will provide better early-season grazing for waterfowl, increasing nutrient mineralization resulting in increased forage quality, and fostering a change in albedo that permits earlier spring growth.

SOUTHERN DEEPWATER SWAMPS

Southern deepwater swamps are found throughout the coastal plain and coastal bays of the southeastern United States. These wetlands are generally characterized by standing water (1 to 2 m) throughout most of the year. The primary productivity in these wetlands is generally high. Major types of deepwater swamps include cypress domes, dwarf cypress, cypress strands, shrub bogs, lake edge, bay, gum, and alluvial river swamps.

Fire occurs in all of these wetlands, although it is more infrequent in the swamp types, which have longer periods of saturated conditions and deeper water environments. While much of the fire in freshwater swamps is low-intensity ground fire, catastrophic crown fire can occur in association with severe burning of vegetation on surrounding hillslopes. Tree species not normally indigenous to wetlands often invade these swamps and become established during the recurring periods of drought and fire that characterize these ecosystems. ■

Physical responses of wetlands to burning depend on the prefire fuel types and accumulations, and the intensity, severity, and duration of the fire. The greatest physical impact of fire on wetlands is likely to come from the transport of sediments and nutrients from burned upland watersheds into the wetland systems. While movement of sediments into wetlands can adversely affect indigenous biota, nutrients associated with the transport of the sediments often can provide a long-term enrichment of the wetland ecosystem.

Origin of postfire nutrients can be surface (overland flows of cation-rich ash material) or subsurface (leaching of mobile anions). A mobile element like N can be leached, immobilized in soil organic matter, denitrified to N_2 gas, or taken up by postfire vegetation and, as a result, not enter wetlands of the aquatic ecosystem. These phenomena are often the case on low-gradient wetlands, where water movement is slow. In higher-gradient systems, nutrients are more likely to be transported off burned sites to wetlands and aquatic ecosystems.

How wetland vegetation responds to burning varies with the severity of the burning and the adaptability of individual plant species to fire. Trees and shrubs in freshwater swamps are often suppressed or killed by fire, allowing wet prairies of aquatic or emergent plants to become reestablished (Glasser 1985). Plant succession in cypress *(Taxodium distichum)* swamps of the Florida peninsula and southern Georgia is generally influenced by the occurrence of fire (Ewel and Mitsch 1978). The entire successional cycle in these cypress swamps can be altered by the occurrence of severe fire. Light or moderate fire is needed to maintain pure stands of cypress, or to convert mixed cypress–hardwood communities back to cypress-dominated stands. Some wetland ecosystems are maintained by recurring fire occurrences, because their floristic associations and fauna are adapted to periodic fires.

Fire burning deep holes in the organic soils of a wetland ecosystem can benefit wildlife by providing a water source where the holes intersect the water table, and by opening sites for colonization by herbaceous plants. Fire often increases plant species richness, which also benefits wildlife. As a consequence, adverse impacts are likely to occur on some wildlife habitats by the exclusion of fire from wetland ecosystems. Fire rarely causes large-scale wildlife mortality.

Organic Soils

Organic soils, and mineral soils containing organic surface layers, are common in wetland ecosystems. These organic layers are formed as a result of low temperatures, high acidity, low nutrient supply, and excessive water and deficient oxygen, all of which slow the decomposition of dead biomass, allowing the layers to accumulate over many years (Hungerford et al. 1995). Organic soils, also referred to as *peats* in some systems, have a root mat at the surface over horizons of highly decomposed muck (sapric) material or less decomposed peat (fibric) material.

Fire can ignite the organic layers of these soils. Only rarely, however, will the soil body burn. The primary form of combustion in organic soils is smoldering. Fire typically burns in concentric patterns outward from the ignition point, resulting in a bowl-shaped burned depression being created. Burning generally continues until it either reaches mineral soil or a location where the moisture content of the soil prohibits further combustion (see Chapter 2, Figure 2.3). Fires in organic soil fires do not move fast, but they can persist for weeks, months, or longer (Example 10.2).

Example 10.2: The Kalimantan Fire: One of the Largest and Longest Burning Fires in the World (Kilmaskossu 1988)

This fire began in the autumn of 1983, a period of record drought in East Kalimantan, Indonesia, as a consequence of farmers burning to clear land of logging slash and eliminate competing vegetation before planting agricultural crops. The numerous small fires originally ignited by the farmers quickly penetrated deeply into the peat soils of the region because of the dry conditions, eventually merging into a more-or-less continuous burn that covered over 3.4 million hectares and remained unchecked into the later 1990s.

It is helpful to place the Kalimantan fire into perspective to appreciate its significance. The area burned (to the later 1980s) was 2.5 times larger than that of the Great Idaho fire of 1910, in which 85 people lost their lives and five towns were destroyed. The Kalimantan fire was almost six times larger than the 1871 Peshtigo fire in Wisconsin, which resulted in 1500 people dying. The fire was 50 times larger than the 1894 fire in Hinckley, Minnesota, when 500 people lost their lives and 12 towns were lost. With the Kalimantan fire almost impossible to extinguish, how large the burned area will ultimately become, and how long the fire might burn, is anyone's guess. ■

RIPARIAN ECOSYSTEMS

Riparian ecosystems are situated in the interfaces between terrestrial and aquatic ecosystems. Riparian ecosystems are found along open bodies of water, such as the banks of rivers and ephemeral, intermittent, and perennial streams, and around lakes, ponds, springs, bogs, and meadows. The abruptness and extent of the transitions between these interfaces are generally site-specific (Figure 10.1).

Riparian ecosystems are delineated by soil characteristics and vegetative communities that require free or unbound water (DeBano and Schmidt 1989, Gregory et al. 1989, DeBano and Neary 1996). These ecosystems are representative of ecological sites, or a combination of sites, in which soil moisture

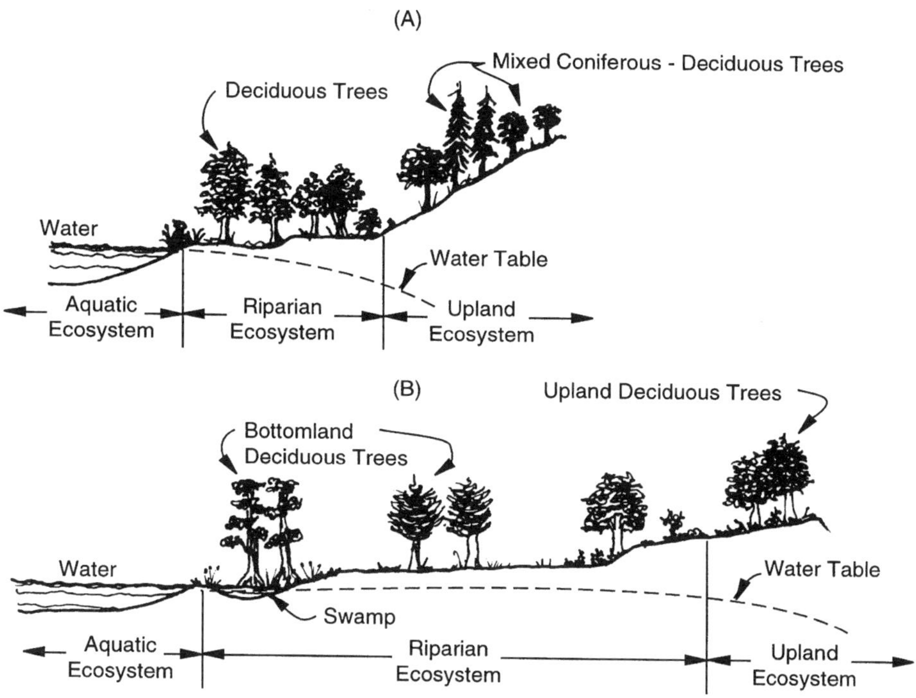

FIGURE 10.1 Transition profiles across hillslope, riparian, and aquatic ecosystems in (A) the southwestern United States (adapted from Johnson and Lowe 1985) and (B) the southeastern United States (adapted from Clark and Benforado 1981). Transitions between these three ecosystems tend to be more abrupt in the dry southwestern United States than in the more humid southeastern United States.

exceeds that otherwise available due to water flowing into and through the ecosystems, or due to subsurface seepage. The excess water that occurs results in establishment of soil–vegetation habitats that reflect the influence of the extra soil moisture (Anderson 1987). The characteristics of these habitats are the differentiating criteria that help to characterize riparian ecosystems.

Riparian ecosystems can furnish habitats for a variety of wildlife species, are grazed by livestock in many instances, and provide areas for fishing, bird watching, picnicking, and camping. Maintaining the integrity of these fragile ecosystems requires careful management and, in some cases, protection from logging activities, livestock grazing, recreational use, and other types of exploitation.

Ecological Significance

Riparian vegetation is adapted to well-drained but periodically flooded soils. This vegetation often consists of deep-rooted trees and shrubs, enabling the plants to survive the frequently occurring flooding events common to riparian ecosystems (Hendricksen and Minkley 1984). Root systems of these plants bind the soil together, provide streambank stability, and promote high infiltration rates. Herbaceous plants also help in armoring streambanks against large, channel-forming flood flows. Trees, shrubs, and herbaceous plants also trap sediments, building a soil body leading to a floodplain and then community succession (Mitsch and Gosselink 1993). Ecosystems located along stream systems include a diversity of plant, insect, mollusk, and crustacean species that are key resources in the aquatic food chain.

The streambanks themselves occupy a crucial boundary position between terrestrial and aquatic habitats in riparian ecosystems (Bohn 1986). Streambanks and stream channel configurations affect the streamflow velocity, sedimentation processes, and stream water temperatures. Streambank condition and the general quality of aquatic habitats are closely linked, with deterioration in the former also resulting in deterioration of the latter.

Riparian–Watershed Linkages

Erosion and surface runoff are key factors affecting the stability of both riparian ecosystems and surrounding hillslopes. A response to disturbance such as fire, therefore, must be considered within this context (DeBano and Schmidt 1989, DeBano et al. 1996). Water flow and sediment movement through a riparian ecosystem are controlled by vegetation, topography, and geologic formations within this linked system.

Riparian ecosystems that are characterized by an equilibrium between

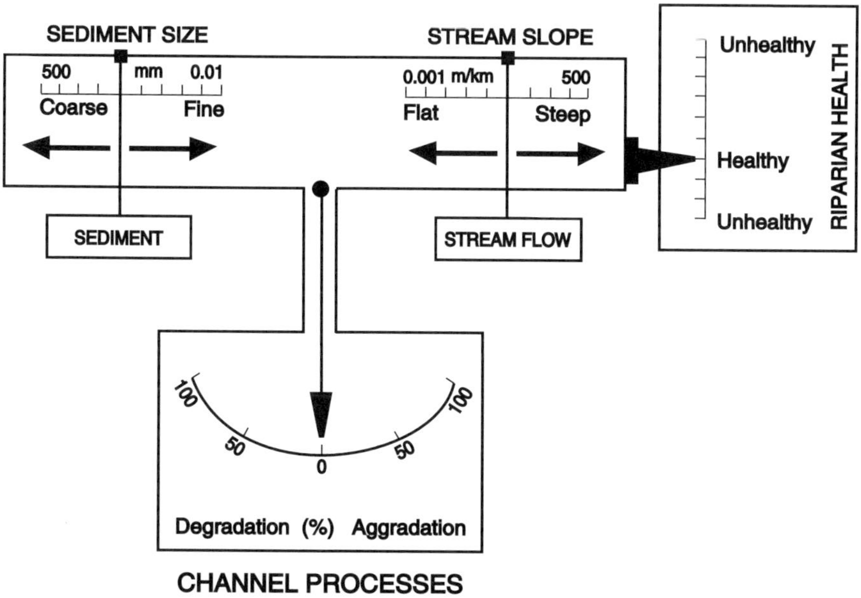

FIGURE 10.2 Dynamic equilibrium between stream channel aggradation and degradation in a healthy riparian ecosystem. (Adapted from Rosgen 1980, based on Lane 1955.)

stream channel deposition (aggradation) and downcutting (degradation) by erosion in stream channels are considered healthy. A *healthy riparian ecosystem* is one that maintains a dynamic equilibrium between the streamflow forces acting to produce change and the resistance of vegetative, geomorphic, and structural features (DeBano and Schmidt 1989). This dynamic equilibrium between stream channel aggregation and degradation is sufficiently stable so that compensations to the external disturbances on surrounding hillslopes occur without altering the equilibrium (Figure 10.2). This resistance to change results largely from adjustments among several factors (such as vegetation, channel depth, and stream morphology) operating simultaneously within the riparian system to increase flow or sediment movement.

Large increases in short-term surface runoff from disturbances on upland watersheds can also increase channel flow volume and velocity, which, in turn, causes channel erosion and increased deposition rates in a downstream riparian community. Under these conditions, a riparian ecosystem oscillates back and forth, so that only little change occurs in the dynamic equilibrium of the riparian system. An equilibrium condition can be reestablished to support a healthy riparian ecosystem when the resiliency of the system is not altered by the disturbance.

Responses to Fire

Fire is a common disturbance in both riparian ecosystems and the surrounding hillslopes. Fire scenarios in riparian ecosystems can consist of burning only on the upland watershed, or burning both the riparian ecosystem and upland watershed. Fire rarely occurs only in a riparian ecosystem, however, unless a controlled burn is prescribed to do so. When only the upland watershed is burned, the riparian ecosystem acts as a buffer between the burned hillslopes and stream. However, when upland watersheds and riparian ecosystems are severely burned together, devastating effects can occur. Of most significance, the buffering effect of the riparian ecosystem can be lost, leading to a disruption of the dynamic equilibrium described previously.

Debris flows and small landslides off surrounding hillslopes are also a frequent postfire occurrence on steep, erodible topography. However, the most obvious fire effects are changes in the character of riparian vegetation (Example 10.3) and consumption of the underlying protective litter layer. The severity of this damage depends largely on the fire intensity (Cox and Morton 1985, Davis et al. 1989, DeBano and Neary 1996). Intense wildfire can cause severe damage to vegetative covers, while a low-intensity cool-burning prescribed fire is likely to have less severe consequences. The degree and speed of recovery of postfire riparian vegetation to prefire conditions are largely dictated by the nature of the combined disturbance of both the fire and postfire flooding, if any.

Example 10.3: Effects of Fire on Riparian Vegetation on the Lower Colorado River Floodplain (Busch 1994, 1995)

Approximately 35 percent of the riparian vegetation between Davis Dam on the Colorado River and the international boundary with Mexico burned in the 12–year period evaluated. Fire extent was largely associated with the area of saltcedar *(Tamarix ramosissima)* habitat burned, while fire disproportionally affected the relatively small areas dominated by mesquite *(Prosopis glandulosa* and *P. pubescens)*. Saltcedar and arrowweed *(Tessaria sericea)* shrubs dominated the postfire riparian communities. Arrowweed replaced riparian trees in response to fire in habitats dominated by cottonwood–willow *(Populus fremontii–Salix gooddingii)* and mesquite habitats. Species turnover was relatively low through the postfire period studied. Therefore, it is likely that the replacement of riparian forests by water- or salinity-stress shrubs is long-term. ■

Destruction of *only* the vegetation on upland watersheds by burning affects downslope riparian ecosystems indirectly through changes in the fluvial processes on surrounding hillslopes and, as a result of these changes, in the dynamic equilibrium situation described previously. The most obvious change in this case also involves the removal of vegetative and litter covers on the surrounding hillslopes. Soil erosion is likely to increase when the burned soil surface becomes exposed to erosive agents. Material that is eroded from the hillslopes can be deposited in stream channels to cause channel aggradation. The frequent occurrence of increased postfire streamflow can lead to accelerated channel downcutting. Channel downcutting generally acts more rapidly than does channel aggradation in changing the equilibrium of a healthy riparian ecosystem. However, both of these processes disrupt the dynamic equilibrium, resulting in the system shifting to a different equilibrium (Figure 10.2).

Intense wildfire can have large impacts by increasing streamflow velocity, sedimentation rates, and stream water temperatures, while low-intensity prescribed burning normally has relatively little impact on these hydrologic processes. The duration of these effects is affected by the rate of revegetation and postfire rehabilitation efforts.

COARSE WOODY DEBRIS

Coarse woody debris plays a role in the hydraulics, sediment routing, and channel morphology of streams that flow through riparian ecosystems (Smith et al. 1993, DeBano and Neary 1996). Through time, entire live and dead trees and shrubs, or parts of them, are likely to fall into stream channels within riparian ecosystems. This debris moves into stream channels by streambank undercutting and collapse, tree blowdown, mass soil movements, and snow avalanches. The larger debris materials can become wedged in the streambanks, increasing channel roughness that (in turn) reduces streamflow velocities and retards the downstream movement of boulders, rocks, and large stones. Accumulations of coarse woody debris in stream channels can also cause a dissipation of the energy that is generated by flowing water. The larger the components of these accumulations are, the longer they function in this manner.

Accumulations of coarse woody debris in stream channels often increase after fire. While this added debris can have a beneficial effect on aquatic habitats in the short-term by providing structure to the streams, long-term impacts can be disruptive to stream morphology and the consequent streamflow and sediment transport regimes. Coarse woody debris accumulations are usually larger in the first few years following a wildfire than in prefire years, but they generally decline to below prefire conditions thereafter (Figure 10.3). The recovery time to prefire conditions ranges from 25 to 300 years depending on the stability of the ecosystem burned. Carefully prescribed fire should not affect the accumulations of coarse woody debris.

Coarse woody debris in the stream channels of burned riparian ecosystems is made up of components of larger individual size, and forms larger accumulations, than that of unburned systems. Postfire debris is also likely to be moved more frequently over longer distances in subsequent streamflow events (Young 1994, Alexander and Rinne 1995). Managers frequently debate the

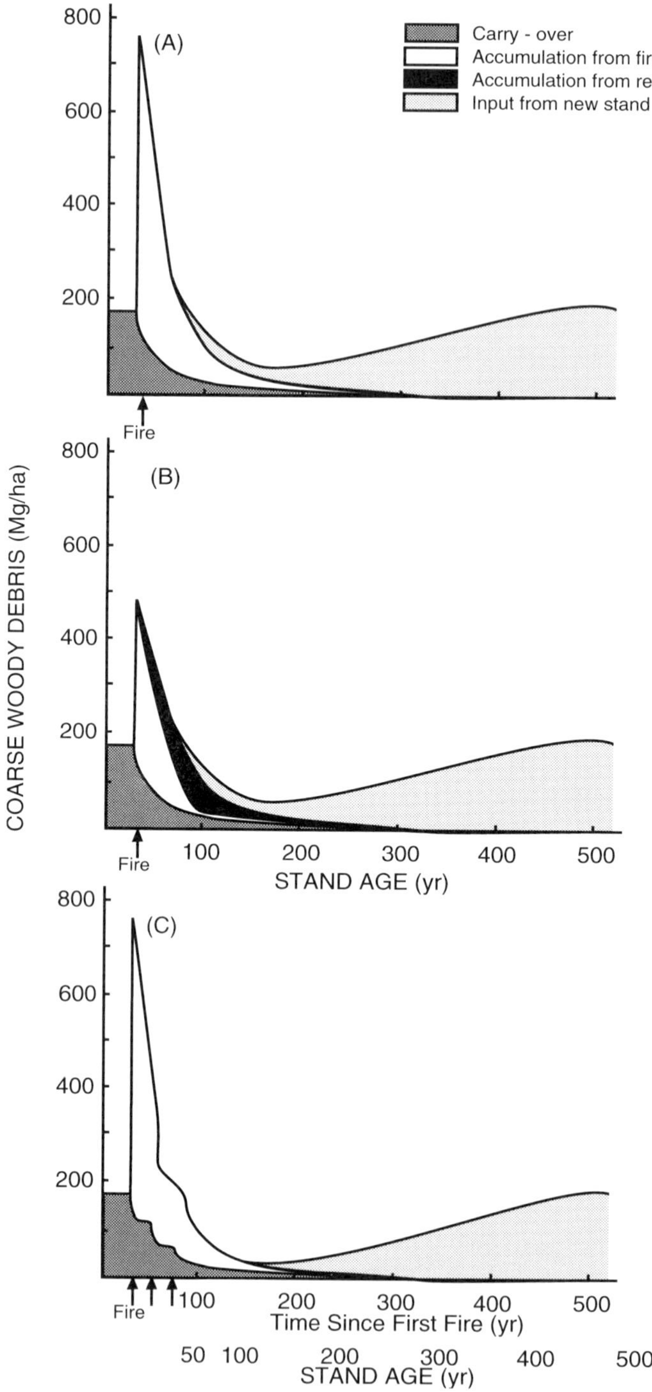

FIGURE 10.3 Predicted changes in coarse woody debris accumulations following fire in a 450-year-old Douglas fir-western hemlock *(Tsuga heterophylla)* forest after (A) a catastrophic fire, (B) a partial burn, and (C) a catastrophic fire and two subsequent fires 25 and 50 years later. Coarse woody debris is separated into that present before the fire, that created by the fire, that produced by the remnant forests, and that produced by the forest growing after the fire. (Adapted from Spies et al. 1988.)

merits of leaving burned dead debris in stream channels following fire. Some managers feel that this material could jam culverts and bridges, causing these structures to wash out and cause flooding, and, as a consequence, feel it should be removed (Barro et al. 1989). However, removal of this debris can also result in changes in channel morphology, a scouring of the channel bed, increases in streamflow velocities and sediment loads, an export of nutrients out of the ecosystem, and a deterioration of biotic habitats. Removal of post-fire accumulations of coarse woody debris, therefore, is likely to best be decided on a site-specific basis.

BUFFER STRIPS

Buffer strips are vegetated bands along streams, or around water bodies, that are left intact or that have been established to filter sediments and nutrients from upland watersheds; to enhance bank stability; to provide shade, shelter, and food for wildlife and for fish and other aquatic organisms; and to create visually diversified landscapes (Figure 10.4). Maintaining buffer strips is an effective way of reducing the velocity of surface runoff from upland watersheds into the stream channel and, as a consequence, decreasing the amount of sediment and associated nutrients being deposited into a stream (Neary et al. 1993, Schultz et al. 1995). Nutrients that are attached to sediments are kept from entering water bodies by reducing sedimentation. Economic products are also provided by buffer strips.

When riparian buffer strips are burned by wildfire, the filtering of sedi-

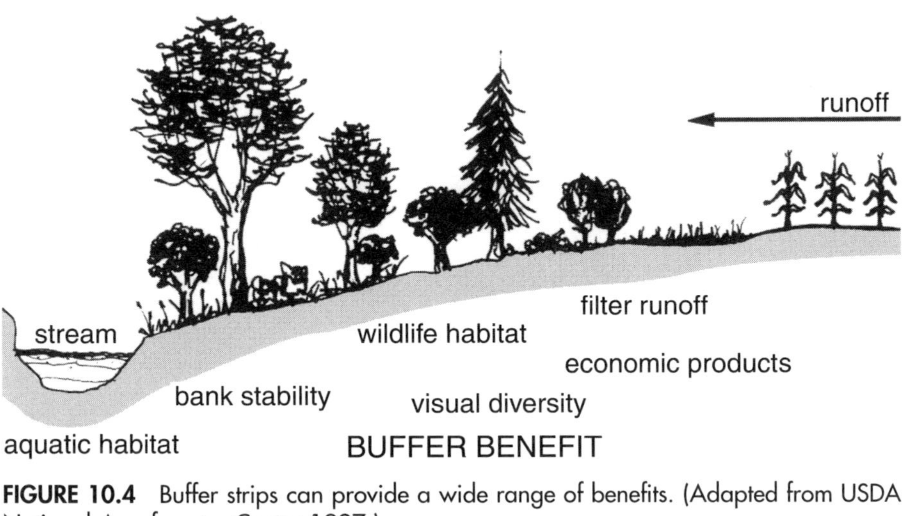

FIGURE 10.4 Buffer strips can provide a wide range of benefits. (Adapted from USDA National Agroforestry Center 1997.)

ments and nutrients from a burned-over upland watershed is lost, and, as a consequence, these materials are likely to enter the stream. However, these impacts can be mitigated when implementing a prescribed burn by recognizing the importance of buffer strips to the stream system and planning accordingly (Cooper et al. 1987, USDA Forest Service 1989, Neary et al. 1993). Recommended buffer strip widths to retain in prescribed burning are available for riparian ecosystems in some parts of the United States.

INSTREAM FLOW

The concept of instream flow has become an important consideration in management of watersheds in the western United States. *Instream flow* is the streamflow regime required to satisfy the conjunctive demands being placed on water while the water remains in the stream channel (Helm 1985). An *instream flow right* protects a designated flow through a specified segment of a stream from depletion by newer water users. This right is especially important where newer upstream uses, cultural developments, diversions, or transfers could threaten existing flows. Benefits of instream flow rights include protection of the diversity of riparian plants and animals that live in or along the water.

Increased surface runoff and sediment movement resulting from fire can change instream flows. Unfortunately, relatively little can be done to assure the retention of instream flows after a severe wildfire has occurred on upland watershed lands. However, when using fire as a management tool, prescriptions that retain the integrity of riparian buffer strips will minimize increased streamflow through and deposition of sediment in stream channels and, as a consequence, will minimize changes in instream flow regimes. Burning a watershed in a mosaic pattern can further decrease the effect of prescribed fire on instream values.

SUMMARY

Wetlands and riparian ecosystems are unique and, in many instances, fragile ecosystems. The effects of fire on these ecosystems, therefore, warrant the special attention paid to this topic. After reading this chapter, you should be able to:

1. Distinguish wetlands from riparian ecosystems in terms of their ecological significance and responses to fire.
2. Understand how a fire can affect water flow and sediment movement through a riparian ecosystem and the stability of both the riparian system and surrounding hillslopes.
3. Describe the role of coarse woody debris in the hydraulics, sediment

routing, and channel morphology of streams that flow through riparian ecosystems.
4. Discuss the importance of buffer strips in the filtering of sediments and nutrients from a burned-over upland watershed.
5. Appreciate the ways in which increased surface runoff and sediment movement resulting from fire can change instream flow.

REFERENCES

Alexander, M. M., and J. N. Rinne. 1995. Preliminary observations on the transportation of large woody organic debris in burned and unburned headwater streams, Tonto National Forest, Arizona. *Hydrology and Water Resources in Arizona and the Southwest* 22–25:56–59.

Anderson, E. W. 1987. Riparian area definition—a viewpoint. *Rangelands* 9:70.

Barro, S. C., P. M. Wohlemuth, and A. G. Campbell. 1989. Postfire interactions between riparian vegetation and channel morphology and the implications for stream channel rehabilitation. In: Abell, D. L., technical coordinator. *Proceedings of the California Riparian Conference: Protection, Management, and Restoration for the 1990s.* USDA Forest Service, General Technical Report PSW-100, pp. 51–53.

Bohn, C. 1986. Biological importance of streambank stability. *Rangelands* 8:55–56.

Busch, D. E. 1994. Fire in southwestern riparian habitats: functional and community responses. In: Covington, W. W., and L. F. DeBano, technical coordinators. *Sustainable Ecological Systems: Implementing an Ecological Approach to Lands Management.* USDA Forest Service, General Technical Report RM-247, pp. 304–305.

Busch, D. E. 1995. Effects of fire on southwestern riparian plant community structure. *The Southwestern Naturalist* 40:259–267.

Caljouw, C. 1995. Role of fire in wetland and riparian systems. *Fire in Ecosystem Management Notes: Unit II-C.* USDA Forest Service, National Advanced Resources Technology Center, Marana, Arizona.

Christensen, N. L. 1981. Fire regimes in southeastern ecosystems. In: Mooney, H. A., T. M. Bonnicksen, N. L. Christensen, J. E. Lotan, and W. A. Reiners, technical coordinators. *Proceedings of the Conference: Fire Regimes and Ecosystem Properties.* USDA Forest Service, General Technical Report WO-26, pp. 112–136.

Clark, J. R., and J. Benforado. 1981. Introduction. In: Clark, J. R., and J. Benforado, editors. *Wetlands of Bottomland Hardwood Forests.* Elsevier Scientific Publishing Company, Amsterdam, The Netherlands, pp. 1–9.

Cooper, J. R., J. W. Gilliam, and R. B. Daniels. 1987. Riparian areas as filters for agricultural sediment. *Soil Science Society of America Journal* 51:416–420.

Cox, J. R., and H. L. Morton. 1985. Above-ground biomass quantities and livestock production at big sacaton riparian areas in southeastern Arizona. In: Johnson, R. R., C. D. Ziebell, D. R. Patton, P. F. Ffolliott, and R. H. Hamre, technical coordinators. *Riparian Ecosystems and Their Management: Reconciling Conflicting Uses.* USDA Forest Service, General Technical Report RM-120, pp. 305–309.

Davis, F. W., E. A. Keller, A. Parikh, and J. Florsheim. 1989. Recovery of the chaparral riparian zone after wildfire. In: Abell, D. L., technical coordinator. *Proceedings of the California Riparian Conference: Protection, Management, and Restoration for the 1990s.* USDA Forest Service, General Technical Report PSW-100, pp. 194–203.

DeBano, L. F., and D. G. Neary. 1996. Effects of fire on riparian systems. In: Ffolliott, P. F., L. F. DeBano, M. B. Baker, Jr., G. J. Gottfried, G. Solis-Garza, C. B. Edminster, D. G. Neary, L. S. Allen, and R. H. Hamre, technical coordinators. *Effects of Fire on Madrean Province Ecosystems: A Symposium Proceedings.* USDA Forest Service, General Technical Report RM-GTR-289, pp. 69–76.

DeBano, L. F., and L. J. Schmidt. 1989. *Improving Riparian Areas Through Watershed Management.* USDA Forest Service, General Technical Report RM-182.

DeBano, L. F., P. F. Ffolliott, and K. N. Brooks. 1996. Flow of water and sediments through southwestern riparian systems. In: Shaw, D. W., and D. M. Finch, technical coordinators. *Desired Future Conditions for Southwestern Riparian Ecosystems: Brings Interests and Concerns Together.* USDA Forest Service, General Technical Report RM-GTR-272, pp. 128–134.

Ewel, K. C., and W. J. Mitsch. 1978. The effects of fire on species composition in cypress dome ecosystems. *Florida Scientist* 41:25–31.

Glasser, J. E. 1985. Successional trends in the Okefenokee Swamp as determined by interspecific association analysis. *The American Midland Naturalist* 113:287–293.

Gregory, S. V., G. A. Lamberti, and K. M. S. Moore. 1989. Influence of valley floor landforms on stream ecosystems. In: Abell, D. L., technical coordinator. *Proceedings of the California Riparian Systems Conference: Protection, Management, and Restoration for the 1990s.* USDA Forest Service, General Technical Report PSW-110, pp. 3–8.

Helm, W. T., editor. 1985. *Aquatic Habitat Inventory: Glossary and Standard Methods.* Habitat Inventory Committee, Western Division, American Fisheries Society, Bethesda, Maryland.

Hendrickson, D. A., and W. L. Minkley. 1984. Cienegas—vanishing climax communities of the American Southwest. *Desert Plants* 6:131–175.

Hungerford, R. D., W. H. Frandsen, and K. C. Ryan. 1995. Ignition and burning characteristics of organic soils. In: Cerulean, S. I., and R. T. Engstrom, editors. *Fire in Wetlands: A Management Perspective.* Proceedings of the Tall Timbers Fire Ecology Conference No. 19, Tall Timbers Research Station, Tallahassee, Florida, pp. 78–91.

Johnson, R. R., and C. H. Lowe. 1985. On the development of riparian ecology. In: Johnson, R. R., C. D. Ziebell, D. R. Patton, P. F. Ffolliott, and R. H. Hamre, technical coordinators. *Riparian Ecosystems and Their Management: Reconciling Conflicting Uses.* USDA Forest Service, General Technical Report RM-120, pp. 112–116.

Kilmaskossu, M. S. E. 1988. Fire as a management tool to improve the renewable natural resources of Indonesia. Master's Thesis, University of Arizona, Tucson, Arizona.

Kirby, R. E., S. J. Lewis, and T. N. Sexson. 1988. *Fire in North American Wetland Ecosystems and Fire–Wildlife Relations: An Annotated Bibliography.* USDI Fish and Wildlife Service, Biological Report 88(1), Washington, DC.

Lane, E. W. 1955. The importance of fluvial morphology in hydraulic engineering. *Journal of Hydraulics Division* 81:745-1 to 745-17.

Linde, A. F. 1985. Vegetation in water impoundments: alternatives and supplements to water level control. In: Knighton, M. D., compiler. *Water Impoundments for Wildlife: A Habitat Management Workshop.* USDA Forest Service, General Technical Report NC-100, pp. 51–60.

Little, J. S., Jr. 1950. Ecology and silviculture of white cedar and associated hardwoods in southern New Jersey. *Yale University School of Forestry Bulletin* 56.

Mitsch, W. J., and J. G. Gosselink. 1993. *Wetlands.* Van Nostrand Reinhold, New York.

Neary, D. G., N. B. Comerford, and L. W. Swift, Jr. 1993. Land and riparian interactions with sediment in the southern United States. In: *Proceedings of a Technical Workshop on Sediments.* Terrene Institute, Washington, DC, pp. 51–60.

Niering, W. A. 1988. Endangered, threatened, and rare wetland plants and animals of the continental United States. In: Hook, D. D., editor. *The Ecology and Management of Wetlands. Volume I: Ecology of Wetlands.* Timber Press, Portland, OR, pp. 227–238.

Rosgen, D. L. 1980. Total potential sediment. In: Mulkey, L. A., coordinator. An approach to water resources evaluation of non-point silvicultural sources: A procedural handbook. EPA-600/8-80-012, Environmental Research, U.S. Environmental Protection Agency, Environmental Research Laboratory, Athens, Georgia, pp. VI.1–VI.43.

Schultz, R. C., T. M. Isenhart, and J. P. Colletti. 1995. Riparian buffer systems in crop and rangelands. In: Rietveld, technical coordinator. *Agroforestry and Sustainable Systems: Symposium Proceedings.* USDA Forest Service, General Technical Report RM-GTR-261, pp. 13–27.

Sharitz, R. R., and J. W. Gibbons. 1982. The ecology of southeastern shrub bogs (pocosins) and Carolina bays: a community profile. U.S. Fish and Wildlife Service, Publication FWS/OBS-82/04.

Smith, R. D., R. C. Sidle, P. E. Porter, and J. R. Noel. 1993. Effects of experimental removal of woody debris on the channel morphology of a forest, gravel-bed stream. *Journal of Hydrology* 152:153–178.

Spies, T. A., J. F. Franklin, and T. B. Thomas. 1988. Coarse woody debris in Douglas-fir forests of western Oregon and Washington. *Ecology* 69:1689–1702.

USDA Forest Service. 1989. Final environmental impact statement: vegetation management in the Piedmont and Coastal Plain. USDA Forest Service, Southern Region, Management Bulletin R8–MB-23, Atlanta, Georgia.

USDA National Agroforestry Center. 1997. Riparian buffers for agricultural land. USDA Forest Service and USDA Natural Resources Conservation Service, National Agroforestry Center, Agroforestry Notes, AF Note 3, Lincoln, Nebraska.

Wharton, C. H., W. M. Kitchens, E. C. Pendleton, and T. W. Sipe. 1982. The ecology of bottomland hardwood swamps of the southeast: a community profile. USDI Fish and Wildlife Service, Biological Services Program, FWS/OBS-81/37.

Young, M. K. 1994. Movement and characteristics of stream-borne coarse woody debris in adjacent burned and undisturbed watersheds in Wyoming. *Canadian Journal of Forest Research* 24:1933–1938.

Chapter

11

Air

INTRODUCTION

Impacts of fire on the atmosphere occur at all scales. Visibility effects and the changes in chemical constituents in the atmosphere (such as CO_2, water vapor, CO, and particulate matter) are a concern at the scale of a small fire of less than a hectare. At a landscape and regional scale, changes in atmospheric processes and chemical deposition emerge as issues, although visibility and the chemical content of air are still important. The major environmental concern on a global scale is the response of atmospheric processes to the chemical constituents added to the atmosphere from the burning of fuels in forests, shrublands, and grasslands. Chemical products of combustion, effects of smoke on the atmosphere, and the resulting considerations for the management of fire in natural ecosystems are the topics of this chapter.

SMOKE

Flames are the spectacular part of a fire, but the universal symbol of fire is smoke. Hundreds of compounds are produced during the exothermic reactions of combustion. These compounds are either oxidized further or released into the atmosphere. Many of these compounds have high boiling temperatures and, as a consequence, readily condense into soot, tar droplets, and water droplets in the cooling columns of air above a fire, producing clouds of visible smoke. The effects of smoke and its constituents on the environment can range from localized visibility impairment to possible changes in the global climate. These effects are largely a result of the primary chemical products and secondary emissions of combustion.

Primary Chemical Products

Primary chemical products of combustion include chemical products volatilized (but not oxidized) during the combustion process, compounds formed by partial or complete oxidation of organic fuels, and compounds pyrosynthesized during combustion. Some of these products, for example, CO_2 and water vapor, are considered to be normal constituents of the atmosphere, but others are often air pollutants.

Air pollutant regulations are established by individual countries. In the United States, the Environmental Protection Agency has issued a set of National Ambient Air Quality Standards for acceptable levels of air pollutants that are potentially harmful to human health. These pollutants are known as *criteria air pollutants.* Criteria air pollutants in smoke include particulates, CO, SO_2, N_2O, and ozone (O_3). Most of the other chemical components of smoke do not have specified ambient air standards, although some can have harmful effects on human health. These emission products are classed as *noncriteria air pollutants.*

One measure often used to assess the amount of combustion products produced by fire, and the impact of fire on air quality, is the *emission rate,* which is the mass of smoke produced per unit of time. The emission rate (ER) is used as an input to air quality models to predict air pollutant concentrations and potential impacts on urban areas and transportation corridors. The ER is obtained by multiplying together the *available fuels,* the *combustion rate,* and the appropriate *emission factor.* Available fuels are the total amount of fuel (Mg/ha) that will be oxidized in combustion under a given set of conditions. The combustion rate is the average speed (ha/min) at which fuels are being burned. Emission factor (EF) values and the combustion process, including the flaming and smoldering phases where much of the primary chemical products and secondary emissions of combustion occur, have been presented in Chapter 2.

Emission factor values have been determined for many chemical products and combustion conditions to help in calculation of the emission rate. For example, EF values relative to different types of fire and combustion phases, different types of fuel, and different geographic areas have been reported for CO_2 (Ward and Hardy 1984, Sandberg and Dost 1990), water vapor (Haddow 1995), CO (Ward 1990, Haddow 1995), particulate matter (Johansen et al. 1985, Haddow 1995), hydrocarbons (Ryan and McMahon 1976), N and S oxides, and aldehydes (Ward and Hardy 1984, Dost 1991).

CARBON DIOXIDE

The largest single atmospheric emission from fire is CO_2. Along with water vapor, CO_2 makes up about 90 percent of the atmospheric emissions from fire, and 80 to almost 90 percent of the C emissions (Figure 11.1). Carbon dioxide can account for 99 percent of the C emissions in *efficient burns*, that is, burns consuming most, if not all, of the available fuels, but account for only 50 percent in low-intensity smoldering fire. Although this chemical compound is not an air pollutant, it is a *greenhouse gas* and, therefore, has a potential to impact the global climate by warming the earth's atmosphere.

All organic matter ultimately decomposes into CO_2 by microbial action, or becomes fixed in soil horizons or organic deposits such as peat with long residence times. Fire accelerates the rate of CO_2 return to the atmosphere. This addition of CO_2 into the air is a concern where large tracts of forests are being cleared, burned, and converted into agricultural crop production. Prescribed fire alone, or slash reduction fire in combination with timber harvesting, does not produce the CO_2 releases that are characteristic of large wildfires.

WATER VAPOR

The second largest chemical product of fire is water vapor. Since water vapor is a normal atmospheric constituent, it is not considered an air pollutant. The main concern about emissions of water vapor from fire is generally their contribution to smoke and visibility reductions in areas near to fire.

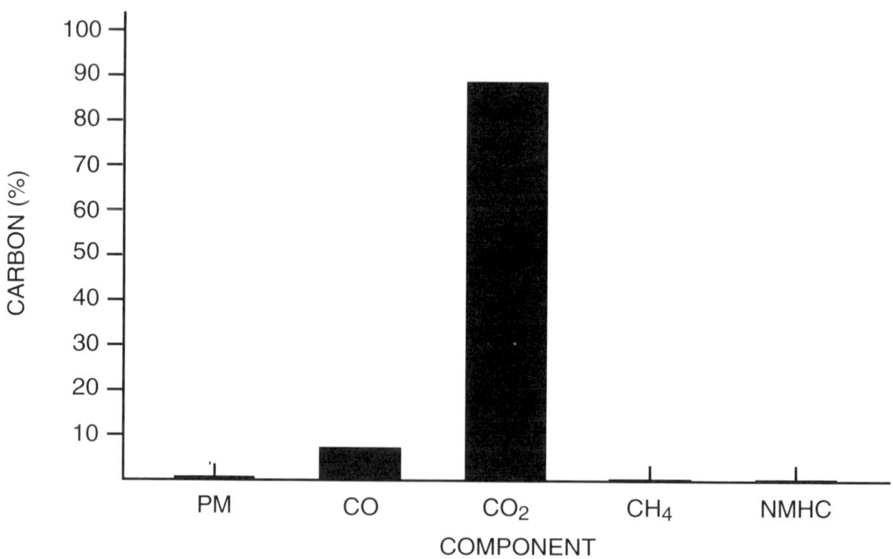

FIGURE 11.1 Total C released by burning in emission components of particulate matter (PM), CO, CO_2, methane (CH_4), and nonmethane hydrocarbons (NMHC). (Adapted from Ward 1990. © Springer-Verlag, New York, NY.)

CARBON MONOXIDE

Carbon monoxide, commonly produced with incomplete combustion of moist fuels, is a criteria air pollutant. High emissions of CO (400 kg/Mg of fuel) have been measured on the fireline of relatively low-intensity fire (Sandberg et al. 1975). The amount of CO emitted by fire is a function of combustion efficiency, increasing as efficiencies drop (see Chapter 2). For example, CO emissions in California chaparral shrubs climbed from 60 kg/Mg in the flaming phase of a prescribed fire to 99 kg/Mg during the smoldering phase of combustion, and averaged nearly 80 kg/Mg for the burn (Hardy et al. 1996). Human health effects from CO exposure in fire situations depend mostly on the duration of exposure, CO concentrations, and level of physical activity.

Firefighters involved in the direct attack of a fire, or constructing firelines close to the burning, need to be aware of possible CO effects—dizziness, weariness, and loss of concentration and orientation. Moving several meters away from the fireline, where CO levels are diminished, will often quickly lead to recovery.

PARTICULATE MATTER

Particulate matter is one of the more important groups of chemical products emitted by fire because of its importance to human health and its effect on visibility. Particulate emissions degrade air quality by reducing visibility, and, in some instances, they affect human health by aggravating respiratory problems. Particulates often increase by a factor of 8 as fire evolves from the flaming phase to smoldering conditions (McMahon and Tsoukalas 1978).

The size distribution of particulates in a smoke plume varies, depending largely on the chemistry of the type of fuel consumed and the heat release characteristics of the individual fires. Almost 80 percent of the particulate mass was reported to be <1 μm, and about 70 percent was <0.3 μm, in smoke of a slash fire in the Pacific Northwest (Sandberg and Martin 1975). Airborne monitoring of smoke plumes from slash disposal fires has detected bimodal distributions (0.15 μm and >43 μm) of particulate mass (Radke et al. 1990). The larger particulates found in smoke plumes are created by air turbulence processes, not through combustion.

The small sizes of many particulates found in smoke plumes allow some of the particulate matter to be carried deep into lung tissue, a human health concern. Some particulates function as light scatterers, since their diameters are close to the wavelength of light.

HYDROCARBONS

Hydrocarbons released by combustion number in the hundreds, are a diverse group of compounds, and differ in their atmospheric reactivity depending on their chemical structure. *Aliphatic hydrocarbons,* which contain the maximum number of hydrogen atoms, are relatively nonreactive in the air. These hydrogen-saturated compounds make up about 15 percent of hydrocarbon emissions in smoldering fire, and 30 percent during the flaming phase of a fire (Sandberg et al. 1975). *Alkenes,* with one or more C=C double bonds, and *alkynes,* containing one or more C≡C triple bonds, are unsaturated compounds with respect to hydrogen atoms and, therefore, atmospherically reactive. Unsaturated hydrocarbons are produced by incomplete combustion of fuels. *Aromatic hydrocarbons* contain benzene ring structures that are also reactive in the air.

Hydrocarbons as a group are not considered to be criteria air pollutants, although trace hydrocarbons can impact air quality (lowering visibility) and human health. Hydrocarbons that are a concern to air quality and human health are aldehydes and polynuclear aromatic hydrocarbons (PAHs). Aldehydes found in smoke that can cause eye, throat, and nose irritations of mucosal membranes are formaldehyde and acrolein (Lipari et al. 1984). Aldehydes can also interfere with the performance of firefighters by depressing breathing rates. Aldehyde production in fire ranges from 0.6 to about 2.5 percent of the fuel weight.

Polynuclear aromatic hydrocarbons are polymers of benzene that are joined in patterns that form in the reducing zone of burning fuels with an optimum range of 700 to 850°C (see Chapter 2). PAH formation is controlled more by the combustion process than the fuel type (McMahon and Tsoukalas 1978). When combustion efficiencies drop during smoldering combustion, or when a *backing fire* is set against an advancing fire to derive it of fuels, PAH formation increases since gaseous fuels have longer residence times in these types of combustion conditions. A backing fire can have 6 to 100 times the emissions of individual PAH chemicals as other fire. Smoldering fire also forms more PAHs but in less variable amounts.

Although the PAH compounds are recognized as potentially having impacts on human health, they are considered noncriteria air pollutants, and, as a consequence, no ambient air standards have been specified. Human exposure to these compounds is rare, however, because of their low emission factors and high dilution in smoke plumes. Several PAHs are carcinogens in animals.

Methane is the third most abundant greenhouse gas contributing to global warming. About 10 percent of the methane released into the atmosphere annually comes from the combustion of biomass (Andreae 1991). All of the other hydrocarbons are released into the atmosphere at the rate of about 70 percent of the methane emissions.

NITROGEN AND SULFUR OXIDES

Production of NO_x by burning is largely dependent on the N content of the fuel consumed. Both NO and NO_2 are reactive gases that are emitted from combustion (Lobart and Warnatz 1993). NO is a thermally stable product of combustion. Although NO_2 is less stable than NO, its abundance increases in smoldering fire. Conversion of fuel N to NO_x can vary from about 5 percent for wood to nearly 40 percent for organic soil; there is a linear relationship between fuel N and NO_x emissions into the atmosphere in smoke plumes (Clements and McMahon 1980). Nitrogen oxides are classified as criteria air pollutants and, in addition, are precursors to the formation of O_3, another criteria air pollutant. Other nitrogenous compounds emitted from fire include ammonia, low-molecular-weight nitrites, hydrogen cyanide, acetonitrile, nitric acid, amines, amino acids, and heterocyclic N compounds.

Sulfur dioxide makes up the primary S emission from fire. While SO_2 is a criteria air pollutant, it is not normally considered a concern, largely because it makes up a relatively small fraction of the smoke constituents. Between 40 and 60 percent of the S in the fuels consumed remains in ash after a fire has burned itself out.

FREE RADICALS

Free radicals are molecules that do not have a balanced charge due to excess electrons. As a consequence, they are highly reactive with other contaminants in the atmosphere and biological tissues. Most free radicals recombine within a few seconds of exiting a fire front, although some can persist for up to 20 minutes (Sandberg and Dost 1990). Free radicals can be damaging to proteins and nucleic acids within cells and sensitive lung tissues. Little is known about the effects of exposure of people to free radicals in smoke from the combustion of fuel accumulations in natural ecosystems.

INORGANIC FRACTIONS

Nutrient-rich ash materials that are produced by a fire can be moved off-site or redistributed within the burned area by convection in a smoke column or by surface wind transport. These processes account for a small portion of atmosphere-related nutrient losses during, and immediately after, a low-intensity fire. The processes are a more important consideration in large wildfires and in windy regions.

Ash that is lost in convective smoke plumes can be 1 to almost 5 percent of the fuel mass burned in low-intensity fire (Raison et al. 1985b). With higher-intensity fire, however, ash losses can exceed 10 percent. Magnitudes of these losses also depend on the type of fuel burned. Direct loss of nutrients to the atmosphere occurs if air temperatures are high enough (Example 11.1). Nitrogen is the element most prone to this form of loss, since it starts to vaporize at 300°C (see Chapter 6).

Example 11.1: Nutrient Release to the Atmosphere (Raison et al. 1985a)

Nutrient releases to the atmosphere from a fire set in an Australian eucalypt forest followed, as expected, the order of N>K>P>Ca, with the losses increasing linearly with increasing fuel combustion. Losses of P ranged from 2.0 to 3.0 kg/ha. Where the fire was cooler, combustion of the organic litter was almost 75 percent and a fine black ash was produced. Fuel combustion was >98 percent and a fine grey ash was produced in hotter parts of the fire. When black ash resulted, particulate and nonparticulate losses (volatilization) to the atmosphere were about equal. In the hotter parts of the fire with grey ash, nonparticulate losses to the atmosphere were three to four times greater than particulate.

Phosphorus is volatilized from burning plant material only as phosphorus pentoxide (P_2O_5) when oxygen is not limiting (cooler fire). However, as fire intensity increases and oxygen becomes more limiting, other compounds like P_4O_6 can form. This volatile form of P probably accounts for the differences in atmospheric P loss after the hotter fire. This example indicates that P transfer to the atmosphere is likely much greater than previously recognized and more significant for the productivity of a postfire forest ecosystem. The increased loss of P is not significant when P fertilization is done during the rotation in high-intensity production forests. However, P can become limiting to successional plants on sites with marginal P reserves in ecosystems where P fertilization is not routinely done. ■

Secondary Emissions

Secondary emissions are air pollutants emitted into the atmosphere as a result of photochemical transformation of primary emissions from fuels, or as exotic chemicals driven off fuels during the preignition phase of combustion. One example of the former is *ozone,* a criteria air pollutant, while chemical compounds released into the atmosphere through the combustion of fuels treated with pesticides represent an example of the latter. The emission factors for secondary emissions, while not well known, are presumed to be low. Secondary emissions are generally not considered to be important for that reason.

OZONE

The atmospheric chemistry of O_3 in relation to its emission by burning is complicated (Evans et al. 1974). Formation of O_3 occurs high in a smoke plume, at elevations where ultraviolet light can penetrate, and levels of hydrocarbons can scavenge NO, reducing the ability of NO to remove O_3. Westberg et al. (1981) measured O_3 concentrations of 55, 40, and 40 parts per billion (ppb) above ambient levels in a smoke plume 3, 24, and 48 km downwind of a forest fire. Concentrations of up to 75 ppb have been measured 30 km downwind of fire plumes. Ozone is a respiratory irritant that has the potential to aggravate air quality problems in urban areas.

PESTICIDES

Perceptions about the combustion of fuels treated with pesticides have led to some concerns about human exposure to compounds that are released into the atmosphere through combustion of the treated fuels. For example, "brown and burn" techniques have been, and continue to be, used to control or kill undesirable vegetation prior to the planting of trees in parts of the Pacific Northwest and southern United States. These techniques usually involve the application of a herbicide from either the ground or air, and then burning the treated vegetation after the herbicide has killed and desiccated the targeted plants. People have occasionally become concerned about exposure from smoke drift away from the vegetation being burned after herbicide application.

Concerns about compounds released through the combustion of fuels treated with pesticides might not always be warranted, however. Herbicidal compounds were not detected in samples of air mass or rain collected in 130 studies throughout the United States (Majewski and Cadel 1995). Also, risk analysis studies conducted in the Pacific Northwest (Sandberg and Dost 1990) and the southern United States (Bush et al. 1987) demonstrated that exposures to pollutants would approach no-effect levels under assumptions that unrealistically produce complete volatilization of pesticide residues on wood. It is likely best to consider concerns about exposure to compounds that are released into the atmosphere through combustion of fuels treated with herbicides or pesticides on a case-by-case basis.

SMOKE MANAGEMENT

Smoke management consists of applying practices needed to meet air quality regulations in managing fuels and developing fire prescriptions. The goals of smoke management practices are reducing fire emissions, enhancing the dispersion of smoke plumes, steering smoke plumes away from smoke-sensitive areas, and coordinating the ignitions of prescribed burning programs (National Wildfire Coordinating Group 1985). Smoke management practices include reducing fuel loads, improving combustion efficiency, igniting fire effectively, and mopping-up efficiently. The specific practices used in a locale or region depend largely on the fire management objectives to be satisfied and the types of fuel to be burned.

Smoke management can range from a simple guideline of not igniting a fire when the wind is blowing from a certain direction to a holistic management program that prioritizes fire occurrences, monitors meteorological and fuel conditions, predicts fire behavior and subsequent impacts, and coordinates fire ignition and suppression with multiple agencies. Questions that can guide the formulation of a smoke management program, or the selection of a smoke management practice, include the following:

- Are air quality standards being affected?
- Are public health and safety affected?
- Are there smoke-sensitive areas in the vicinity?
- Does the public complain about prescribed fire?
- Does the local meteorology or terrain interfere with smoke dispersal plans?
- Are the goals of the current prescribed fire program changing?
- Is the number of burning days limited?

Atmospheric conditions and topography often interact to diminish or aggravate smoke problems. A fire manager should possess an understanding of these interactions in developing a smoke management plan for a particular area (Example 11.2). Unstable air promotes the rise of smoke to great heights and the dispersion of smoke at altitudes that will not affect local visibility (Pyne et al. 1996), but it can affect haze. On the other hand, stable air produces inversions that hinder smoke rise and reduce visibility at the lower levels of the atmosphere, where it can affect human activities and health. (An *inversion* is a stable layer of air in which temperature increases with altitude rather than decreases.) When a smoke plume hits an inversion, it will cease lifting and spread horizontally, substantially reducing its mixing with unaffected air and decreasing visibility over a larger area.

Example 11.2: Interactions of Atmospheric Conditions and Topography on Smoke: An Illustration of Local Effects (Lahm et al. 1990)

The need to consider the interactions between atmospheric conditions and topography can be illustrated by an incident in Sedona, Arizona, 160 km north of Phoenix. In September of 1988, about 60 ha of ponderosa pine slash (a fuel loading of approximately 40 Mg/ha) was burned along the Mogollon Rim in central Arizona, about 600 m above and 20 km north-by-northwest of Sedona. Knowing that down-canyon nighttime airflows would be likely to transport smoke into the town, the fire manager initiated a smoke management plan that specified only 80 ignitions a day. Still, the people of Sedona suffered an impairment of visibility on three successive mornings after the fire was ignited. Elevated levels of particulates disturbed those suffering from asthma and allergies. ■

Although inversions can form almost anywhere, they happen more frequently over terrain with high mountain ranges and deep valleys. Open terrain, such as the Great Plains in the western United States, tends to have less stable air and fewer inversions, because geographic features do not exist to block the flow of air or move air flows to higher altitudes. Unstable air promotes the dispersion of smoke over much larger landscapes, effectively reducing visibility problems at any one location.

Visibility

Visibility is a primary focus of smoke management. Impaired visibility, in general, can adversely affect the safety of people during fire control operations, public safety along roadways, recreational values, and the reaction of people to prescribed burning programs. Visibility can be 200 km or more in mountainous terrain of the western United States, or 40 km or more in the eastern part of the country with good visibility. Visibility becomes poorer when particulate matter or gaseous pollutants in the air scatter or absorb light; it worsens still more with the condensation of atmospheric moisture on smoke particulates, a condition in which the number of particles that block light transmission is increased. A result of this condensation is the creation of a *smoke fog,* which can reduce visibility to near zero in the vicinity of a fireline and, as a consequence, increase the danger associated with fire control efforts.

Particulate matter concentrations can increase to 5000 $\mu g/m^3$ at ground level within close proximity to an active fire, and visibility can drop to 0.2 km at 100 m above the ground, or near zero within a smoke plume.

Visibility Risk Indices

Some regions such as the southern United States are prone to the development of smoke fog. Smoke fog can develop in any environment when particulate matter in the air adsorbs moisture at high relative humidities. The resulting decrease in visibility can have drastic consequences along roadways. Sections of the interstate highways have been equipped with lights because of the frequent occurrence of this problem in parts of the southern United States. Some managers have stopped or reduced the use of prescribed fire because of the risks involved and the liability problems associated with smoke-induced fog along transportation corridors. A need to assess the risk of hazardous visibility prior to human-induced ignition becomes apparent in these instances.

Relative humidity has been used as a crude measure of the risk of hazardous visibility in some instances. Managers involved in fire use and suppression have long known that relative humidities above 70 percent can produce smoke fogs that reduce visibility. However, it is generally difficult to quantify the risk to visibility by using relative humidity by itself. Therefore, a number of visibility risk indices have been developed to more quantitatively assess the risk of hazardous visibility that might result from burning (Example 11.3). These indices are largely based on some estimate of the dispersion of the chemical products of combustion into the atmosphere.

Example 11.3: A Dispersion Index for Prescribed Burning (Lavdas 1986)

The so-called dispersion index (DI) is a derived meteorological measure of mixing in the atmosphere and is greatest when wind speeds are high and the air mass is unstable. The DI, specifically, is the inverse of predicted ground level smoke concentrations immediately downwind of a 50-km^2 smoke source. The DI values are positive integers, with higher numbers indicating greater dispersion in the atmosphere. A doubling of the DI indicates a doubling of dispersion effectiveness. The DI has a diurnal component that tracks changes in windspeeds, air mass stability, and atmospheric mixture that occur daily in many locales. It is common in many locations to have a poor DI at night and fair to good during the day, especially where topographic differences are pronounced. The mathematical basis of the DI, response of the DI to meteorological parameters, and interpretations of DI values are presented by Lavdas (1986). ■

Smoke Management Strategies

Smoke reduction efforts for wildfire consist largely of suppressing the fire. Air quality has not generally been integrated into wildfire management decisions. However, air quality impacts of wildfires have become a concern in the United States as a consequence of the large wildfires in the late 1980s and early 1990s. Some control of smoke from wildfire can be achieved by the management of fuel loadings through prescribed fire programs (Deeming 1990). Active smoke management is more pertinent to planning prescribed fire and prescribed natural fire programs.

Smoke management strategies for prescribed burning consist largely of conducting fire under specified fuel loadings, fuel moisture regimes, and meteorological conditions, and specifying ignition techniques that maintain the smoke's environmental impact within acceptable limits. Strategies used to attain smoke management objectives include, but are not necessarily limited to, emission reduction, avoidance and dilution, alternative management techniques, and education (National Wildfire Coordinating Group 1985, Haddow 1995). These strategies can be employed singly or, when appropriate, in many combinations.

EMISSION REDUCTION

Emission reduction involves the use of fuel management practices to reduce smoke when fire occurs. Smoke and particulate emissions are reduced by burning less fuel. Nearly 50 percent of the emissions from prescribed burning can result from logging residues. Something as simple as lowering the wood-utilization standards and, as a result, harvesting more residues can significantly reduce smoke emissions (Figure 11.2). Isolating large pieces of wood residue (such as stumps) that are prone to smolder for long periods can also lower the level of emissions. Burning smaller parcels of land and keeping soil and forest floor material out of slash piles can also reduce emissions.

Prescribed burns conducted when small fuel moisture is low, large fuel

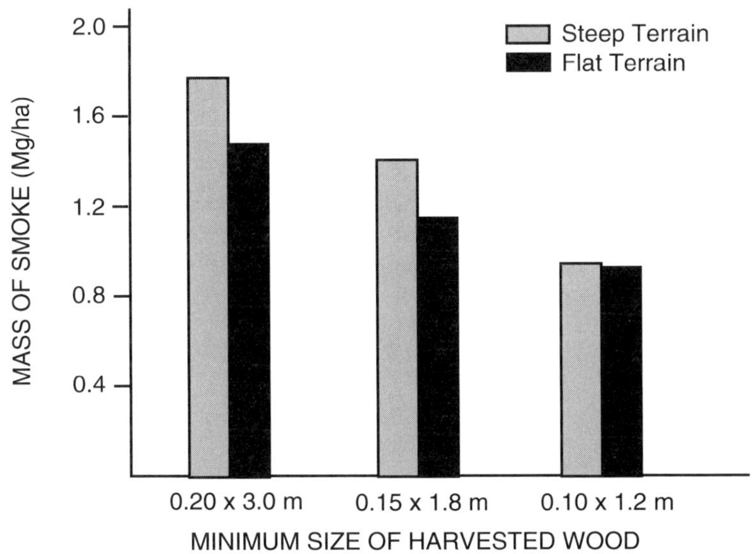

FIGURE 11.2 Smoke emission reductions on steep and flat terrains in the Cascade Mountains of the Pacific Northwest as affected by lowering wood-utilization standards and harvesting more residues. (Adapted from Sandberg and Dost 1990. In: *Natural and Prescribed Fire in Pacific Northwest Forests,* edited by J. D. Walstad, S. R. Radosevich, and D. V. Sandberg. Copyright © 1990 Oregon State University Press. Reproduced by permission.)

moisture is high, and forest floor moisture is high will also substantially reduce emissions by making less fuel available for combustion. Prescribed fire can have a wide range of particulate emission factors, with magnitudes of these factors depending on whether the fire is dominated by flaming combustion or smoldering. Smoldering fires have the greatest emission factors for nearly all combustion products.

AVOIDANCE AND DILUTION

Avoidance and dilution are fire management strategies that take advantage of weather conditions to direct smoke away from smoke-sensitive areas, or dilute the smoke before it reaches unacceptable concentrations. However, it is necessary to evaluate dispersion of particulates, estimate pollutant loads, and quantify total emissions before a fire ignites to formulate these management strategies (Potter and Fox 1996). A number of models have been developed for this purpose to simulate the behavior of smoke that is released from fire (Sandberg and Peterson 1984, Riebau et al. 1988, Ross and Fox 1991). These models vary from relatively simple to complex in their structure, and in relation to how realistically they accomplish these simulations.

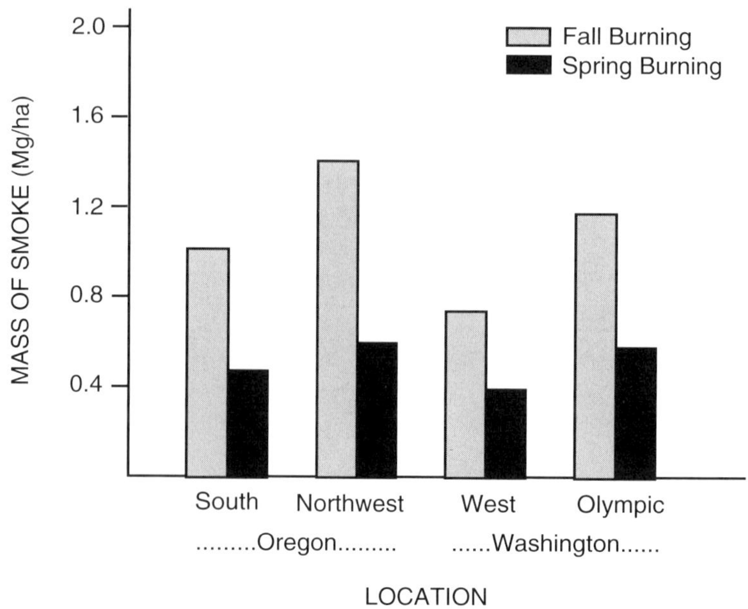

FIGURE 11.3 Smoke emission reductions in the Cascade Mountains of the Pacific Northwest as affected by the season of slash burning. The mass of smoke emitted was reduced by over 50 percent in all four locations by spring burning. (Adapted from Sandberg and Dost 1990. In: *Natural and Prescribed Fire in Pacific Northwest Forests,* edited by J. D. Walstad, S. R. Radosevich, and D. V. Sandberg. Copyright © 1990 Oregon State University Press. Reproduced by permission.

Avoidance of potential smoke problems is achieved by scheduling burning during the season when fuel moisture and meteorological conditions are likely to result in reduced smoke emissions or movement into urban areas (Figure 11.3). Dilution involves controlling emission rates by fire management techniques, or scheduling prescribed fire when atmospheric processes favor unstable conditions and, therefore, maximum mixing of the lower atmosphere. A key to good smoke dilution and dispersal is accurate weather forecasts on wind speeds and direction, atmospheric stability, and mixing heights.

ALTERNATIVE MANAGEMENT TECHNIQUES

Slash management techniques other than prescribed burning include mechanical and chemical treatments. Some of the mechanical treatments used in the past, including crushing, chopping, shearing, raking and piling, and chipping, can be effective but often have significant economic, environmental, and terrain limitations (Neary et al. 1984, Neary and Hornbeck 1994). Chemical treatments also have environmental limitations.

Herbicides have been used to control vegetation and desiccate plants prior to burning in some ecosystems, as mentioned elsewhere in this book. While herbicides do not reduce fuel loads, they can be effective in preparing sites supporting dense vegetation for prescribed fire regimes. Although herbicides remain controversial because of health and sociopolitical considerations, their impacts on the environment are often less than mechanical treatments (Neary and Michael 1997).

EDUCATION

One of the major problems confronting fire managers is a lack of understanding by the general public and some air regulatory agencies of effective smoke management techniques, the roles of fire in ecosystem management, the ecological importance of fire in maintaining the integrity of some ecosystems, and the consequences of complete fire suppression on ecosystem functioning. Education of the public and training of fire suppression personnel are necessary, therefore, to ensure that prescribed burning and prescribed natural fire can remain, or become, an integral tool for managers.

Smoke Management Guidelines

Smoke management guidelines recommended by the National Wildfire Coordinating Group (1985) and Haddow (1995) include *prefire management guidelines* that should be adopted to the extent possible. These guidelines are:

- Have clearly stated fire management objectives.
- Use weather forecasts to predict future fire conditions.
- Do not burn when air stagnation advisories are in effect, during urban pollution episodes, or when relative humidity or visibility risk indices indicate that smoke problems might occur.
- Comply with air pollution control and smoke management regulations.
- Notify local fire-control organizations, air monitoring units, local law enforcement agencies, and nearby residents of proposed burning plans.
- Develop emergency plans.

There are also a number of guidelines that can be used to reduce the amount of smoke generated during *actual burning* operations. These guidelines include:

- Burn when conditions are good for rapid air dispersal to reduce the effects of emissions.
- Burn under favorable moisture conditions.
- Use backing fire when appropriate to increase combustion efficiencies.
- Burn small parcels of land.
- Burn when fine fuels have low moisture contents, large fuels have high moisture contents, and the forest floor has a high moisture content.

- Be cautious when burning at night.
- Mop-up as soon as possible after burning.
- Increase the utilization of harvested wood material to lessen the accumulation of slash.
- Expand the burning season, if possible, to include early spring or late fall to obtain favorable fuel moisture contents.

Additional smoke management guidelines suggested by Mahaffey and Miller (1994) are:

- Use mass ignition techniques that allow burning at higher fuel moisture.
- Consider the volume of smoke already present in a geographic area when making decisions about prescribed fire and prescribed natural fire.
- Burn fuel concentrations (such as slash piles, windrows, and other debris accumulations) outside the prescribed burning season.
- Avoid burning on days when smoke might affect smoke-sensitive areas, heavily used recreation areas, and public holidays.
- Monitor prescribed natural fire on a daily basis to assess smoke dispersal and adherence of the fire to prescription conditions.

REGULATORY CONSIDERATIONS

It should always be remembered that smoke contains air pollutants and, therefore, contains components for which there may be emission standards. Smoke that is generated as a result of human ignitions is subject to governmental regulation in some countries, while it is not in others. Importantly, managers who are planning prescribed fire programs need to appreciate the regulations that can affect them and, also, be prepared to mitigate and accept liability for downwind effects of fire. Air quality regulations often restrict prescribed burning programs because of established air quality standards.

Prescribed natural fire is a management decision and, as such, is generally considered a prescribed fire that is subject to regulatory control.

The United States has some of the most stringent regulations regarding smoke generated from prescribed fire. The Clean Air Act, as amended in 1977 and 1990, identifies air quality standards and legal requirements for prescribed fire that must be met by federal, state, and local agencies and private industry (Mitchell 1995). Before the 1990 amendments, the Clean Air Act did not apply specifically to prescribed burning in natural ecosystems. However, the 1990 amendments incorporated a number of sections that directly and indirectly place prescribed fire within the purview of the Clean Air Act's regulatory controls.

There are also instances in the United States where state and local governments have imposed additional air quality regulations (beyond those specified by the federal Clean Air Act) that pertain to prescribed fire. For example, an air quality district in California has imposed a fee for prescribed burning that is based on the estimated tons of CO, PM, hydrocarbons, and NO_x (which are air pollutants) to be emitted. Oregon has developed a state-wide fee system. Washington bans prescribed fire on weekends during the summer months for the protection of visibility. Wyoming requires fire managers to predict the downwind particulate concentrations for each prescribed fire planned from dispersion models. This trend in state and local actions is likely to continue into the future as demands for air quality improvement increase.

GLOBAL CLIMATE IMPACTS

The observed warming of the earth's atmosphere over the past 150 years has been linked to increases of greenhouse gases (for example, CO_2, CO, methane, and NO_x) produced largely by the burning of fossil fuels during industrialization of many countries. Despite temporary short-term reversals and a leveling of temperatures from 1940 to 1975, the trend of increasing global temperatures has been persistently upward since 1910 (Figure 11.4). Events such as the Mt. Pinatubo (Philippines) eruption of 1991 slowed this trend somewhat due to reflection of radiant energy back into the atmosphere by volcanic dust and S gases (Kerr 1996).

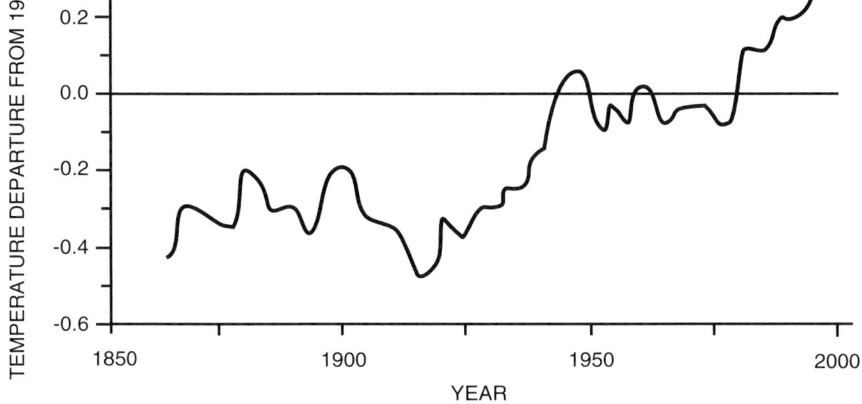

FIGURE 11.4 Global temperature departures from the 1961–1990 average for the years between 1860 and 1996. (Adapted from Kerr 1996. Reproduced by permission of Climatic Research Unit, University of East Anglia and the Meteorological Office, Bracknell, UK.)

There is debate about the role of fire on the warming of the earth's atmosphere, however. Occurrences of fire in natural ecosystems have been common since humans achieved their dominance on the planet, although the frequency and areal extent of fire have increased as human populations have expanded (Pyne et al. 1996). Burning occurs throughout the world for a variety of reasons, including to clear forest and shrubland vegetation for agricultural and urban development, reduce the levels of fuel accumulations in natural ecosystems, control undesirable vegetation, release nutrients onto agricultural lands to improve fertility, and (in some countries) produce charcoal. This burning results in about 8.7×10^9 Mg of dry matter being consumed annually, of which nearly 60 percent of the fuels come from natural ecosystems largely in the tropics (Andreae 1991).

At the present rate of burning in the tropics, C (in the form of CO_2, CO, methane, and hydrocarbons) is being released into the atmosphere at almost twice the rate it is being fixed in the mid- and high-latitude regions of the world. Burning at mid- and high-latitudes is a minor component of C inputs into the atmosphere. In the absence of sufficient cooling feedback mechanisms, such as smoke, volcanic ash, or S gas inputs into the upper atmosphere, continuing inputs of greenhouse gases into the atmosphere from the burning in the tropics and elsewhere are likely to affect the course of global climate change into the future.

SUMMARY

Smoke and its constituents can impact the environment on scales ranging from the impairment of local visibility to possible changes in the global climate. These effects are caused largely by the primary chemical products and secondary emissions of combustion. After completing this chapter, you should have a better understanding of these impacts of fire. Specifically, you should be able to:

1. Describe the primary chemical products and secondary emissions that are recognized as criteria air pollutants and those that are noncriteria air pollutants.
2. Answer the following questions:
   a. What factors should be considered in the formation of a smoke management program and the selection of a smoke management practice?
   b. What strategies can be used to attain smoke management objectives?
   c. What prefire smoke management guidelines can be followed to mitigate the effects of smoke on the environment? What are some additional management guidelines to consider in attempting to reduce the amount of smoke generated during actual burning operations?

3. Discuss the salient aspects of the Clear Air Act enacted by the U.S. Congress to regulate smoke from prescribed burning.
4. Discuss the possible role of fire in the warming of the earth's atmosphere.

REFERENCES

Andreae, M. O. 1991. Biomass burning: its history, use, and distribution and its impact on environmental quality and global climate. In: Levine, J. S., editor. *Global Biomass Burning: Atmospheric, Climatic, and Biospheric Implications.* MIT Press, Cambridge, Massachusetts, pp. 1–21.

Bush, P. B., D. G. Neary, C. K. McMahon, and J. W. Taylor, Jr. 1987. Suitability of hardwoods treated with phenoxy and pyridine herbicides for use as firewood. *Archives of Environmental Contamination and Toxicology* 16:333–341.

Clements, H. B., and C. K. McMahon. 1980. Nitrogen oxides from burning forest fuels examined by thermogravimetry and evolved gas analysis. *Thermochimica Acta* 35:133–139.

Deeming, J. E. 1990. Effects of prescribed fire on wildfire occurrence and severity. In: Walstad, J.D., S. R. Radosevich, and D. V. Sandberg, editors. *Natural and Prescribed Fire in Pacific Northwest Forests.* Oregon State University Press, Corvallis, Oregon, pp. 95–104.

Dost, F. N. 1991. Acute toxicology of components of vegetation smoke. *Reviews of Environmental Contamination and Toxicology* 119:1–46.

Evans, L. F., N. K. King, D. R. Packham, and E. T. Stephens. 1974. Ozone measurements in smoke from forest fires. *Environmental Science and Technology* 8:75–76.

Haddow, D. 1995. Air resources. *Fire in Ecosystem Management Notes: Unit II-J.* USDA Forest Service, National Advanced Resources Technology Center, Marana, Arizona.

Hardy, C. C., S. G. Conard, J. C. Regelbrugge, and D. R. Teesdale. 1996. Smoke emissions from prescribed burning of southern California chaparral. USDA Forest Service, Research Paper PNW-RP-486.

Johansen, R., W. Deeming, Jr., M. Long, and D. E. Ward. 1985. Smoke production characteristics and effects. In: Gardner, F., F. Abeita, L. Bancroft, W. L. McCleese, H. N. Miller, R. Hutch, H. Ryan, J. W. van Wagtensonk, R. Wilson, and F. Zontek, coordinators. *Prescribed Fire Smoke Management Guide: PMS 420–1.* National Wildfire Coordinating Group, National Interagency Fire Center, NFES Publication 1279, Boise, Idaho, pp. 5–10.

Kerr, R. A. 1996. 1995 the warmest year? Yes and no. *Science* 271:137–138.

Lahm, P., D. Fox, and A. Riebau. 1990. Smoke management: an emerging profession. In: Krammes, J. S., technical coordinator. *Effects of Fire Management of Southwestern Natural Resources.* USDA Forest Service, General Technical Report RM-191, pp. 139–145.

Lavdas, L. G. 1986. An atmospheric dispersion index for prescribed burning. USDA Forest Service, Research Paper SE-256.

Lipari, F. W., J. M. Dasch, and W. F. Scruggs. 1984. Aldehyde emissions from woodburning fireplaces. *Environmental Science and Technology* 18:326–330.

Lobart, J. M., and J. Warnatz. 1993. Emissions from the combustion process in vegetation. In: Crutzen, P. J., and J. G. Goldammer, editors. *Fire in the Environment:*

The Ecological, Atmospheric, and Climatic Importance of Vegetation Fires. John Wiley & Sons, Chichester, England, pp. 15–36.

Mahaffey, L., and M. Miller. 1994. Air quality. In: Anderson, L. D., R. G. Clark, J. Findley, R. C. Hanes, L. Mahaffey, M. Miller, K. Stinson, and G. T. Zimmerman, contributors. In: *Fire Effects Guide.* Prescribed Fire and Fire Effects Working Team, National Wildfire Coordinating Group, National Interagency Fire Center, NFES Publication 2394, Boise, Idaho, pp. III-1 to III-13.

Majewski, M. S., and P. D. Cadel. 1995. *Pesticides in the Atmosphere: Distribution, Trends, and Governing Factors.* Ann Arbor Press, Chelsea, Michigan.

McMahon, C. K., and S. N. Tsoukalas. 1978. Polyaromatic hydrocarbons in forest fire smoke. In: Jones, D. W., and R. I. Freundenthal, editors. *Volume 3–Carcinogenesis: A Comprehensive Survey.* Raven Press, New York, pp. 61–73.

Mitchell, B. 1995. The federal Clean Air Act and its impact on fire management programs. In: Brown, J. K., R. W. Mutch, C. W. Spoon, and R. H. Wakimoto, technical coordinators. *Proceedings: Symposium on Fire in Wilderness and Park Management.* USDA Forest Service, General Technical Report INT-GTR-320, pp. 74–76.

National Wildfire Coordinating Group. 1985. *Prescribed Fire Management Guide: PMS 420–1.* National Wildfire Coordinating Group, National Interagency Fire Center, NFES Publication 1279, Boise, Idaho.

Neary, D. G., and J. W. Hornbeck. 1994. Impacts of harvesting and associated practices on off-site environmental quality. In: Dyck, W. J., D. W. Cole, and N. B. Comerford, editors. *Impacts of Forest Harvesting on Long-Term Site Productivity.* Chapman and Hall, London, pp. 41–80.

Neary, D. G., and J. L. Michael. 1997. Herbicides—protecting long-term sustainability and water quality in forest ecosystems. *New Zealand Journal of Forestry Science:* 26:241–264.

Neary, D. G., L. A. Morris, and B. F. Swindel. 1984. Site preparation and nutrient management in southern pine forests. In: Stone, E. L., editor. *Forest Soils & Treatment Impacts. Proceedings of the 6th North American Forest Soils Conference.* University of Tennessee, Knoxville, Tennessee, pp. 121–144.

Potter, D. U., and D. G. Fox. 1996. Clean air and healthy ecosystems: managing emissions from fires. In: Ffolliott, P. F., L. F. DeBano, M. B. Baker, Jr., G. J. Gottfried, G. Solis-Garza, C. B. Edminster, D. G. Neary, L. S. Allen, and R. H. Hamre, technical coordinators. *Effects of Fire on Madrean Province Ecosystems: A Symposium Proceedings.* USDA Forest Service, General Technical Report RM-GTR-289, pp. 205–216.

Pyne, S. J., P. L. Andrews, and R. D. Laven. 1996. *Introduction to Wildland Fire.* John Wiley & Sons, New York.

Radke, L. F., J. H. Lyongs, and D. A. Hegg. 1990. *Airborne Monitoring and Smoke Characterization of Prescribed Fires on Forest Lands in Western Washington and Oregon.* USDA Forest Service, General Technical Report PNW-251.

Raison, R. J., P. K. Khanna, and P. V. Woods. 1985a. Mechanisms of element transfer to the atmosphere during vegetation fires. *Canadian Journal of Forest Research* 15:132–140.

Raison, R. J., P. K. Khanna, and P. V. Woods. 1985b. Transfer of elements to the atmosphere during low-intensity prescribed fires in three Australian subalpine eucalypt forests. *Canadian Journal of Forest Research* 15:657–664.

Riebau, A. L., D. G. Fox, M. L. Sestak, B. Dailey, and S. E. Archer. 1988. Simple approach smoke estimation model. *Atmospheric Environment* 22:783–788.

Ross, D. G., and D. G. Fox. 1991. Evaluation of an air pollution analysis system for complex terrain. *Journal of Applied Meteorology* 30:909–923.

Ryan, P. W., and C. K. McMahon. 1976. Some chemical and physical characteristics of emissions from forest fires. In: Proceedings of the 69th Annual Meeting of the Air Pollution Control Association, Paper 76–2.3, Portland, Oregon.

Sandberg, D. V., and F. N. Dost. 1990. Effects of prescribed fire on air quality and human health. In: Walstad, J. D., S. R. Radosevich, and D. V. Sandberg, editors. *Natural and Prescribed Fire in Pacific Northwest Forests*. Oregon State University Press, Corvallis, Oregon, pp. 191–218.

Sandberg, D. V., and R. E. Martin. 1975. Particle sizes in slash fire and smoke. USDA Forest Service, Research Paper PNW-199.

Sandberg, D. V., and J. Peterson. 1984. A source strength model for prescribed fires in coniferous logging slash. In: Proceedings of the 21st Annual Meeting of the Northwest International Section of the Air Pollution Control Association, Paper 84–20, Portland, Oregon.

Sandberg, D. V., S. G. Pickford, and E. F. Darley. 1975. Emissions from slash burning and the influence of flame retardant chemicals. *Journal of Air Pollution Control Association* 25:278–281.

Ward, D. E. 1990. Factors influencing the emissions of gases and particulate matter from biomass burning. In: Goldammer, J. G., editor. *Fire in the Tropical Biota: Ecosystem Processes and Global Challenges*. Springer-Verlag, Berlin, pp. 418–436.

Ward, D. E., and C. C. Hardy. 1984. Advances in the characterization and control of emissions from prescribed fires. In: Proceedings of the 78th Annual Meeting of the Air Pollution Control Association, Paper 84–36.3, San Francisco, California.

Westberg, H., K. Sexton, and D. Flyckt. 1981. Hydrocarbon production and photochemical ozone formation in forest burn plumes. *Journal of the Air Pollution Control Association* 31:661–676.

Chapter

12

Cultural Resources

INTRODUCTION

Fire is related to cultural resources from two perspectives (Spoerl 1996). First, it is known that past societies affected their environments through prehistoric use of fire. It has been shown, for example, that prehistoric settlements and the surrounding landscapes have been burned repeatedly (Deneven 1992). The second perspective deals with the effect that postoccupational and postdepositional fires (prehistoric, historic, and modern) have on the materials remaining from previous occupants. Archaeologists recognize that wildfire has been a recurrent phenomenon in many parts of the world and, therefore, need to include any modification or damage by wildfire and prescribed burning in their interpretations and assessments of cultural resources (Rowlett 1991). This second perspective is the focus of this chapter.

ARCHAEOLOGICAL PERSPECTIVE

Cultural resources are considered here in an archaeological perspective. The term "culture" refers to all those patterns of human behavior present in the modern world, or which can be detected in the past through archaeological records based on the interpretation of *artifacts,* that is, objects made or modified by humans that reflect human behavior, their use, or their discard. The purpose of archaeology is reconstructing prehistoric human behavior and lifestyles, and understanding the processes of cultural change through time

(Spoerl 1996). This interpretation is based partly on the analysis of artifacts. *Cultural resources management* aims to ensure the conservation of these irreplaceable resources into perpetuity, thus retaining their value for interpretation and cultural appreciation by future generations.

Archaeologists have assumed an increasingly greater role in fire-related activities from both a management and research perspective since the 1970s. This management involvement has focused on reviewing site and survey records while a wildfire is progressing to assist in locating firelines and routes for heavy firefighting equipment, and on assisting in mop-up and postfire rehabilitation activities (Spoerl 1996). Limited research efforts have centered on determining what effect specific heating regimes (different fire severities) have on artifact stability and subsequent interpretation. Planning for the protection of cultural resources during prescribed fires involves both archaeological and fire management specialists and scientists to minimize the possibility of fire-related damage.

REGULATIONS AND LEGISLATION

The nonrenewable nature of cultural resources has prompted interest and led to the enactment of regulations covering the study, protection, and preservation of cultural resources. The first "formal" interest in the cultural resources of the United States was expressed in 1846 by the newly founded Smithsonian, which focused investigations on the origin of prehistoric "mound builders" in the eastern and midwestern United States. Formation of the U.S. Geological Survey and the Smithsonian's Bureau of American Ethnology in 1879, and the founding of the Anthropological Society of Washington and the Archeological Institute of America, led to the study and exploration of Native American ruins in the southwestern United States. This activity led to historic preservation through formal withdrawal, beginning in 1889, of specific pueblo sites such as Casa Grande and Chaco Canyon in the southwestern United States.

Federal legislation was first passed in 1906, when the Antiquities Act was passed as part of President Theodore Roosevelt's conservation effort. The Antiquities Act was a transition from individual site protection by Congress to protection as a material class through federal ownership. The law authorized the President to designate historic and prehistoric landmarks in federal ownership as national monuments. The National Park Service, created in 1916, began managing cultural resources as part of its charter. The Historic Sites Act of 1935 established a national policy to preserve historic (including prehistoric) sites, buildings, and objects of national significance. This legislation also created a national register for historic structures and buildings.

The post-World War II building boom in the 1950s, and a growing conservation concern in the 1960s, led to the passage of the National Historic Preservation Act in 1966, a major transition from protection through federal ownership to protection through regulation of government activities. This Act further established the National Register of Historic Places, which includes any heritage resource sites found to be "significant" in American history, architecture, archeology, and culture. The Reservoir Salvage Act of 1960 required that before any agency of the United States shall undertake the construction of a dam or issue a license for construction of a dam, it shall provide written notice to the Secretary of the Interior concerning the consideration of heritage resources within the project area.

Further legislation included the Archaeological and Historic Preservation Act of 1974, which amended the Reservoir Salvage Act of 1960 to include all federal or federally assisted construction projects rather than just dam projects. The Archaeological Resources Protection Act was passed in 1979 to increase protection of cultural resources by prohibiting the unauthorized excavation and removal of archaeological resources from federal and Native American lands. Other acts (such as the National Environmental Policy Act of 1969) also contained provisions for cultural preservation and protection.

CATEGORIES OF CULTURAL RESOURCES

Categories of cultural resources in the United States include prehistoric sites, historical locations, and cultural areas. All of these cultural resources are important in interpreting and understanding the nature and activities of previous cultures. Prehistoric sites are those locations created by peoples prior to European arrival in the Americas. These sites include (Bureau of Land Management 1995):

- Artifact scatters that may indicate locations where tools were made
- A series of depressions in the soil surface made by a pit house village
- A mound of rocks of a pueblo ruin
- A rock art panel with figures carved into rock

Prehistoric sites can represent religious, ceremonial, or habitation activities of early inhabitants. Although these sites may have been perceived initially as archaeological sites, they are important to Native Americans and other indigenous groups for other reasons.

Historical locations represent a later presence of the Euro-American occupation of the Americas. These locations, many of which are less than a century old, are represented by cabins, trails, battlegrounds, smelters, logging camps, and ghost towns, which represent more recent settlement, resource exploitation, and development of the American continent. Historic locations are important because they may add to the historical record of a site. Many of these locations were identified during, or just before, the period of aggressive fire suppression policies in the United States. As a consequence, many of the locations have not been subjected to any form of fire and, in many cases, are likely to be in good condition (Bureau of Land Management 1995). Old buildings on historic locations are frequently constructed of wood and, therefore, vulnerable to destruction by fire. Prescribed burns need to be designed to avoid damage to this type of cultural resource.

Cultural areas are places that commemorate an event of some kind, even though culturally important materials associated with the event may not be present. Examples of cultural areas are a meadow where a treaty was signed or a place of religious importance (Bureau of Land Management 1995). Cultural areas can also be areas where plants are traditionally collected, animals are hunted, or other secular activities of importance to people have occurred or continue to occur. Sacred places of many Native Americans include constructed shrines, which help identify them, but many are simply a rock, mountain, tree, or view that may be recognizable to someone of a particular culture.

ARTIFACT DAMAGE BY FIRE

Fire can have one or several effects on artifacts that are found on prehistoric sites, historical locations, and cultural areas. The amount of alteration that occurs to artifacts with burning is largely dependent on the type of materials present, their location (surface or subsurface), and the fire severity (Lissoway and Propper 1990). The amount of heating is also related to the burn severity, which (in turn) is dependent on the soil and fuel moisture content, amount of woody fuels, litter and duff accumulations, duration of combustion, humidity, and topography (see Chapter 3).

Long fire durations and smoldering litter and thick duff layers that produce extensive subsurface heating (see Chapter 2) can be damaging to both surface and subsurface artifacts. Fuel loadings are heavier than in the past in many regions of the United States because of increased fire suppression (Lissoway and Propper 1990). Therefore, the potential threat to cultural resources can be far greater than previously thought.

Direct effects of fire on artifacts include complete combustion, structural change, melting, a breaking off of a piece of artifact as a result of heating, generally called *spalling*, changes in the artifact's chemical nature, and rearrangement of materials at the site. Artifact materials affected by burning include ceramics, grinding stones, lithic artifacts and glass, architectural material, rock art panels, and wood and other organics. Heating can also destroy the usefulness of research dating techniques, including those involving carbon-14 ratio analysis, obsidian hydration, archaeomagnetism, and dendrochronology.

When structures are built or artifacts are deposited, they bear spatial relationships to each other, allowing a cultural resource specialist to interpret more than the individual pieces themselves reflect. Damage to the integrity of these historical arrangements also involves less obvious indirect effects. Indirect causes of fire effects on archeological sites include:

- Overland flow and erosion following fire causing damage to the integrity of the cultural resources
- Trees falling on aboveground structures or penetrating the soil profile and exposing cultural sites
- Compaction, rutting, and other disturbances caused by the use of mechanized equipment
- Collection of artifacts by fire crews and visitors to the postfire site occurring due to a higher visibility of artifacts as a result of the fire
- Fire retardants, foams, and petrochemicals (gas spills from vehicles) discoloring the surface of artifacts
- Postfire rehabilitation impacting on scenic beauty (see below)
- Disruption of ongoing cultural activities and ceremonies due to smoke, the presence of fire management personnel, and postfire rehabilitative activities

Ceramics

Ceramic artifacts include whole vessels and, more commonly, *sherds,* the latter being fragments of broken ceramics. Time, temperature, and mineralogy of the clay used in constructing the object have a major effect on the point at which chemically bound water is driven from clay. Once fired at the minimum critical firing temperature of 600°C, prolonged exposure to water will not cause it to rehydrate (Knudsen 1996). No more chemically bound water will be lost if the firing temperature is not again reached or exceeded.

Damage to ceramic artifacts from fire includes sooting, spalling, oxidation, pigment alterations, vitrification, adhesions, and cracking (Bureau of Land Management 1995). *Vitrification* results in a glassy, glossy quality occurring on the surface of a sherd accompanied by an overall brittleness. Adhesions are sticky black substances (probably organic) produced by the combustion of pine pitch or substances released from needle combustion. Prehistoric pottery is usually structurally safe at temperatures below 600°C, although organic paints burn away completely at temperatures of 350°C (Knudsen 1996). Altered paint, discoloration, smudging, or deposition of surface carbon residue can create problems with identification of the artifact and ceramic dating (Jones and Euler 1986). Carbonaceous residues produced by smoke can often be removed from ceramic artifacts by a rinsing with dilute hydrochloric acid followed by a washing with water.

Fire alterations are limited mostly to ceramic pieces located on the soil surface. Buried sherds are likely not to be altered by low-severity burns because of the minimal soil heating that would be expected, while sherds on the soil surface can exhibit light-to-medium postfire sooting, some light spalling, and light-to-medium oxidation. Moderate-severity fire is still not likely to alter subsurface artifacts. However, surface artifacts can undergo light-to-heavy sooting, light-to-high oxidation and spalling, adhesions, and pigment alteration. Fire of high-severity can refire ceramics, recombining the constituents, oxidizing certain elements, burning out carbon paints, and causing a brittleness of those materials found on the soil surface.

Grinding Stones

Fire effects on grinding stones vary according to the condition and type of material. Common effects include discoloration, severe cracking, spalling, and disintegration of the stone. Temperatures of about 375°C cause heavy smudging of grinding stones. Low fire severity minimally affects the surface of the stone artifacts, while high-severity burns can alter the stones to a large degree. Again, if the artifact is buried, it is subjected to much less damage.

Lithic Artifacts and Glass

The effect of fire on lithic artifacts is generally limited to discoloration and development of a dull, dark patina. Other damage can include sooting, potlids, and crazing. *Potlids* are small pieces of lithic material that are exfoliated from the surface, causing a depression in the artifact from the size of a pinhead to larger. *Exfoliation* occurs as a result of the buildup of vapor and gases within the artifact during heating. *Crazing* is a cracking pattern that has numerous fine asymmetrical fissures. Breaking and spalling of chert and obsidian artifacts can also occur. Intense heating over 500°C can result in water loss, causing the artifact to become brittle and break easily. No adverse effects are likely to occur below 500°C.

With the exception of some sooting, little damage to lithic artifacts is likely to occur during low-severity fires. Most of the damage to these artifacts occurs where there are surface accumulations of fuels, such as downed logs, that are burned. The resulting high and long-duration temperatures can cause sooting, potlids, and some crazing of the artifacts.

Caution must be exercised in interpreting the appearance of lithic materials, however, because heat treatment might have occurred prehistorically (Spoerl 1996). Raw lithic materials are sometimes deliberately subjected to thermal alteration to make them easier to work while constructing the desired tool.

Glass commonly found on historic sites is mostly inert, with a relatively high lead content. While little damage occurs to these artifacts, window glass can crizzle, scum, or melt at temperatures above 500°C (Knudsen 1996).

Architectural Materials

Architectural materials consist mainly of tuff elements that are used in masonry structures. *Tuff* is a rock formed of compacted volcanic fragments of varied composition. Information from lightly, moderately, and severely burned sites reveals that tuff elements can be affected by fire in different ways. Rocks can be spalled, blackened, reddened, broken, or disintegrated. For example, a study of a wildfire in the Jemez Mountains in New Mexico showed that about 85 to 100 percent of the building blocks were affected by the fire on moderately and severely burned sites, while lightly burned sites only showed 0 to 50 percent fire damage (Lentz et al. 1996). The degree of impact by the fire was related to fire severity. If there were logs or branches lying across the structure that burned (producing long residence burning times), or if other heat-conducting material was near the structure, the resultant damage was greatest.

The most apparent and widespread damage to architectural materials is the spalling of the tuff elements (Lentz et al. 1996). A high degree of spalling along with exploded rock and blackening is indicative of moderate-to-severe burning conditions, and indicates that the tuff has been exposed to a "hot" fire. Tuff that is exploded or cracked by a hot fire is usually also blackened. Blackened tuff can result from smoke released during the burning of a thick bed of coniferous needles and branches that cover the structure.

Over time and after many fires, tuff begins to undergo chemical and physical changes. There can be enough heat generated during a moderately severe fire to alter between 85 and 100 percent of the material (Lentz et al. 1996). Therefore, the threshold at which substantial physical and chemical changes occur is normally between light and moderate fire severities. The chemical and physical alterations of the tuff can also cause sites to become susceptible to damage by increased erosion. Heating causes tuff to spall and crack and become increasingly friable. The tuff can be reduced to powder in some cases.

Rock Art Panels

Rock art panels can also be damaged by fire, although only a limited number of observations have been made on potential damage to them. A wildfire in the Canyonlands National Park in Utah demonstrated that potential damage to rock art panels can occur as a consequence of extensive exfoliation of cliff faces exposed to fire (Noxon and Marcus 1983). Such impacts on cliff faces and other cultural resources having rock art writings and rock shelter structures can be significant. It has been hypothesized that the extensive exfoliation results from variable expansion–contraction ratios between the parent rock formation and the cliff face surfaces, elevated cliff face temperature cooling, or moisture trapped between the cliff face surface and the parent rock that may have been converted to steam by the fire and, therefore, developed high pressures. Carbon blackening as a result of fire can also damage these rock panels.

Damage to rock art panels can be mitigated by clearing potential fuels from areas around the rock face. Removal of this combustible material minimizes both the exfoliation and carbon blackening (Romme et al. 1993). Stone-strengthening compounds, such as organosilicon compounds, can also be used to help prevent exfoliation of porous sandstones bearing petroglyphs.

Wood and Other Organics

Exposed wood and other organics (including hides, fibers, bones, shells, and plant residues such as pollen) are susceptible to damage by fire, even during low-severity fires. Wood is altered at relatively low temperatures (see Chapter 2), and other organics can also be destroyed at low temperatures. Many sites contain shells and bones that become calcined and fragile upon moderate heating. Bone will become charred at temperatures of 400°C and severely altered at higher temperatures (Example 12.1). Surface pollen can be destroyed at temperatures above 300°C.

Example 12.1: Damage to Bone Fragments by Fire: A Classification System (Picha et al. 1991)

Damage to bone fragments by burning can be classified according to their appearance after being exposed to different temperatures of heating. Classes recognized are:

- No color change—unburned
- Surface charred (charred less than 1 mm deep), totally charred (charred black throughout), or reddish brown (altered throughout)—temperature reached 285°C
- Gray-blue-white-black (colors often combined, with lighter colors at the surface)—temperatures reached 525°C or higher ■

Organics that are buried tend to be undamaged by fire, particularly during low-severity fires. Complete fire exclusion from areas containing wood and other organics of cultural value is necessary to assure the integrity of these resources, because they would probably be damaged with even a low-severity fire.

It is possible to separate more recently burned plant material from prehistoric plant evidence. For example, archaeological charcoal was thoroughly carbonized and no textural or color differences were evident after the La Mesa fire in New Mexico (Ford 1990). The archaeological charcoal was more friable than were the more recently burned materials. Conversely, recently burned material was harder and varied in degree of combustion even within a single specimen. Fire of moderately high severities in forests with high fuel loadings can have a detrimental effect on the preservation of surface pollen (Scott 1990). The effect of less severe fires is not known. Pollen buried in soil is usually not affected by fire.

Heat Effects on Dating Techniques

Dating of artifacts is important for the interpretation of cultural resources. Dating techniques involve carbon-14, obsidian hydration, archaeomagnetic properties, and dendrochronology. Fire can affect the results that are obtained from the use of these techniques, leading to incorrect interpretation of the artifact's age.

Artifacts that are contaminated by ash and charcoal produced during a fire often indicate more recent dates than the true date of the sample when using carbon-14 techniques. Obsidian hydration is based on the amount of moisture in the external surface of an obsidian artifact. Its interpretation is centered on the assumption that moisture is absorbed by the freshly flaked obsidian surface at a constant rate. However, heat from a fire can alter the moisture content of the artifact and, as a consequence, lead to an incorrect estimation of its date (Example 12.2).

Example 12.2: Damage to Obsidian Artifacts by Fire (Trembour 1990)

A study after the La Mesa, New Mexico, burn indicated that surface-collected obsidian artifacts were damaged by the fire, while subsurface obsidian showed little damage. Heating showed the following responses of obsidian artifacts to different levels of heating:

- *Unburned.* A uniform rind of about 10.7 μm depth of white color, with a sharp interface line separating hydrated from unhydrated obsidian.
- *Heating from 170 to 350°C.* Progressively more tinting of the rind to gray or violet in plain light, lessening of rind brilliance in polarized light, increasing broadening and blurring of the interface line. The interface line was no longer measurable at 350°C.
- *Heating to 430°C.* Thermal obliteration of all traces of rind and its inner boundary.
- *Heating to 540°C.* Initial appearance of thermal crazing on the obsidian surface. The cracks are shallow and tend to form an intersecting network visible with a magnifying handglass.
- *Heating to 760°C.* Beginning of melting and conversion to a frothy mass of escaping volatiles. ■

The archaeomagnetic technique of dating is based on the orientation of electrons in stones from prehistoric hearths. The magnetic orientation is fixed at the original time of heating (prehistoric). Reheating these artifacts to above

675°C produces a reorientation of the electrons, indicating a more recent date (Rowlett 1991). *Thermoluminescence,* a phosphorescence developed on a substance by gentle heating, can also be affected by heating to temperatures of 400°C.

Dendrochronological techniques, involving the dating of tree rings in standing live and dead trees and wooden beams, are sensitive to heating. The record to be obtained through the dendrochronology technique can be lost completely upon ignition and combustion of the artifact.

Archaeological Implications

Many artifacts have important value in dating and further interpreting past events using modern-day research techniques. However, these artifacts are currently being exposed to both wildfire (Example 12.3) and prescribed burning. Therefore, effects of heating on these artifacts need to be considered not only from the standpoint of archaeological interpretation, but also in terms of fire suppression and fuel management activities.

Example 12.3: An Assessment of Wildfire Damages to Cultural Resources at the Mesa Verde National Park, Colorado (Fiero 1995)

A wildfire burned 1215 ha in the Mesa Verde National Park in 1989. About 200 mapped archaeological sites were located in the area burned. Firelines were constructed and earth-moving activities were undertaken by hand, with an archaeologist assigned to each crew; because of the high concern for cultural resources, no heavy equipment was used in the fire suppression. This approach represented a new and untested strategy for fire suppression in areas rich in cultural resources, and it was the first time archaeologists were deployed with firefighters to help avoid damage to archaeological sites. Archaeological sites in the burned area included prehistoric to historic materials, and the artifacts ranged from isolated hearths to surface pueblos and cliff dwellings. A postfire survey of the sites was made to assess the fire damage on different types of cultural materials.

Direct effects of the fire ranged from minimal impact on buried sites to the complete loss of historic wood structures. In general, the most vulnerable materials were wood structures such as hogans, cabins, and fence posts. Other vulnerable artifacts were rock art panels, because of rock spalling, and cliff dwellings that contained combustible materials such as wood. Pack rat debris in two cliff dwellings caught fire, producing an intense fire that destroyed the combustible materials in these rooms. Less vulnerable to damage were lithic scatters and water control structures such as check dams. Sites with the least vulnerability to fire were buried unexcavated pueblos and pithouses. Rehabilitation work following the fire included erosion control, water diversion, and stabilization. It was concluded that the absence of heavy equipment, and the presence of an archaeologist on the firelines, reduced the amount of damage to the cultural resources. ■

Evidence of fire must be recognized in the archaeological context to determine whether changes have resulted from natural processes or cultural activities (Bellomo 1991).

DIRECT EFFECTS

Direct effects of fire on artifact materials depend largely on the type and size of the artifact, its location with respect to heating, and severity of the fire (Figure 12.1). Wood and other organics are most susceptible to destruction

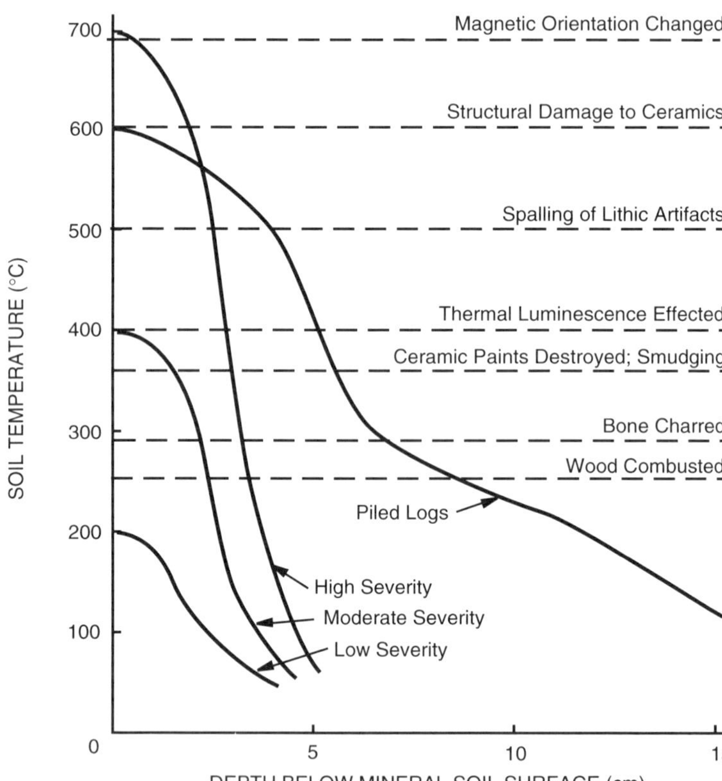

FIGURE 12.1 Temperatures, fire severities, and soil depths at which archaeological interpretations of buried artifact materials are likely to be damaged. An indication of the effect of burning a large fuel accumulation on top of artifacts buried in soil is also illustrated by the burning of a pile of eucalyptus logs in Australia (see Chapter 5).

by fire at low severities and will ignite at 280°C. Cool fires below 400°C are not likely to cause permanent damage to most inorganic materials.

Fire-related impacts to artifacts on the surface can be substantial (Picha et al. 1991). Virtually all artifacts on the surface will be scorched and smoke-blackened. Artifacts that are buried 5 cm or deeper in soil will not usually be damaged (Traylor et al. 1990), with the exception of those that are near smoldering roots where temperatures can reach 1500°C.

Grassland fire is usually less damaging to artifacts than fires in either forest or shrubland ecosystems (Picha et al. 1991). Only 1 cm of soil is usually sufficient to protect the artifacts from fire damage in grassland ecosystems.

Artifacts can be classified in order of their vulnerability to direct damage by fire (Romme et al. 1993). These classes can be useful in identifying sites of potential loss in cultural resources to the occurrence of fire. The vulnerability classes are:

- *High vulnerability.* These artifacts include historic wooden structures, alcoves and cliff dwellings, rock art panels, and scarred trees (such as scarring caused by stripping the bark for food by Native Americans).
- *Moderate vulnerability.* These artifacts include Euro-American structures (those associated with park management) and surface lithic scatter with shallow hearth.
- *Low vulnerability.* This class includes all artifacts and building materials associated with unexcavated pueblos. Prehistoric check dams would also sustain limited damage. Fire generally has little effect on deeply buried sites.

INDIRECT EFFECTS

Indirect effects of burning vary from erosion with its associated damage following the fire to collection of artifacts that have been exposed by the fire. Planning fire-crew activity to minimize disturbance or destruction of cultural resources is also important in both fire suppression during wildfire and prescribed burning activities. Fire retardants consisting of water and a slurry can cause surface artifacts to be stained pink temporarily. This staining usually washes away after several rainstorms, however. Other activities, such as the construction of firelines and the movement of mechanized equipment, can disturb the integrity of a cultural resource site (see Example 12.3). Rehabilitation activities should be analyzed as carefully for disturbance as those associated with fire suppression.

SCENIC BEAUTY

The scenic beauty of a natural landscape can also represent a cultural resource to many people. Fire and the activities associated with fire management often have an impact on this scenic beauty.

Smoke can obstruct the viewing of scenic landscapes by people, although this obstruction diminishes as the fire producing the smoke is controlled and, eventually, extinguished. A longer-termed impact of fire is the change in the vegetation on a postfire landscape (see Chapter 9). Burned, or totally consumed, vegetation is likely to lower the scenic beauty of a landscape to many people. On the other hand, however, a fire might remove some vegetation that has become too dense and, in doing so, return the landscape to its original setting.

Opportunities to mediate the impacts of burning on the scenic beauty of a landscape during and after a wildfire are generally limited. However, scenic beauty can be more fully protected when planning a prescribed burn. Quantitative techniques are available to evaluate people's perceived values of different scenes representing landscapes following prescribed burning. These techniques involve photographing landscape scenes or reconstructing them through computer simulations (Daniel and Boster 1976). Modifications to the scenes, such as those that might be caused by burning, can then be viewed by people, who are asked to evaluate the scenes in terms of their personal values.

Evaluation of these values helps to determine the people's perceptions of different severities of prescribed burning on the landscape (Example 12.4). Results of the evaluation allow the most preferred scenes to be identified and incorporated into the planning process for the prescribed burn.

Example 12.4: People's Perceptions of Prescribed Fire (Taylor and Daniel 1984)

An evaluation was made of a public's response to scenic quality, and its acceptance of different severities of prescribed burning as they affected various recreational uses in ponderosa pine forests of the southwestern United States. Scenes were shown of ponderosa pine forests that were unburned, and that had been subjected to light or severe fires 1 to 5 years previously. These scenes were represented by a series of selected slides that were randomly ordered for presentation. Participants in the evaluation studied brochures on fire effects beforehand, and then undertook a scenic beauty estimation rating that measured both their fire knowledge and attitude relative to a proposed prescribed burn.

The ratings that the participants assigned to the slides were translated into values that indicate scenic beauty was improved by low-severity fire, but diminished by more severe burnings. Low-severity fire improved scenic quality for a 3- to 4-year postfire period, while the more severe fires detracted from the scenic quality for an unknown length of time exceeding the 5–year study period. While the brochures and education increased the participants' knowledge and tolerance of fire, the knowledge obtained did not significantly affect ratings of scenic or recreational quality. Overall, the participants supported prescribed burning. Recreational activities affected the most by the more severe burnings were camping, picnicking, hiking, backpacking, and nature study. Picnicking, hiking, and nature study were not affected by low-severity fire. Results of this evaluation showed managers that prescribed burning was generally acceptable in southwestern ponderosa pine forests, and that the participants supported it.

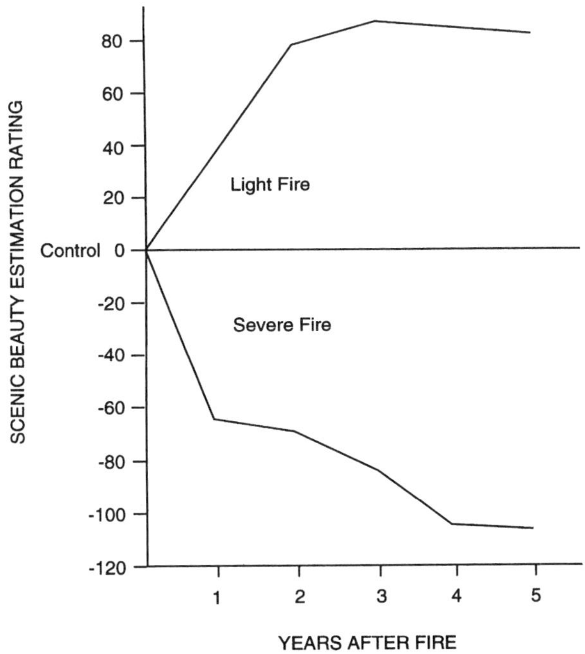

Scenic beauty estimation for a public response over time to severe and light prescribed fires. (Adapted from Taylor and Daniel 1984. Reprinted from the *Journal of Forestry,* vol. 82, no. 6, p. 365; published by the Society of American Foresters, 5400 Grosvenor Lane, Bethesda, MD 20814-2198. Not for further reproduction.) ■

CULTURAL RESOURCES PROTECTION

Mitigating the damaging effects of fire on cultural resources has become a critical part of fire management in the United States. The irreplaceable values of these resources must be protected to the extent possible. Cultural resources protection involves a prefire inventory of the resources and planning for the possible impacts of fire suppression, minimizing the impacts of wildfire, and thorough planning of prescribed burning.

Prefire Inventory and Planning of Fire Suppression

Prefire inventory and the planning of fire suppression to minimize the effects of fire present the most opportunities for protecting cultural resources. Activites that are considered in this regard include (Lissoway and Propper 1990):

- Conducting a prefire inventory in coordination with an archaeologist to identify and map the locations of cultural resources at risk. The National Register needs to be consulted for the status and potential listing of cultural resources. On-site clearances should be obtained where possible.
- Determining local fire history, natural vegetative types, and fuel loadings to ascertain fire danger.
- Identifying the potential heating and fire severities generated under different fuel moisture regimes and weather conditions.
- Classifying fire management units in terms of fire management objectives, fuel type, and cultural resource values at risk.
- Estimating the ways in which fire is likely to affect the identified cultural material and incorporation of these estimates into the fire management plan.
- Considering ways to reduce the risks to vulnerable cultural sites by reducing fuel loads, constructing firelines, and removing fuels by hand.
- Identifying and documenting Native American concerns, including the location of ceremonial sites or areas. Conflicts with traditional ceremonies should be considered in the case of prescribed burns. Identifying the locations of trails and roads used for ceremonies is also important.
- Presenting training in cultural site identification and protection to fire suppression personnel.
- Training and fire-qualifying archaeologists as fireline locators, heavy equipment supervisors, resource advisors, or for other tasks needed by a fire manager.

Minimizing the Impacts of Wildfire

Two major impacts on cultural resources usually result from a wildfire. The wildfire itself can directly damage existing cultural resources when it burns over a site. Most likely, the severity will be high, so surface-situated materials could sustain some damage. However, most damage from a wildfire occurs as a result of suppression activities, including the use of fire retardants and other activities associated with fighting the fire. Among the activities for minimizing the impacts of wildfire suppression are (Lissoway and Propper 1990):

- Having a cultural resource specialist available to brief suppression crews and other field personnel.
- Considering the location of cultural resources in determining appropriate suppression responses (including confinement, containment, or control strategy).
- Restricting the use of heavy equipment in culturally sensitive areas. If heavy equipment must be used, a fire-qualified archaeologist should be

available to flag the sensitive areas and other cultural features during suppression and rehabilitation.

Planning Prescribed Burns

Incorporation of the need to protect cultural resources into the planning of a prescribed burn can help to mitigate adverse effects. Prescribed burning is a way to minimize damages from wildfire by reducing high fuel loadings in the vicinity of cultural sites (Bureau of Land Management 1995). Prescribed fire is also useful on cultural sites where the encroachment of trees and shrubs has allowed deep litter and duff layers to accumulate and, as a consequence, increased the potential for longer duration and hotter wildfire. Important considerations when planning a prescribed fire include (Lissoway and Propper 1990, Spoerl 1996):

- Obtaining on-site inspection and archaeological clearance of the proposed burn area to determine the likelihood of flammable cultural resources.
- Assisting the fire manager in formulating the burning prescription to ensure that cultural resources are protected.
- Knowing the temperatures and durations of the flaming front, the type and loading of fuels, and how the combination of the two might affect artifact materials.
- Considering the use of chemical foam or other special protective measures on sensitive burn sites.
- Listing all burn-preparation needs in the plan and ensuring these needs are met before burning. This action could result in physically removing or reducing the fuels on and around cultural sites.
- Briefing all personnel on the objectives of the burn and on the identification and proper protection of cultural features.
- Having an archaeologist remove and document cultural materials from a burn site according to regulations established by the Archaeological and Historic Preservation Act (1974). This activity should be done according to an established research design because the site is being destroyed by collection.

Other Considerations

Rehabilitation activities following the burning should receive the same level of planning concerning the protection of cultural resources as those used to design the prescribed fire. A cultural resource specialist should be involved in the development of the rehabilitation plans to specify mitigation measures for cultural materials. Mapping the locations of postfire treatment areas, and of rehabilitative activities as they are related to cultural sites, needs to be done to avoid damage to any cultural features. Site records should be updated to reflect any changes in cultural resources that occurred as a result of the rehabilitation activities.

SUMMARY

The recognition of the possibility that burning can detrimentally affect archaeological interpretations of cultural resources has become an important part of fire management in the United States. Effects of fire on artifact materials depend largely on the type of artifact, its location on the site, and severity of the fire. It is important, therefore, that the effects of fire be mitigated to the extent possible by protecting cultural resources that are likely to be exposed to the heat produced during a fire. After reading this chapter, you should be able to:

1. Describe the categories of cultural resources.
2. Explain the ways in which artifact materials can be damaged by fire.
3. Explain how heating by fire affects the dating of artifact materials.
4. Appreciate the impacts that fire might have on scenic beauty of a natural landscape.
5. Discuss the ways by which cultural resources can be protected by a prefire inventory and planning of fire suppression, minimizing the impacts of wildfire, and careful planning of prescribed burns.

REFERENCES

Bellomo, R. V. 1991. Identifying traces of natural and humanly-controlled fire in the archaeological record: the role of actualistic studies. *Archaeology in Montana* 32:75–93.

Bureau of Land Management. 1995. Human interface—Lesson 4; Cultural Resources: Unit 4. *Training Manual for Introduction to Fire Effects.* National Interagency Fire Center, Boise, Idaho, pp. 4A.1–4A.12.

Daniel, T. C., and R. S. Boster. 1976. Measuring landscape esthetics: the scenic beauty estimation method. USDA Forest Service, Research Paper RM-167.

Deneven, W. 1992. The pristine myth: the landscape of the Americas in 1492. *Annuals of the Association of American Geographers* 83:369–385.

Fiero, K. 1995. The effects of fire on cultural resources of Mesa Verde National Park, Colorado. In: Shaw, D. W., E. F. Aldon, and C. LoSapio, technical coordinators. *Desired Future Conditions for Pinon-Juniper Ecosystems.* USDA Forest Service, General Technical Report RM-258, pp. 176–178.

Ford, R. I. 1990. Appendix B — Ethnobotanical consequences of the La Mesa fire, Bandelier National Monument. In: Traylor, D., L. Hubbell, N. Wood, and B. Fiedler, coordinators. *The 1977 La Mesa Fire Study: An Investigation of Fire and Fire Suppression Impact on Cultural Resources in Bandelier National Monument.* Branch of Cultural Resources Management, Division of Anthropology, National

Park Service, Southwest Cultural Resources Center Professional Paper 28, Santa Fe, New Mexico, pp. 147–152.

Jones, A. T., and R. C. Euler. 1986. Effects of forest fires on archaeological resources at Grand Canyon National Park. *North American Archaeologist* 7:243–254.

Knudsen, G. D. 1996. Cultural resources considerations. *Fire in Ecosystem Management: Unit II-H.* USDA Forest Service, National Advanced Resource Technology Center, Marana, Arizona.

Lentz, S. C., J. K. Gaunt, and A. J. Willmer. 1996. *Fire Effects on Archaeological Resources, Phase 1: The Henry Fire, Holiday Mesa, Jemez Mountains, New Mexico.* USDA Forest Service, General Technical Report RM-273.

Lissoway, J., and J. Propper. 1990. Effects of fire on cultural resources. In: Krammes, J. S., technical coordinator. *Effects of Fire Management of Southwestern Natural Resources: Proceedings of the Symposium.* USDA Forest Service, General Technical Report RM-191, pp. 25–30.

Noxon, J. S., and D. A. Marcus. 1983. Wildfire-induced cliff face exfoliation and potential effects on cultural resources in the Needles District of Canyonlands National Park, Utah. *Southwestern Lore* 49:1–8.

Picha, P. R., S. A. Ahler, R. D. Sayler, and R. W. Seabloom. 1991. Effect of prairie fire on selected artifact classes. *Archaeology in Montana* 32:15–28.

Romme, W. H., L. Floyd-Hanna, and M. Connor. 1993. Effects of fire on cultural resources at Mesa Verde NP. *Park Science* 13:28–30.

Rowlett, R. M. 1991. Ceramic thermoluminescence response effects after the La Mesa forest fire, Bandelier National Monument. *Archaeology in Montana* 32:49–56.

Scott, L. 1990. Appendix C—Pollen analysis of three sites in the La Mesa study area. In: Traylor, D., L. Hubbell, N. Wood, and B. Fiedler, coordinators. *The 1977 La Mesa Fire Study: An Investigation of Fire and Fire Suppression Impact on Cultural Resources in Bandelier National Monument.* Branch of Cultural Resources Management, Division of Anthropology, National Park Service, Southwest Cultural Resources Center Professional Paper 28, Santa Fe, New Mexico, pp. 153–164.

Spoerl, P. M. 1996. The effects of fire on interpretations of the past. In: Ffolliott, P. F., L. F. DeBano, M. B. Baker, Jr., G. J. Gottfried, G. Solis-Garza, C. B. Edminster, D. G. Neary, L. S. Allen, and R. H. Hamre, technical coordinators. *Effects of Fire in Madrean Province Ecosystems: A Symposium Proceedings.* USDA Forest Service, General Technical Report RM-GTR-289, pp. 118–124.

Taylor, J. G., and T. C. Daniel. 1984. Prescribed fire: public education and perception. *Journal of Forestry* 82:361–365.

Traylor, D., L. Hubbell, N. Wood, and B. Fiedler, coordinators. 1990. *The 1977 La Mesa Fire Study: An Investigation of Fire and Fire Suppression Impact on Cultural Resources in Bandelier National Monument.* Branch of Cultural Resources Management, Division of Anthropology, National Park Service, Southwest Cultural Resources Center Professional Paper 28, Santa Fe, New Mexico.

Trembour, F. N. 1990. A hydration study of obsidian artifacts, burnt vs. unburnt by the La Mesa forest fire. In: Traylor, D., L. Hubbell, N. Wood, and B. Fiedler, coordinators. *The 1977 La Mesa Fire Study: An Investigation of Fire and Fire Suppression Impact on Cultural Resources in Bandelier National Monument.* Branch of Cultural Resources Management, Division of Anthropology, National Park Service, Southwest Cultural Resources Center Professional Paper 28, Santa Fe, New Mexico, pp. 174–179.

PART

IV

Management Implications

Part IV places the earlier parts of the book into a management perspective. Economic considerations, including the assessment of postfire changes, their economic valuation, and their role in fire management decisions, are discussed in Chapter 13. These considerations are requisite to evaluations of fire effects on ecosystem resources. Results of such evaluations (in turn) often lead to decisions relative to the use of fire to meet ecosystem management objectives. Chapter 14, therefore, discusses the possible benefits from the use of fire in ecosystem management along with related concerns and constraints. The role of fire in ecosystem restoration is also addressed. A description of the Fire Effects Information System, a database by which further information on site-specific fire effects can be acquired, concludes the book.

Chapter

13

Economic Considerations

INTRODUCTION

The main emphasis of this book has been placed on descriptions of the effects of wildfire, prescribed burning, and prescribed natural fire on soil, water, vegetative, air, and cultural resources found in natural ecosystems. Management implications relative to these effects are considered in this chapter with respect to the assessment of postfire changes, their economic valuation, and their role in fire management decisions.

ASSESSMENT OF POSTFIRE CHANGES

Postfire changes in soil properties, water flows, vegetative communities, air quality, and cultural resources usually take place at the time of, or shortly after, a fire occurrence. When quantitative data are available, the magnitudes of these changes can often be described (Example 13.1). When only qualitative information on resource change is available, however, a narrative statement of expected changes is generally presented within the context of fire severities. Some assessment of postfire changes is necessary in either case.

Example 13.1: Assessing Physical Effects of Fire

NUTRIENT LOSSES

It was estimated in Chapter 6 that the total N lost from plants and litter in Arizona chaparral stands that were burned at low, moderate, and high fire severities would be about 160, 345, and 550 kg/ha, respectively. Additional losses of soil N might also occur at moderate- and high-severity burns (DeBano 1990). P losses from plants and litter would be 8, 18, and 29 kg/ha under low-, moderate-, and high-severity burns, respectively. It is important to remember that the estimates represent only changes under the plant canopy and, therefore, must be reduced in proportion to the amount of interspace area between shrubs when estimating total nutrient losses on an area basis. However, the calculations of N and P losses illustrate the importance of burning chaparral shrubs as coolly as possible to reduce consumption of surface litter and minimize soil heating. It is also important to be aware of these nutrient losses so that N restoration to a site will be considered as part of postfire management.

WATER QUALITY CHANGES

Sample plots located in ponderosa pine forests near Tucson, Arizona, were burned at low intensities (approximately 35 to 345 kJ/m/s) to determine the postfire effects on surface water quality. Mean concentrations of Ca, Mg, and fluoride (F), and the average pH of runoff water from the plots, increased as a result of the controlled burning treatments (Sims et al. 1981). There were no changes in the concentrations of Na, chloride (Cl), SO_4, or bicarbonate (HCO_3), while the sodium adsorption ratio (SAR) decreased. The observed changes in water quality were short-lived, however, with concentrations of the chemical constituents returning to prefire conditions within a year or two of the burns.

Empirical relations quantify the postfire changes in water quality as functions of fire intensity, forest overstory density, physiography of the site, storm-event runoff water, and time since burning. The fire intensities of the controlled burns were almost 10 times less than those of a severe wildfire in the ponderosa pine forests in north-central Arizona (Campbell et al. 1977).

LITHIC ARTIFACT DAMAGE

Damage to lithic artifacts was assessed on sites exposed to varying fire severities of burning following a wildfire in northern New Mexico. Most of the observed damages were qualitatively described (Lentz et al. 1996). As discussed in Chapter 12, intense heating over 500°C can result in water loss, causing the artifact to become brittle and break easily. No adverse effects are likely to occur at temperatures below 500°C. It was observed that lithic artifacts lying on the surface or those that were excavated from deep soil depths showed no significant fire effects on sites burned at low severity. Only the top layer of the duff was burned on these sites, and, fortunately, most of the artifacts were found in the humus layer and were undamaged.

Sites experiencing a moderately severe burn showed increased damage to lithic artifacts, however. About 30 to 45 percent of the surface lithics exhibited some signs of burning, such as light sooting and crazing on their surfaces. Lithics in subsurface layers showed no signs of burning damage. The humus layer was completely burned on these sites, charring the upper 2 to 3 cm of the soil. Severely burned sites showed the greatest artifact damage, with adhesions and other burning effects on 70 to 100 percent of the total lithic surface. This damage included light sooting, potlids, and medium crazing. The top 10 cm of one pit had a lithic artifact that had adhesions and crazing. Almost the entire area containing artifacts was affected on sites where partially buried logs, which burned with long residence times, were adjacent to artifacts. ■

Observations of postfire conditions in past years mostly focused on damaging fire effects only and were largely limited to impacts on a few primary resources. Now, *fire control* has evolved to *fire management,* with its recognition of both damaging and beneficial effects of fire on resource systems and its insistence on a more symmetric treatment of benefits and damages on all fires, whether wildfires or prescribed burns. Emphasis on *ecosystem management* on many public lands in the United States has resulted in newer and more complex management objectives, necessitating the examination of a wider range of fire effects on an increasing number of resources and resource attributes.

Damages and Benefits

Damages are unfavorable effects of fire-caused changes in a resource system, while *benefits* are represented by favorable effects of fire-caused changes in the ecosystem. *Favorable effects* are those effects that contribute to the attainment of management objectives; *unfavorable effects* make their attainment more difficult (Crosby 1977). However, neither damages nor benefits can be defined without reference to overall management objectives for the resource systems. Even the definition of the relevant resources to be observed postfire will depend on the stated management objectives. Relevant fire effects in an ecosystem being managed to optimize wilderness or habitat objectives may be different, or be assessed in a different way, from those occurring in a primarily timber-producing system, for instance. That is not to say that seemingly "nonrelevant" observations of fire effects cannot provide feedback leading to a revision of the management objectives. But, definitions of damages and benefits are tied to management objectives, and a realistic picture of fire effects requires the assessment of both benefits and damages, so defined, through time.

Methods of assessing postfire changes in terms of biophysical effects are many, and they can be complex, as previous chapters of this book have indicated. Long-term effects must be considered as well as those apparent immediately following the fire. Weather and other unpredictable factors often make assessments of such outcomes over time uncertain, suggesting that a full description of the effects would involve a probability distribution of outcomes through time in the absence of fire and another such distribution after fire has occurred.

Unfortunately, the knowledge necessary to implement this kind of full description of effects is seldom available. Therefore, simpler assessment approaches have been devised to permit the use of the kind of data that are available, or can be made available, at relatively low cost.

Time-Trend Response Functions

One relatively simple approach to assessing postfire changes, and a method that recognizes the necessity of a long-term perspective, is the use of *time-trend response functions* (Lowe et al. 1978, Ffolliott et al. 1988). In this approach, postfire changes are viewed as flows of damages or benefits through time. Time-trend response functions can be developed to represent either:

- Postfire changes in a resource in relation to time since a fire of a specified (known) severity
- Postfire changes in a particular resource in relation to time for a range of fire severities; that is, a set of functions such as that defined above

These functions can then become a basis for subsequent valuations of the effects of fire on a resource system in terms of volume of timber, production of forage, concentrations of suspended sediment in the resulting streamflow, or other appropriate biophysical units. A framework for the development of time-trend response functions is as follows:

- Resource values for postfire conditions are obtained by either (a) sampling of the attribute at different points in time after a fire has occurred, or (b) sampling the attribute on a series of burned sites representing fires of similar severity but varying fire histories.
- These postfire values are then compared to corresponding resource values obtained by sampling an unburned (control) area. Assuming that the resource responds in the same manner to fluctuations in weather conditions, time of year, and cyclic alterations, differences in the two sets of values are considered to be indicators of changes in the resource due to fire only.
- The differences in the two sets of values are shown as either (a) a *stream of ratios* over time between the resource values obtained for the burned and unburned (control) areas (Example 13.2), or (b) *absolute changes* (either increases or decreases) over time in resource value following fire in relation to the unburned area (Example 13.3). Applying the ratio

Example 13.2: Time-Trend Response Functions Shown as a Stream of Ratios Between Resource Values (Ffolliott et al. 1988)

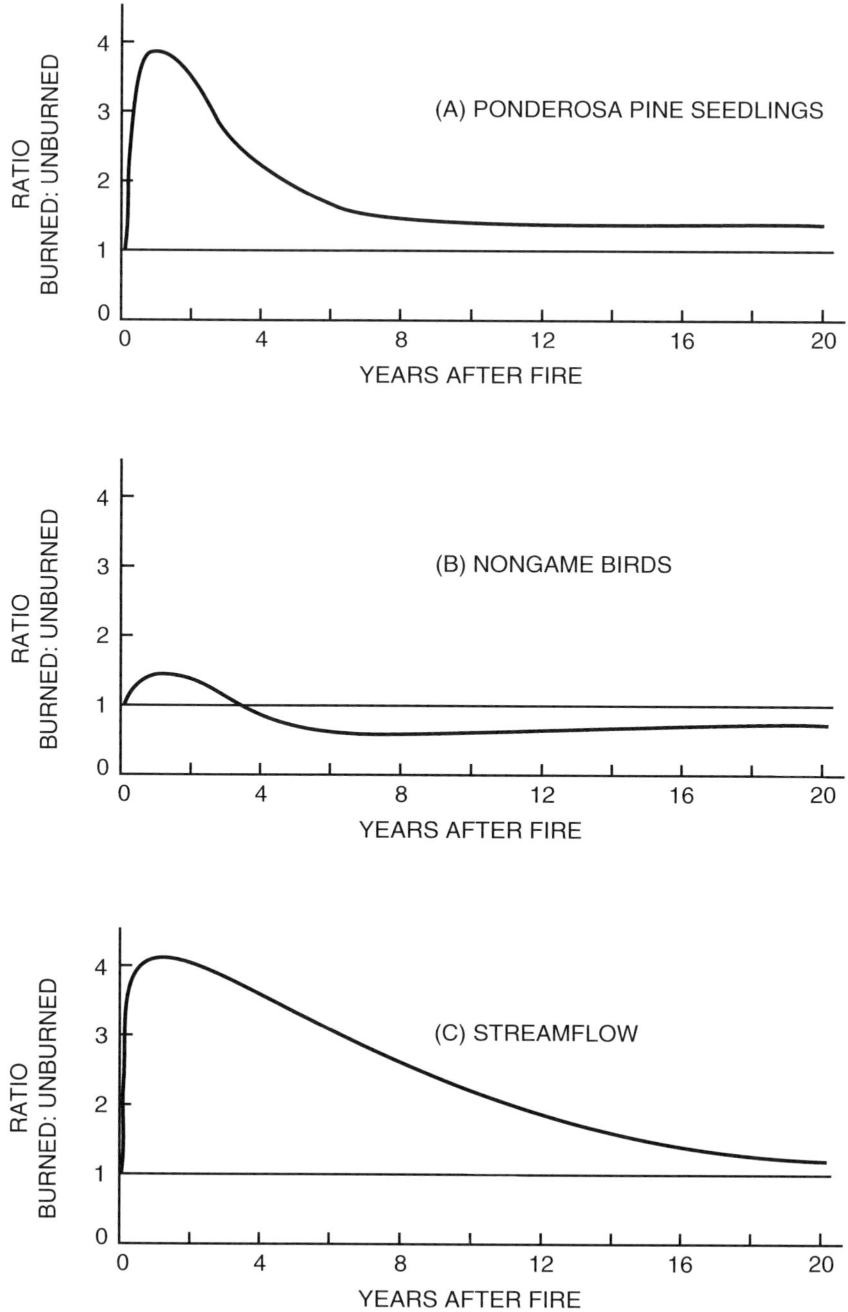

Time-trend response functions for the (A) establishment and initial survival of ponderosa pine seedlings, (B) changes in nongame bird populations, and (C) annual streamflow amounts in relation to time since the occurrence of fire of a known severity in north-central Arizona. The functions were developed by plotting ratios of the resource values obtained by sampling on burned and unburned (control) sites. (Adapted from Ffolliott et al. 1988.) ■

Example 13.3 Time-Trend Response Functions Shown as Absolute Changes (Increases or Decreases) in Resource Values (DeBano et al. 1996)

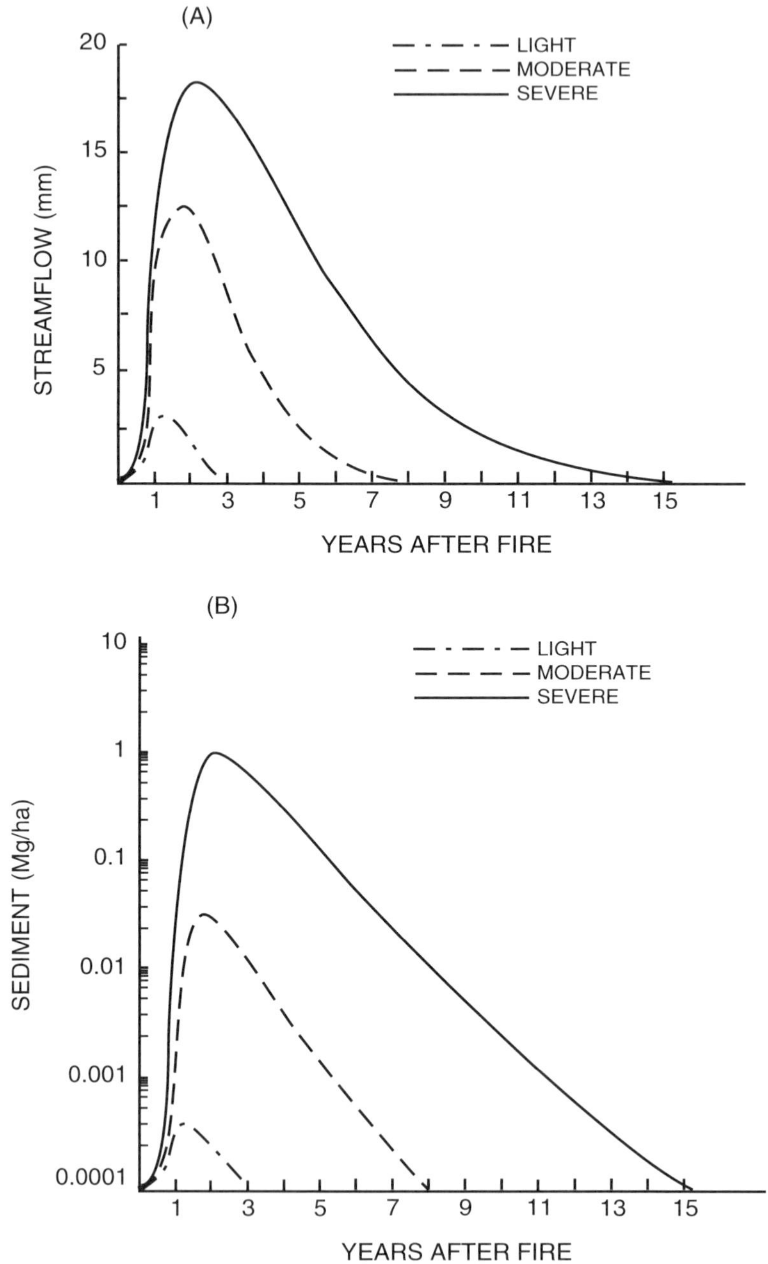

Time-trend response function for (A) annual streamflow and (B) sediment production in relation to time since the occurrence of fire of a known severity in north-central Arizona. ■

from function (a) to unburned values will enable the manager to define a time-stream of postfire resource values for a particular resource. This application provides a basis for calculating the sum of discounted fire-caused damages, benefits, or both, which are expected to occur over time. The changes reflected in response function (b) permit the same kind of calculation for the resource to which it is applicable. Time-trend response functions developed for different fire severities incorporate additional information, further sharpening the manager's ability to estimate fire impact over time.

It is seldom that a manager has all of the "data points" necessary to statistically define a time-trend response function for a resource; more likely, only a few observations of postfire change will be available. However, personal experience and professional judgment should not be discounted in describing how resource values might change as a result of fire. Some combination of on-site measurements and the experience and judgment of the manager will often suffice for estimating the form of a time-trend response function for a resource and fire in many cases.

ECONOMIC EVALUATION AND FIRE MANAGEMENT DECISIONS

Economics deals with the choices that must be made when a manager is confronted by limitations on available resources. Fire management personnel have faced such decisionmaking from the beginning. Long before functional lines of management became blurred and interdisciplinary management teams became commonplace, a fire control officer was responsible for working with multiple resources in considering the preparation of fire control plans and assessing the impacts of (actual or potential) fire on resource values. Priorities had to be established in developing fire suppression strategies and in the required long-term planning, so it became necessary to assign certain "weights" to resources. It is not surprising, therefore, that economic guidelines for allocating resources were considered early in the history of fire management (Sparhawk 1925, Flint 1928), since decisions had to be made "across resources" relative to some management objective, or set of objectives, that was broader than the individual resource objectives of that time. In some ways, this "past world" of fire control was a microcosm that foreshadowed the broader and often stickier issues of today's ecosystem management.

Present Net Worth Criterion

Fire managers, therefore, have turned to the economists' criterion of *present net worth maximization*, with its normative connotations, as a resource allocation guide. When comparing fire management alternatives, the *maximization* of net worth (present values of revenues minus the present values of costs) corresponds to a *minimization* of the sum of discounted net losses (damages) due to fire under a particular alternative plus the fire management costs that are associated with that alternative, where both losses and costs are assigned positive signs. (Minimization of a positive number seems to have been a more intuitively acceptable process than the maximization of a negative number.) Now, as prescribed fire and its benefits are discussed more frequently, managers no longer refer to losses but to net value change (NVC). If fire management costs are represented by (C), then the expression to be minimized becomes (C + NVC) (Althaus and Mills 1982). Net value change is found by subtracting the present value of the resource outputs associated with the outcomes of a particular fire management alternative from the present value of the resource outputs in the absence of fire. In this formulation, net value change will be *negative* in the case of *net benefits* and *positive* in the case of *net losses*, an anomaly reflecting its historical development in the "least cost plus loss" context.

Fire Management Inputs and Fire Outcomes

Difficulties in applying the (C + NVC) criterion have been numerous (Zivnuska 1972). Relationships between fire management inputs and the resulting fire outcomes have been hard to identify (at either a scientific level or through statistical analysis), either at the level of an individual fire or when predicting fire patterns through time and space in evaluating alternative fire management programs. Fire economics models such as the *National Fire Management Analysis System*, currently used by the USDA Forest Service, and the proposed *Fire Economics Evaluation System* use computer simulation to predict fire outcomes. These two models have been compared and discussed by Omi et al. (1986).

Even when fire outcomes can be specified adequately, the task of tracing their resource effects over time remains. (Readers of this book should appreciate the complexities involved in this step.) Knowledge of management objectives will determine what resource changes are relevant, and how they must be measured. For example, timber burned in a wilderness area will not be considered or valued as a merchantable timber loss. Weather patterns following a fire will be important in many cases, and postfire changes in the management program, including (but not limited to) rehabilitation measures, must also be considered. Both on-site and off-site fire effects must also be considered.

There is general acceptance of the notion that *production functions* are most completely specified in the case of fire-caused resource change when probability distributions of outputs are developed for each alternative considered (Zivnuska 1977, Mills and Bratten 1982). Expected values of the economic efficiency measure (C + NVC) can then guide a risk-neutral choice being made among alternatives, while knowledge of the underlying probability distributions of outputs will aid in making choices based on other risk preferences. The low-probability but high-impact outcomes that are associated with some alternatives can be particularly important when weighing fire management alternatives. Examples of such outcomes are a possible escape of major proportions in a prescribed burning plan or the failure of initial attack to suppress a small fire in a high-hazard and a high-value situation.

Value Weights

With these steps completed, the fire manager turns to market clues to provide *value weights* for the inputs and outcomes of the alternatives being considered. These weights are *monetary measures,* which reflect a society's expressed preferences to a greater or lesser degree. The great virtue of monetary measures to the fire manager is that these measures tend to make the disparate quantities of various inputs and outcomes commensurable and, in doing so, permit choices to made on that basis.

Markets do not provide monetary measures for many important resources, however. Consider the problems of pricing the habitats of endangered plant and animal species, losses of human life and health, the availability of wilderness areas for both users and nonusers, air quality deterioration, and damage to archaeological resources. Continuing efforts to develop monetary values in these and similar cases reflect the usefulness of *economic models* in simplifying allocation decisions. Sometimes these efforts also raise questions as to when the apparent comparability permitted by such values is more illusory than real (Diamond and Hausman 1994). Despite any reservations, these efforts continue. For example, *contingent market valuations* are increasingly used and accepted in "noncommodity valuations" of wildlife and recreational resources (Vaux et al. 1984, Loomis and Gonzalez-Caban 1994), and in the general area of attaching values to "environmental damage" such as that resulting from oil spills (Smith 1993, Portney 1994).

There are fewer difficulties in attaching monetary values to inputs since most enter into the market process. Cost problems are most often those of making sure that inputs (including appropriate overhead inputs) are adequately specified and those associated with identifying relevant costs as distinct from "bookkeeping" costs. Until the middle 1970s, fire control agencies in the United States were provided suppression funds with few questions asked. Since that time, however, the U.S. Congress and other legislative bodies have pressured these agencies to answer numerous questions about the size of the fire management budget and the cost-effectiveness of programs, the allocation of expenditures to different management functions, and the costs and benefits of fire management in general.

In the context of the need for valuing inputs and outputs of fire management alternatives, and knowing how such values are used, a discussion of valuation procedures can proceed. To this end, Marty and Barney (1981) outlined a practical and relatively comprehensive approach to identifying economic values involved in a fire, whether a wildfire, prescribed burn, or prescribed natural burn. Originally proposed for use by personnel of the USDA Forest Service for application on public lands, this general valuation procedure has wider applications and, therefore, is outlined below.

A Valuation Procedure

Valuation of fire effects, as proposed by Marty and Barney (1981), is accomplished by tracing the chain of events that occur as a result of fire's impact to obtain an economic valuation of the fire. The original development of the procedure had ten valuation components. However, components can be added or deleted as necessary to meet the needs of valuation. The ten components in the original procedure were:

Costs
- Presuppression costs
- Suppression costs
- Rehabilitation costs

Losses
- Property losses
- Injury and illness losses

Resource Effects (Either Damages or Benefits)
- Timber effects
- Forage effects
- Wildlife effects
- Water effects
- Recreation effects

Important attributes of the valuation procedure that a user should understand include:

- The procedure is formulated in a way that will provide estimates of the economic impacts of wildfire, prescribed burning, or prescribed natural fire. All of these three kinds of fires can cause costs, damages, and benefits. Costs and losses often outweigh benefits when wildfire occurs. However, benefits should exceed costs and losses for prescribed burns or prescribed natural fire.
- The procedure, developed in a public agency context, considers costs, damages, and benefits to all stakeholders affected by fire, including land-

owners (public or private), resource managers, loggers, ranchers, and recreationists. The means of accomplishing this centers on (a) including the costs of all aspects of fire management to all organizations and individuals; and (b) valuing the changes in timber, forage, wildlife, water, and recreation on the basis of their value to the user, rather than from the standpoint of receipts that they might generate to the landowner or responsible management organization, such as the federal government in the case of fire on public lands in the United States. Unit values of the resources are based largely on local market conditions, personal experience, and RPA assessments when fire occurs on public lands. Unit values of amenity resources must be considered on a judgmental basis in many instances.

- While much of the economic impact of fire is often concentrated in the short-term, other costs, losses, and benefits can occur in the long-term, as discussed previously. Economic impacts occurring in the long-term are discounted to the present, using an appropriate discount factor. The discount factor selected in the valuation determines how much weight is given to values at a specified point in time. The greater the discount factor, the more heavily future changes are discounted. If a 5 percent discount factor is used, values for 1 year after a fire are weighted 2.5 times as heavily as values for 20 years following the fire. If a 10 percent discount factor is used, however, values for 1 year after a fire are weighted more than six times as heavily as values for 20 years following the fire.

Valuation Analysis

A basis for estimating the biophysical effects of fire must exist to apply this or any other valuation procedure. One method is that outlined earlier—that is, to use *time-trend response functions* to determine the physical effects of fire. However, the valuation procedure presented by Marty and Barney (1981) is concerned largely with valuing postfire changes once they have been identified explicitly, regardless of the method used to estimate the physical effects.

Marty and Barney (1981) suggested a reporting form, to be used on large fires, for their valuation procedure. This form is flexible and adaptable to many situations (Figure 13.1). Items to be completed in the report, when and where necessary, are listed below:

- *Identification items.* Name of area, name of fire, date of fire, date of valuation report, name of person preparing the report, and so on.
- *Control information items.* Area burned, unreimbursed suppression effort when appropriate, and so on.

1. Identification

Area ____________
Fire name ____________
Date of fire ____________
Date of report ____________
Prepared by ____________
Area burned ____________
Unreimbursed suppression effort ____________

2. Injuries and Illnesses

Name ____________
Length of hospitalization(days) ____________
Worktime loss ____________

3. Property Loss

Kind of property ____________
Original cost or cost of repair ____________
Age ____________

4. Rehabilitation Treatments

Treatment type ____________
Area (ha) ____________
Cost ____________

5. Timber Effects

Merchantable stumpage destroyed ____________
Volume of stumpage salvaged ____________
Growing stock losses and gains ____________
Volume of timber products destroyed ____________

6. Forage Effects

Volume of forage destroyed ____________
Carrying capacity lost ____________
Area burned ____________
Duration of effect (years) ____________

7. Wildlife Effects

Habitat destroyed ____________
Loss of wildlife carrying capacity ____________
Short- and long- term changes in habitat ____________
Changes in unit values ____________

8. Water Effects

Short- and long- term changes in streamflow ____________
Short- and long- term changes in sediment ____________
Changes in unit values ____________

9. Recreation Effects

Loss of recreational facilities and opportunities ____________
Short- and long- term changes in recreational opportunities ____________
Changes in unit values ____________

10. Other Effects

Description of effects ____________

Cost, Loss, and Benefit Summary

	5% Present Values		10% Present Values	
	Cost or Loss	Benefit	Cost or Loss	Benefit
Presuppression cost	____	____	____	____
Suppression cost	____	____	____	____
Life and health loss	____	____	____	____
Property loss	____	____	____	____
Rehabilitation cost	____	____	____	____
Timber effects	____	____	____	____
Forage effects	____	____	____	____
Wildlife effects	____	____	____	____
Water effects	____	____	____	____
Recreation effects	____	____	____	____
Totals	____	____	____	____
Net present values	____		____	

FIGURE 13.1 A reporting form for the fire valuation procedure. (Adapted from Marty and Barney 1981.)

- *Injury and illness items.* Name of person injured or who became ill, length of hospitalization when appropriate, worktime lost.
- *Property loss items.* Kind of property, original and replacement costs or cost of repair, age of equipment or structure.
- *Rehabilitation treatment items.* Treatment type, area, costs.
- *Timber effect items.* Volume of the timber products destroyed, short- and long-term growing stock losses and (in cases where fire causes increases in increment) growing stock gains, unit values of timber products, and so on.
- *Forage effect items.* Forage resources destroyed and immediate losses in livestock carrying capacity, short- and long-term changes in forage production and livestock carrying capacities, and area burned.
- *Wildlife effect items.* Habitats destroyed and (when appropriate and identifiable) losses in wildlife carrying capacities, short- and long-term changes in habitats and wildlife carrying capacities, and unit values of wildlife species.
- *Water effect items.* Short- and long-term changes in streamflow, sedimentation, and water quality; and unit values of water resources.
- *Recreation effect items.* Losses of recreational facilities and opportunities, short- and long-term changes in recreational opportunities, and unit values of recreational facilities and opportunities.
- *Other resource effects.* Any fire effects not included above, for example, air pollution, archaeological and cultural sites, other unique resource losses, and unit values.

Costs, losses, and benefits that result from a fire are summarized and then translated into economic values in the valuation procedure. The net estimated present values of items considered in the valuation are obtained by subtracting the total present values of the costs and losses from the total present values for benefits (Figure 13.1). An appropriate discount factor must be applied to obtain the present values of the items. For illustrative purposes only, 5 and 10 percent discount factors are indicated on the valuation reporting form. The information entered on the fire valuation form includes:

- *Presuppression costs.* All prefire protection expenditures.
- *Suppression costs.* All direct fire suppression expenditures.
- *Rehabilitation costs.* Immediate and short- and long-term rehabilitation expenditures. (Marty and Barney recognized that rehabilitation costs should result in reductions in damages due to fire, with the reductions reflected in the relevant categories. From an economic viewpoint, if the reduction in losses does not exceed rehabilitation costs, the rehabilitative measure should not be undertaken. In any case, rehabilitation costs should not be added on to original, unchanged estimates of damage.)
- *Property losses.* Costs for destroyed property (original cost adjusted by

age and useful life) and repair costs for damaged but repairable property.
- *Injury and illness losses.* Costs for medical treatment and lost worktime.
- *Timber effects.* Costs, losses, and benefits of timber products destroyed, or damaged but salvaged, and short- and long-term growing stock losses and gains.
- *Forage effect items.* Costs, losses, and benefits of forage resources destroyed and livestock carrying capacity lost, and short- and long-term changes in forage production and livestock carrying capacities.
- *Wildlife effect items.* Costs, losses, and benefits of habitats destroyed and wildlife carrying capacities lost, and short- and long-term changes in habitats and wildlife carrying capacities.
- *Water effect items.* Costs, losses, and benefits of short- and long-term changes in streamflow, sedimentation, and water quality.
- *Recreation effect items.* Costs, losses, and benefits of recreational facilities and opportunities lost, and short- and long-term changes in recreational opportunities.

Net present values for the items considered in a postfire valuation can be positive, indicating that benefits occur, or negative, reflecting damages. Summation of the net present values determines whether the total net present values for the fire effects are positive (benefits) or negative (damages).

Use of Valuation Data

The fire valuation procedure outlined by Marty and Barney (1981), or any other valuation approach, provides estimates of costs, losses, and benefits that are associated with wildfires, prescribed burns, or prescribed natural fires. As valuation data accumulate in a region, it often becomes possible to group fires of similar size, intensity, and severity, providing a basis to estimate benefits and damages from fire for planning purposes.

One use of the valuation data obtained is assessing fire management policies and programs. It is important that assessments be made continually throughout the developmental process for these policies and programs. Assessments that occur before a policy or program is implemented are called *appraisals,* while assessments that look at a past policy or program are referred to as *evaluations* (Gregersen et al. 1987). Appraisals should be made to determine the feasibility of proposed prescribed burning programs or prescribed natural fire policies (Example 13.4). These appraisals are often based on evaluations of the effects of earlier wildfires, prescribed burns, or prescribed natural fires. Information obtained from the evaluation of postfire changes represents a critical database in the assessment in both instances.

Example 13.4: Appraisal of a Proposed Prescribed Burning Program or Prescribed Natural Fire Policy (Gregersen et al. 1987)

The appraisal process should include (at least) the following steps:

1. Specification of the objectives of the program or policy: What is the program or policy trying to achieve? Toward what objectives is it aimed? What problems is it attempting to overcome?
2. Identification of the alternatives for achieving the objectives. The following steps are carried out for each of the alternatives identified:
   a. Definition and quantification of the physical inputs and outputs involved. The necessary input–output relationships are obtained from an assessment of physical effects.
   b. Determination of unit values (actual market values when possible and economic values) for the physical inputs and outputs. These values are determined in a fire evaluation procedure.
   c. Development of a value flow table showing values of benefits and costs estimated to occur over the life of the program. This table is obtained from the assessment and physical effects and a fire evaluation procedure.
   d. Calculation of relevant measures of net present values or other relevant measures of program worth and other indices needed to answer the following:
      - Are economic benefits of program or policy greater than costs?
      - What are the distribution impacts of the program or policy, that is, who will gain and who will lose?
      - What are the budget impacts likely to be for the agencies and stakeholders involved?
   e. Development of a sensitivity analysis to determine how the net present values, and other measures and indices in the analysis, might change with changes in input–output relationships, unit values, size of burns, and other features of the program or policy. This analysis provides a measure of the "latitude" that might be built into the program or policy to achieve its desired outcome. ■

SUMMARY

The objective of assessing postfire changes in ecosystem resources, the primary focus of this chapter, is analyzing the changes that can take place at the time of, or shortly after, the occurrence of fire. Damages and benefits associated with these changes can sometimes be analyzed by the use of time-trend response functions, in which postfire changes are viewed as flows of damages or benefits through time. Questions concerning fire management inputs and fire outcomes, and the value weights assigned to these inputs and outcomes, must be answered in conducting a valuation of fire effects. With this in mind, you should be able to:

1. Discuss the alternative approaches that can be used to develop time-trend response functions
2. Explain the meaning and importance of the "present net worth criterion" in comparing fire management alternatives
3. Design a general approach to the valuation of fire effects using the procedure proposed by Marty and Barney (1981) as a basis

REFERENCES

Althaus, I. A., and T. J. Mills. 1982. *Resource Values in Analyzing Fire Management Programs for Economic Efficiency.* USDA Forest Service, General Technical Report PSW-57.

Campbell, R. E., M. B. Baker, Jr., P. F. Ffolliott, F. R. Larson, and C. C. Avery. 1977. Wildfire effects on a ponderosa pine ecosystem: an Arizona case study. USDA Forest Service, Research Paper RM-191.

Crosby, J. S. 1977. A guide to the appraisal of wildfire damages, benefits, and resource values protected. USDA Forest Service, Research Paper NC-142.

DeBano, L. F. 1990. Effects of fire on the soil resource in Arizona chaparral. In: Krammes, J. S., technical coordinator. *Effects of Fire Management of Southwestern Natural Resources.* USDA Forest Service, General Technical Report RM-191, pp. 65–77.

DeBano, L. F., M. B. Baker, Jr., P. F. Ffolliott, and D. G. Neary. 1996. Fire severity and water resource response in the Southwest. *Hydrology and Water Resources in Arizona and the Southwest* 26:26–39.

Diamond, P. A., and J. A. Hausman. 1994. Contingent valuation: Is some number better than no number? *Journal of Economic Perspectives* 8(4):45–64.

Ffolliott, P. F., D. P. Guertin, and W. D. Rasmussen. 1988. Simulating the impacts of fire: a computer program. *Environmental Management* 12:809–814.

Flint, H. R. 1928. Adequate fire control. *Journal of Forestry* 26:624–638.

Gregersen, H. M., K. N. Brooks, J. A. Dixon, and L. S. Hamilton. 1987. Guidelines for economic appraisal of watershed management projects. FAO Conservation Guide 16, Rome.

Lentz, S. C., J. K. Gaunt, and A. J. Willmer. 1996. *Fire Effects on Archaeological Resources, Phase 1: The Henry Fire, Holiday Mesa, Jemez Mountains, New Mexico.* USDA Forest Service, General Technical Report RM-273.

Loomis, J., and A. Gonzalez-Caban. 1994. Estimating the value of reducing fire hazards to old growth forests in the Pacific Northwest: a contingent valuation approach. *International Journal of Wildland Fire* 4:209–216.

Lowe, P. O., P. F. Ffolliott, J. H. Dieterich, and D. R. Patton. 1978. *Determining Potential Wildlife Benefits from Wildfire in Arizona Ponderosa Pine Forests.* USDA Forest Service, General Technical Report RM-52.

Marty, R. J., and R. J. Barney. 1981. *Fire Costs, Losses, and Benefits: An Economic Valuation Procedure.* USDA Forest Service, General Technical Report INT-108.

Mills, T. J., and F. W. Bratten. 1982. *FEES: Design of a Fire Economics Evaluation System.* USDA Forest Service, General Technical Report PSW-65.

Omi, P. N., L. D. Teeter, and A. A. Dyer. 1986. The national fire management analysis system and fire economics evaluation system: a comparative analysis. In: Long, J. N., editor. *The Challenge of Protection and Use: Proceedings of Symposium on Fire Management.* Utah State University, Logan, Utah, pp. 63–70.

Portney, P. R. 1994. The contingent valuation debate: why economists should care. *Journal of Economic Perspectives* 8(4):3–17.

Sims, B. D., G. S. Lehman, and P. F. Ffolliott. 1981. Some effects of controlled burning on surface water quality. *Hydrology and Water Resources in Arizona and the Southwest* 11:87–90.

Smith, V. K. 1993. Nonmarket valuation of environmental resources: an interpretive appraisal. *Land Economics* 69:1–26.

Sparhawk, W. N. 1925. The use of liability ratings in planning forest fire protection. *Journal of Agricultural Research* 30:693–762.

Vaux, H. J., Jr., P. D. Gardner, and T. J. Mills. 1984. *Methods for Assessing the Impact of Fire on Forest Recreation.* USDA Forest Service, General Technical Report PSW-79.

Zivnuska, J. A. 1972. Economic tradeoffs in fire management. In: *Fire in the Environment: Symposium Proceedings.* USDA Forest Service, Denver, Colorado, pp. 69–74.

Zivnuska, J. A. 1977. Fire exclusion practice—cost and benefits. In: Mooney, H. A., and C. E. Conrad, technical coordinators. *Proceedings of the Symposium on the Environmental Consequences of Fire and Fuel Management in Mediterranean Ecosystems.* USDA Forest Service, General Technical Report WO-3, pp. 146–150.

Chapter

14

Fire in Ecosystem Management

INTRODUCTION

Effects of fire on ecosystem resources can represent damages, benefits, or some combination of both, depending largely on the characteristics of the fire site, the severity of the fire, the time period of valuation, and the values placed on the resources affected by the fire. When it is planned to use fire as a tool in ecosystem management, which is a common goal of a prescribed burning program or of a policy on prescribed natural fire, the (hoped for) benefits of fire must be balanced against concerns and constraints before a comprehensive fire management program is formulated.

BENEFITS OF FIRE

The use of fire for beneficial purposes is considered, where appropriate, in terms of reducing fuel loads, disposing of slash, preparing seedbeds, thinning overstocked stands, increasing forage plant production, improving wildlife habitats, changing hydrologic processes, and improving aesthetic environments. While the focus below is placed largely on opportunities for the use of prescribed burning, it is also true that vegetation-modifying and vegetation-replacing wildfire and prescribed natural fire can be important in obtaining these benefits.

Reducing Fuel Loads

The importance of fire in reducing the potential flammability of forests and other woody ecosystems through the consumption of excessive fuel accumulations is widely recognized (Martin et al. 1979, Pyne et al. 1996, Sackett et al. 1996). Knowledge of fire intensities necessary to meet fuel reduction objectives and techniques for controlling fire intensities remain incomplete in many instances, however. It is likely that periodically prescribed fire might be necessary to maintain the achieved reductions in litter amounts.

Predictive equations and tables are available to estimate the consumption of naturally occurring fuel loads from readily available or easily obtained variables in some ecosystems (Anderson 1982, Brown and Simmerman 1986, Harrington 1987). The magnitude of fuel consumption can often be predicted from knowledge of the moisture content of the humus layer and of either prefire litter depth or tree density in some ecosystems. These equations, once they are validated, enable managers to estimate the amounts of fuel that could be consumed by prescribed fire or through unplanned ignition.

Disposing of Slash

The additional fuels created by timber harvesting or precommercial thinning can be a major concern to fire managers (Brown et al. 1977, Snell and Brown 1980, Freeman et al. 1982). Piling and then burning of slash is a preferred fuel treatment in many forest ecosystems. However, the continuity of the fuel loads can be insufficient to support fire spread on some sites.

Planning for the use of fire to dispose of slash should include considerations of the effects of the fire on future timber production, wildlife habitats, soil characteristics, the level of water yield and quality, and recreation values (Hirsch et al. 1979). It is also important that a planned fire be hot enough to significantly reduce fuel loads but not kill the trees at the same time, unless their removal is also an objective of the burning.

Preparation of Seedbeds

Conditions affected through burning that favor germination of seeds and early growth of seedlings include creation of a more receptive seedbed through the removal of litter accumulations and exposure of mineral soil, increased nutrient availability, and more favorable soil moisture and temperature regimes (Chrosciewicz 1974, Keeley and Zedler 1978, Dieterich 1980, Little et al. 1994, Sackett et al. 1996). Seedbeds can remain receptive to regeneration for several years if prescribed burning is repeated. Tree seedlings have been planted successfully after fire has removed competing vegetation and, in doing so, prepared the sites for planting.

A seed crop is also required for successful natural regeneration. It is important, therefore, that seed trees not be severely damaged in burning treatments to prepare a seedbed (Sackett 1984). Fire-scorched trees, but not trees whose foliage has been totally consumed, can have a relatively high survival rate, depending largely on the time and intensity of the fire and the extent of scorching.

Thinning Overstocked Stands

Use of fire to thin overstocked and often stagnated stands is prescribed by silviculturalists in some forest ecosystems. The effectiveness of fire in the thinning of overstocked stands is largely related to the concentrations of fuels on the ground before burning (McNab 1977, Abbott and Loneragan 1983, Wade 1993, Peterson et al. 1994, Sackett et al. 1996). While fire can be a "good start" in reducing the density of overstocked stands, it is likely that additional prescribed burning will be necessary before these stands reach optimal productivity.

Attention must also be directed to the long-term survival of fire-damaged trees that remain in the stands after burning. Only rarely will all trees escape damage by fire. Therefore, the survival potentials of fire-damaged trees should be incorporated into the planning of prescribed fire to thin stands.

Increasing Forage Plant Production

Forage plant growth can be increased through burning. Competing woody overstories are often reduced in density, and soil fertility is usually increased, at least temporarily, by the release of nutrients, encouraging the growth of herbage plants. Plant vigor is promoted by the removal of senescent shoots and foliage, and, in many cases, burning of the litter prevents the interception of light and water, once again favoring plant growth (Wood 1988b, Ffolliott and Guertin 1990, Riggs et al. 1996). Nutrient values of forage plants, including crude protein, phosphorus, and *in vitro* digestibility, can be enhanced, at least temporarily, by burning. Broad generalizations on the effects of fire on forage production should not be made, however, because of the frequently observed wide variations in the response of herbage plants (Oswald and Covington 1983).

The season of burning is important in terms of its influence on forage plants. In many arid and semiarid environments, winter burning often has little effect on the density or vigor of forage species, unlike the results reported for warm-season burns (Pase 1971, Pase and Knipe 1977). Winter burning might be helpful in maintaining a dominance of forage plants in areas converted from woody overstories to herbaceous cover, however.

Improving Wildlife Habitats

Food and cover for game and nongame wildlife species are modified by the disturbance of fire. That these changes impact on wildlife habitats, often beneficially, has generally been recognized (Lyon et al. 1978, Patton 1992, Ganey et al. 1996, Riggs et al. 1996). When burned areas are kept small and interspersed with trees, an increase in both food and cover is likely to occur for a variety of wildlife species. Furthermore, the edge effect created by the burned openings generally enhances wildlife environments. Recurring fire that burns in irregular patterns, leaving unburned patches, often creates a diversity of habitat conditions.

Small burns within otherwise unburned forests and woodlands create a greater variety of food and cover for deer and other big game species than is available on large areas, either burned or unburned. Opening up the dense overstories by fire provides additional space for deer movements and an increase in the abundance and quality of browse species. Increases in upland game birds are also frequently noted after the burning treatments. Habitats for nongame bird populations are also improved by fire in many instances. This response to fire is attributed to the floristic changes after the fire, resulting in an increase in the diversity and productivity of the birds.

Changing Hydrologic Processes

Fire that reduces litter depths and the density of a forest or woodland overstories is likely to reduce ET losses and, in doing so, make more soil water available on a site, which (in turn) might make more water available for overland flow and, ultimately, streamflow runoff (Anderson et al. 1976, Tiedemann et al. 1979, Wells et al. 1979, Baker 1990, DeBano et al. 1996). Average runoff efficiencies, that is, the ratios of surface runoff to precipitation, can also increase with the severity of burn (Campbell et al. 1977).

Sediment concentrations and the chemical quality of postfire streamflow regimes also can be important to managers. The exposure of soil and subsequent soil movement can be increased by "light" and "moderate" burning, although much of the eroded soil materials often move a short distance downslope and become stabilized shortly after the burning. The movement of eroded soil materials immediately after a fire is often greater than in later years. As a consequence, increases in sediment yields can also occur in the short-term.

Concentrations of Ca, Mg, F, and other chemicals in streamflow can increase after burning. However, the elevated concentrations often decline in subsequent runoff events.

Improving Aesthetic Environments

Controlled burns can improve the aesthetics of a forest landscape by keeping the forest open and "park-like," and favoring the development of large trees that can readily be viewed by the public. Monotonous, dense, debris-littered clumps of saplings are generally uninviting, both visually and physically. Adverse scenic impacts of fire are frequently short-lived and, therefore, might not be a significant deterrent to prescribed burning or prescribed natural fire in certain conditions (Anderson et al. 1982, Taylor and Daniel 1984, Taylor et al. 1986). However, careful attention to these adverse impacts is warranted in most cases.

CONCERNS AND CONSTRAINTS

There are concerns and constraints that should be addressed before burning programs become operational. Environmental factors, cultural impacts, economic considerations, and a need for public support are all important in this respect.

Environmental Factors

The use of fire on sites where fuels have accumulated for a number of years often results in changes in surface organic materials and nutrient storage. From a study in a southwestern ponderosa pine forest, for example, Covington and Sackett (1984) reported that the amount of surface organic materials was reduced by a prescribed fire, with nutrient storage somewhat less affected. This burning also released much of the nutrients bound in the surface organic materials, however, improving the conditions for plant growth. It is important, therefore, that guidelines for prescribed burning or prescribed natural fire consider the impacts of fire on nutrient regimes and, ultimately, the potential loss of nutrients.

Burning of organic materials on a soil surface removes the protective litter layers, volatilizes large amounts of nitrogen and smaller amounts of other elements, and transforms less volatile elements into soluble forms that are more easily absorbed by plants or lost by leaching (Wells et al. 1979, Neary et al. 1996). Heating of the underlying soil layers also alters the physical, chemical, and biological properties of the soil that depend on the organic materials on the soil surface. These effects are generally variable and largely unpredictable.

The use of fire for beneficial purposes on a widespread scale can be constrained by smoke. Smoke management to minimize the impairment of air quality includes fuel management and fire prescriptions that improve combustion efficiencies; firing and "mopping up" techniques to reduce emissions; and the scheduling of burning to enhance convection and dispersion, and to ensure plume trajectories away from "sensitive" areas (Sandberg et al. 1979, Potter and Fox 1996, Pyne et al. 1996). Legislation specifying the conditions when fire can safely be prescribed or be allowed to burn has been passed and is being implemented in the United States to ensure the maintenance of desired atmospheric quality.

Cultural Impacts

Many artifacts have important value in dating and further interpreting past events. These artifacts are continuously being exposed to both wildfire and prescribed burning (Bellomo 1991). Therefore, the effects of heating on these artifacts need to be considered in terms of fire management activities. Fire evidence must be recognized in the archaeological context to determine whether changes have resulted from natural processes or from cultural activities.

Damage to cultural resources occurs not only as a direct effect of the fire, but also from erosion following fire with associated damage to collections of artifacts. Minimizing the disturbance or destruction of cultural resources is important in both prescribed burning and fire suppression during wildfire. Some fire retardants and foams used for fire suppression can cause carbon smudging of sherds. Construction of firelines and the movement of mechanized equipment can also disturb the integrity of a particular cultural resource site. Postfire rehabilitative activities should be analyzed as carefully for potential disturbance as those associated with fire suppression. Research techniques valuable to cultural resource specialists can be compromised by severe fires, particularly with respect to those artifacts that are on the soil surface and exposed to direct heating by fire.

Economic Considerations

Althaus and Mills (1982) indicated that minimization of the sum of fire costs plus the net value change in resource outputs, that is, C + NVC, is a useful selection criterion for analyzing the economic efficiencies of alternative fire management strategies (see Chapter 13). The (C + NVC) criterion considers both beneficial and detrimental effects of fire. However, site-specific data must be available to quantify costs and changes in the resource values for the alternative fire management strategies considered.

It should be remembered that the economic efficiency criterion (C + NVC) is only valid when it is used to decide between two or more alternatives. If a prescribed burn is approached as a single project that is designed to meet specified objectives, the project must be evaluated on a benefit–cost basis, where the benefit will be the present value of resource outputs that are associated with fire outcomes *with the project* minus the present value of resource outputs that are associated with fire outcomes *in the absence of the project*. If the benefit does not exceed the cost, the project should not be undertaken. The descriptions of fire outcomes with and without the project should include likely impacts on future fire patterns. In the case of a prescribed burn, the positive probability that is associated with an escape (or escapes of varying sizes and impacts) should be recognized, quantified as fully as possible, and incorporated in the analysis of benefits and costs. The risk of escape is a major consideration in prescribed burning, and the continuing expansion of the urban–wildland interface into areas where prescribed burning is being considered intensifies the problem (Daniel et al. 1996). When there are substantial liability risks to agencies or landowners contemplating prescribed burns, enthusiasm is dampened. These risks, the problems of meeting air quality requirements, and cost constraints may be the three factors most responsible for the relatively limited use of prescribed fire.

While data sets are available to quantify the costs of burning (Wood 1988a, Cleaves and Brodie 1990, Riggs et al. 1996), adequate information is scarce, unfortunately, in many instances. Therefore, additional information for a range of conditions and sites will likely be necessary to evaluate the economic efficiencies of alternative fire management strategies in most cases.

Public Support

Opportunities to use fire for beneficial purposes in the future are highly dependent on obtaining the necessary public support. That the public recognizes that fire can be both beneficial and detrimental has been shown by Cortner et al. (1984), who also determined that public acceptance and understanding of the potential benefits of fire are often high, in relative terms. Furthermore, knowledge of the effects of fire on natural ecosystems often increases the "tolerance" for fire by the public (Taylor et al. 1986). To be effective in obtaining public support for fire management programs in the future, educational efforts should be oriented to the local conditions and needs, and to local knowledge and acceptance of fire.

ECOSYSTEM RESTORATION

Many managers have begun to support the reintroduction of fire in some form to fire-dependent ecosystems. *Fire-dependent ecosystems* are those where fire plays a vital role in determining the composition, structure, and landscape patterns (Arno 1996). Great strides have been made in learning more about the ecological effects of fire in recent years. Application of this knowledge has been difficult in some instances, however, because of the policies of land management agencies that have assumed that one could largely exclude fire from fire-dependent ecosystems with little adverse effect (Pyne 1982, Agee 1993).

That these policies were not necessarily "always correct" is shown by the often-cited negative effects of fire exclusion on the "forest health" on millions of acres of ponderosa pine and related forest communities in the Inland West (Mutch et al. 1993). Ironically, this problem had been identified more than 50 years ago by Weaver (1943), who traced its roots to fire exclusion. Weaver (1943, 1967) and his colleagues called for the use of prescribed burning to restore fire to these forests, which had been burned historically by frequently occurring low-intensity fires. Their arguments have been accepted eventually by many foresters and ecologists.

As a result of the above arguments and other supporting evidence, many management agencies in the United States, such as the USDA Forest Service, have adopted policies that broadened their mission to more holistic *fire management,* not solely *fire control* (Nelson 1979). These revised policies consider prescribed burning as a management tool as well as recognizing the need for suppression of unwanted fire.

There is considerable knowledge about, and increasing public support for, the reintroducing of fire to fire-dependent ecosystems. Nevertheless, there are also obstacles confronting managers attempting to do so (Arno 1996, Czech 1996, Mutch and Cook 1996). The obstacles include the following:

- Funding for prescribed burning programs and silvicultural treatments to reduce the accumulations of fuels has traditionally been secondary to the funding available to suppress wildfire.
- Fuel treatment is generally financed from an annually appropriated budget, while fire suppression funding is covered by "emergency" money that is perceived by some to be limitless.
- Managers are frequently held responsible for the smoke emissions produced by prescribed burning. Managers can also suffer "career setbacks" when prescribed fire escapes control for any reason. Conversely, the motives of managers who attempt to exclude fire, an action that can eventually result in damaging wildfire, are seldom questioned. Moreover, managers are rarely blamed for the high costs of fighting wildfire occurring in built-up fuels, or for the frequent severe smoke or damages such fire produces.

Management agencies, regulatory institutions, decisionmakers, and society in general have challenging opportunities to implement meaningful ecosystem restoration programs, including reintroducing fire when appropriate. These programs must be at a large enough scale to sustain the health of fire-adapted ecosystems, while benefitting people, their property, and natural resources at the same time (Mutch and Cook 1996). Society must move away from litigation and courtroom strategies as primary tools in managing and sustaining natural resources if these programs are to succeed. Available scientific knowledge and managerial experience need to be employed in harmony with disturbance factors for ecosystem restoration to be effective.

FIRE EFFECTS INFORMATION SYSTEM

A large body of scientific and technical knowledge on the effects of fire on ecosystem resources has accumulated over the years. Unfortunately, this knowledge is fragmented throughout the literature, and, as a consequence, it is unwieldy and time-consuming to gather information for interpretation and management planning. Some of this literature formed a basis for writing this book and has been highlighted in this concluding chapter on the role of fire in ecosystem management. However, additional informational sources are available on species-specific fire effects, but this information is frequently difficult to identify and obtain, especially by field-based fire managers.

The Fire Effects Information System (FEIS) was created by the USDA Forest Service to help in overcoming the problem of efficiently storing and retrieving the large body of knowledge that has accumulated on the effects of fire on ecosystem resources (Fischer et al. 1996). The FEIS represents an easy to use, computerized knowledge management system that stores and retrieves species-specific fire effects information in the form of text organized in encyclopedic format. The system consists of the knowledge base, the query program, the builder program, and the citation retrieval system.

The knowledge base contains summaries of fire effects and related ecological information. The query program allows the user to readily access the knowledge base through on-screen prompts and menu guides. The builder program facilitates the addition of information to, or otherwise editing of, the knowledge base. The builder program is restricted to the system management staff (Fischer et al. 1996). The citation retrieval system is a computerized bibliography of all literature that has been used to assemble the knowledge base summaries; it can be searched by keyword, author, or date.

The knowledge base contains fire effects and related biological, ecological, and management-oriented information in three categories—plant species, wildlife species, and plant communities. The plant community category, in turn, includes three levels of classification—a broad ecosystem level, a potential natural vegetation level (Kuchler 1964), and a cover type level. The FEIS provides information for predetermined topics within several subject areas for each category and level (Figure 14.1). The knowledge base and geographic distribution of the FEIS are worldwide. Some of the species represented also occur beyond North America, as either native, introduced, or naturalized species. Other locations include Australia, New Zealand, India, the British Isles, France, Germany, Spain, Scandinavia, Russia, and South America.

The FEIS was developed originally to meet informational needs related to

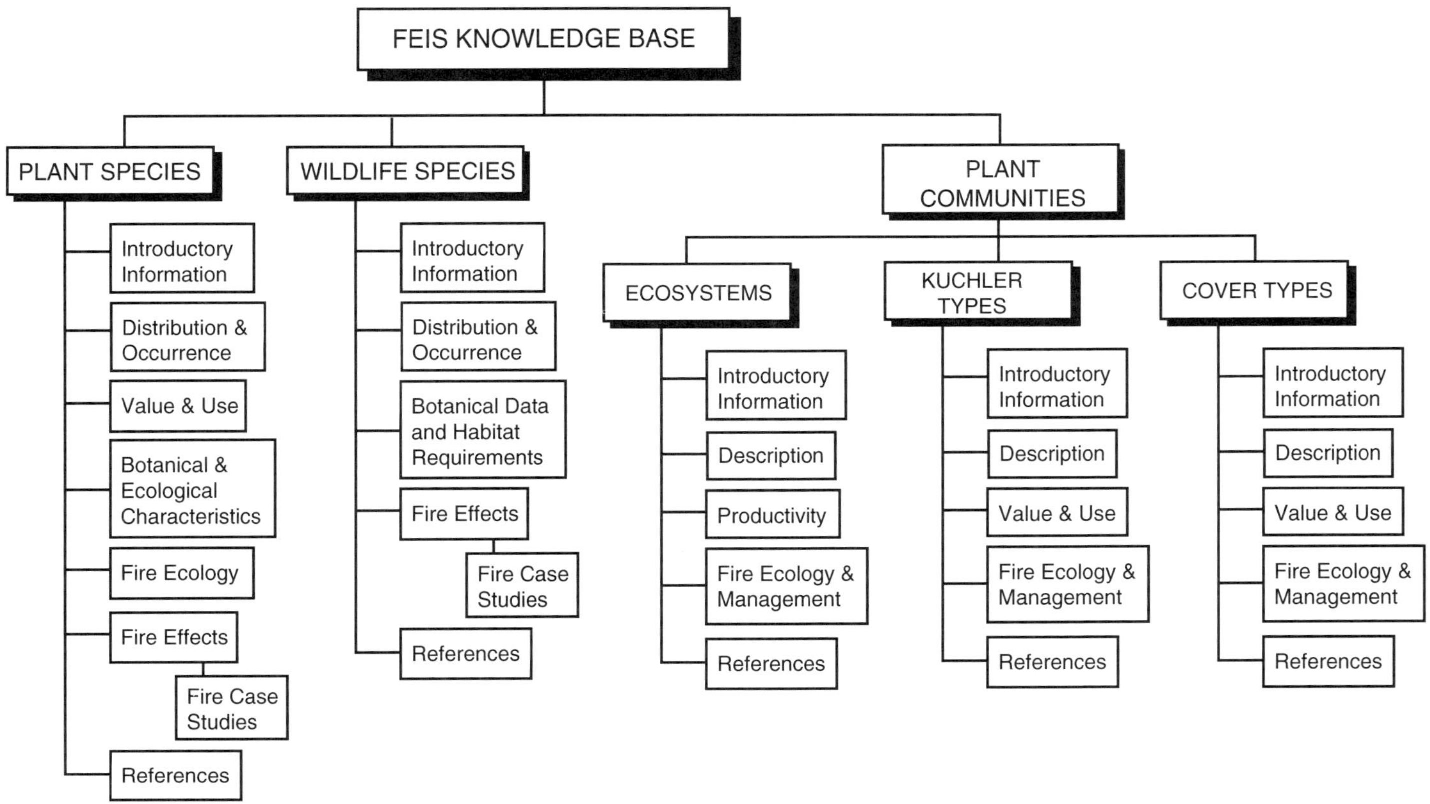

FIGURE 14.1 Subject matter content of the FEIS knowledge base. (Adapted from Fischer et al. 1996.)

prescribed fire. However, the system, which is still being developed, is also recognized as a valuable tool for obtaining information about species ecology for other applications. The FEIS knowledge base is available on the Internet (Fischer et al. 1996). It is accessed through the USDA Forest Service home page on the World Wide Web. Its address, "http://www.fs.fed.us/database/feis," provides a direct connection to the FEIS main menu.

SUMMARY

Fire can be a valuable tool in ecosystem management. It is important, however, that the benefits of fire be weighed against the concerns and constraints related to the use of fire in management before implementing operational management programs. Fortunately, a large body of information on the effects of fire on ecosystems resources is readily available to aid a fire manager in making this determination. After reading this final chapter of the book, you should have an appreciation of how fire might be used in ecosystem management and, as a result, be able to:

1. Discuss the ways in which fire can be used for beneficial purposes in ecosystem management.
2. Identify possible concerns and constraints related to the use of fire.
3. Discuss the role of fire in ecosystem restoration.
4. Explain how the Fire Effects Information System (FEIS) can be accessed on the Internet.

REFERENCES

Abbott, I., and O. Loneragan. 1983. Influence of fire on growth rate, mortality and butt damage in Mediterranean forest of western Australia. *Forest Ecology and Management* 6:139–153.

Agee, J. K. 1993. *Fire Ecology of Pacific Northwest Forests.* Island Press, Washington, DC.

Althaus, I. A., and T. J. Mills. 1982. *Resource Values in Analyzing Fire Management Programs for Economic Efficiency.* USDA Forest Service, General Technical Report PSW-57.

Anderson, H. E. 1982. *Aids to Determining Fuel Models for Estimating Fire Behavior.* USDA Forest Service, General Technical Report INT-122.

Anderson, H. W., M. D. Hoover, and K. G. Reinhart. 1976. *Forest and Water: Effects of Forest Management on Floods, Sedimentation, and Water Supply.* USDA Forest Service, General Technical Report PSW-18.

Anderson, L. M., D. J. Levi, T. C. Daniel, and J. H. Dieterich. 1982. The esthetic effects of prescribed burning: a case study. USDA Forest Service, Research Paper RM-415.

Arno, S. F. 1996. The seminal importance of fire in ecosystem management—impetus for this publication. In: Hardy, C. C., and S. F. Arno, editors. *The Use of Fire in*

Forest Restoration. USDA Forest Service, General Technical Report INT-GTR-341, pp. 3–5.

Baker, M. B. Jr. 1990. Hydrologic and water quality effects of fire. In: Krammes, J. S., technical coordinator. *Effects of Fire Management of Southwestern Natural Resources.* USDA Forest Service, General Technical Report RM-191, pp. 31–42.

Bellomo, R. V. 1991. Identifying traces of natural and humanly-controlled fire in the archaeological record: the role of actualistic studies. *Archaeology in Montana* 32:75–93.

Brown, J. K., and D. G. Simmerman. 1986. *Appraising Fuels and Flammability in Western Aspen: A Prescribed Fire Guide.* USDA Forest Service, General Technical Report INT-205.

Brown, J. K., J. A. K. Snell, and D. L. Bunnell. 1977. *Handbook for Predicting Slash Weight of Western Conifers.* USDA Forest Service, General Technical Report INT-37.

Campbell, R. E., M. B. Baker, Jr., P. F. Ffolliott, F. R. Larson, and C. C. Avery. 1977. Wildfire effects on a ponderosa pine ecosystem: an Arizona case study. USDA Forest Service, Research Paper RM-191.

Chrosciewicz, Z. 1974. Evaluation of fire-produced seedbeds for jack pine regeneration in central Ontario. *Canadian Journal of Forest Research* 4:455–457.

Cleaves, D. A., and J. D. Brodie. 1990. Economic analysis of prescribed burning. In: Walstad, J. D., S. R. Radosevich, and D. V. Sandberg, editors. *Natural and Prescribed Fire in Pacific Northwest Forests.* Oregon State University Press, Corvallis, Oregon, pp. 271–282.

Cortner, H. J., M. J. Zwolinski, E. H. Carpenter, and J. G. Taylor. 1984. Public support for fire-management policies. *Journal of Forestry* 82:359–361.

Covington, W. W., and S. S. Sackett. 1984. The effect of a prescribed burn in southwestern ponderosa pine on organic matter and nutrients in woody debris and forest floor. *Forest Science* 30:183–193.

Czech, B. 1996. Challenges to establishing and implementing sound natural fire policy. *Renewable Resources Journal* 14(2):14–19.

Daniel, T. C., M. Meitner, and E. Weidemann. 1996. Human desires and fears in ecologically rational wildland fire management. In: Ffolliott, P. F., L. F. DeBano, M. B. Baker, Jr., G. J. Gottfried, G. Solis-Garza, C. B. Edminster, D. G. Neary, L. S. Allen, and R. H. Hamre, technical coordinators. *Effects of Fire on Madrean Province Ecosystems: A Symposium Proceedings.* USDA Forest Service, General Technical Report RM-GTR-289, pp. 37–42.

DeBano, L. F., P. F. Ffolliott, and M. B. Baker, Jr. 1996. Fire severity effects on water resources. In: Ffolliott, P. F., L. F. DeBano, M. B. Baker, Jr., G. J. Gottfried, G. Solis-Garza, C. B. Edminster, D. G. Neary, L. S. Allen, and R. H. Hamre, technical coordinators. *Effects of Fire on Madrean Province Ecosystems: A Symposium Proceedings.* USDA Forest Service, General Technical Report RM-GTR-289, pp. 77–84.

Dieterich, J. H. 1980. Chimney Spring forest fire history. USDA Forest Service, Research Paper RM-220.

Ffolliott, P. F., and D. P. Guertin. 1990. Prescribed fire in Arizona ponderosa pine forests: a 24–year case study. In: Krammes, J. S., technical coordinator. *Effects of Fire Management of Southwestern Natural Resources.* USDA Forest Service, General Technical Report RM-191, pp. 250–254.

Fischer, W. C., M. Miller, C. M. Johnston, J. K. Smith, D. G. Simmerman, and J. K.

Brown. 1996. *Fire Effects Information System: User's Guide.* USDA Forest Service, General Technical Report INT-GTR-327.

Freeman, D. R., R. M. Loomis, and P. J. Roussopoulos. 1982. *Handbook for Predicting Slash Weight in the Northeast.* USDA Forest Service, General Technical Report NC-75.

Ganey, J. L., W. M. Block, and P. F. Boucher. 1996. Effects of fire on birds in Madrean forests and woodlands. In: Ffolliott, P. F., L. F. DeBano, M. B. Baker, Jr., G. J. Gottfried, G. Solis-Garza, C. B. Edminster, D. G. Neary, L. S. Allen, and R. H. Hamre, technical coordinators. *Effects of Fire on Madrean Province Ecosystems: A Symposium Proceedings.* USDA Forest Service, General Technical Report RM-GTR-289, pp. 146–154.

Harrington, M. G. 1987. Predicting reduction of natural fuels by prescribed burning under ponderosa pine in southeastern Arizona. USDA Forest Service, Research Note RM-472.

Hirsch, S. N., G. F. Meyer, and D. L. Radloff. 1979. *Choosing an Activity Fuel Treatment for Southwest Ponderosa Pine.* USDA Forest Service, General Technical Report RM-67.

Keeley, J. E., and P. H. Zedler. 1978. Reproduction of chaparral shrubs after fire: a comparison of sprouting and seeding strategies. *The American Midland Naturalist* 99:142–161.

Kuchler, A. W. 1964. *United States: Potential Natural Vegetation of the Conterminous United States* (1:3,168,000 colored). Special Publication 36, American Geographical Society, New York.

Little, R. L., D. L. Peterson, and L. L. Conquest. 1994. Regeneration of subalpine fir following fire. *Canadian Journal of Forest Research* 24:934–944.

Lyon, L. J., H. S. Crawford, E. Czuhai, R. L. Fredriksen, R. F. Harlow, L. J. Metz, and H. A. Pearson. 1978. *Effects of Fire on Fauna: A State-of-Knowledge Review.* USDA Forest Service, General Technical Report WO-6.

Martin, R. E., H. E. Anderson, W. D. Boyer, J. H. Dieterich, S. N. Hirsch, V. J. Johnson, and W. H. McNab. 1979. *Effects of Fire on Fuels: A State-of-Knowledge Review.* USDA Forest Service, General Technical Report WO-13.

McNab, W. H. 1977. An overcrowded loblolly pine stand thinned with fire. *Southern Journal of Applied Forestry* 1:24–26.

Mutch, R. W., and W. A. Cook. 1996. Restoring fire to ecosystems: methods vary with land management goals. In: Hardy, C. C., and S. F. Arno, editors. *The Use of Fire in Forest Restoration.* USDA Forest Service, General Technical Report INT-GTR-341, pp. 9–11.

Mutch, R. W., S. F. Arno, J. K. Brown, C. E. Carlson, R. D. Ottmar, and J. L. Peterson. 1993. *Forest Health in the Blue Mountains: A Management Strategy for Fire-Adapted Ecosystems.* USDA Forest Service, General Technical Report PNW-GTR-310.

Neary, D. G., S. A. Overby, G. J. Gottfried, and H. M. Perry. 1996. Nutrients in fire-dominated ecosystems. In: Ffolliott, P. F., L. F. DeBano, M. B. Baker, Jr., G. J. Gottfried, G. Solis-Garza, C. B. Edminster, D. G. Neary, L. S. Allen, and R. H. Hamre, technical coordinators. *Effects of Fire on Madrean Province Ecosystems: A Symposium Proceedings.* USDA Forest Service, General Technical Report RM-GTR-289, pp. 107–117.

Nelson, T. C. 1979. Fire management policy on national forests—a new era. *Journal of Forestry* 77:723–725.

Oswald, B. P., and W. W. Covington. 1983. Changes in understory production following a wildfire in southwestern ponderosa pine. *Journal of Range Management* 36:507–509.

Pase, C. P. 1971. Effect of a February burn on Lehmann lovegrass. *Journal of Range Management* 24:454–456.

Pase, C. P., and O. D. Knipe. 1977. Effect of winter burning on herbaceous cover on a converted chaparral watershed. *Journal of Range Management* 30:346–348.

Patton, D. R. 1992. *Wildlife Habitat Relationships in Forested Ecosystems.* Timber Press, Portland, Oregon.

Peterson, D. L., S. S. Sackett, L. J. Robinson, and S. M. Haase. 1994. The effects of repeated prescribed burning on *Pinus ponderosa* growth. *International Journal of Wildland Fire* 4:239–247.

Potter, D. U., and D. G. Fox. 1996. Clean air and healthy ecosystems: managing emissions from fires. In: Ffolliott, P. F., L. F. DeBano, M. B. Baker, Jr., G. J. Gottfried, G. Solis-Garza, C. B. Edminster, D. G. Neary, L. S. Allen, and R. H. Hamre, technical coordinators. *Effects of Fire on Madrean Province Ecosystems: A Symposium Proceedings.* USDA Forest Service, General Technical Report RM-GTR-289, pp. 205–216.

Pyne, S. J. 1982. *Fire in America: A Cultural History of Wildland and Rural Fire.* Princeton University Press, Princeton, New Jersey.

Pyne, S. J., P. L. Andrews, and R. D. Laven. 1996. *Introduction to Wildland Fire.* John Wiley & Sons, New York.

Riggs, R. A., S. C. Bunting, and S. E. Daniels. 1996. Prescribed fire. In: Krausman, P. R., editor. *Rangeland Wildlife.* Society for Range Management, Denver, Colorado, pp. 295–319.

Sackett, S. S. 1984. Observations on natural regeneration in ponderosa pine following a prescribed fire in Arizona. USDA Forest Service, Research Note RM-435.

Sackett, S. S., S. M. Haase, and M. G. Harrington. 1996. Prescribed burning in southwestern ponderosa pine. In: Ffolliott, P. F., L. F. DeBano, M. B. Baker, Jr., G. J. Gottfried, G. Solis-Garza, C. B. Edminster, D. G. Neary, L. S. Allen, and R. H. Hamre, technical coordinators. *Effects of Fire on Madrean Province Ecosystems: A Symposium Proceedings.* USDA Forest Service, General Technical Report RM-GTR-289, pp. 178–186.

Sandberg, D. V., J. M. Pierovich, D. G. Fox, and E. W. Ross. 1979. *Effects of Fire on Air: A State-of-Knowledge Review.* USDA Forest Service, General Technical Report WO-9.

Snell, J. A. K., and J. K. Brown. 1980. *Handbook for Predicting Residue Weights of Pacific Northwest Conifers.* USDA Forest Service, General Technical Report PSW-103.

Taylor, J. G., and T. C. Daniel. 1984. Prescribed fire: public education and perception. *Journal of Forestry* 82:361–365.

Taylor, J. G., H. J. Cortner, P. D. Gardner, T. C. Daniel, M. J. Zwolinski, and E. H. Carpenter. 1986. Recreation and fire management: public concerns, attitudes, and perceptions. *Leisure Sciences* 8:167–187.

Tiedemann, A. R., C. E. Conrad, J. H. Dieterich, J. W. Hornbeck, W. F. Megahan, and D. D. Wade. 1979. *Effects of Fire on Water: A State-of-Knowledge Review.* USDA Forest Service, General Technical Report WO-10.

Wade, D. D. 1993. Thinning young loblolly pine stands with fire. *International Journal of Wildland Fire* 3:169–178.

Weaver, H. 1943. Fire as an ecological and silvicultural factor in the ponderosa pine region of the Pacific slope. *Journal of Forestry* 41:7–14.

Weaver, H. 1967. Fire and its relationship to ponderosa pine. *Proceedings of the Tall Timbers Fire Ecology Conference* 7:127–149.

Wells, C. G., R. E. Campbell, L. F. DeBano, C. E. Lewis, R. L. Fredriksen, R. C. Froelich, and P. H. Dunn. 1979. *Effects of Fire on Soil: A State-of-Knowledge Review.* USDA Forest Service, General Technical Report WO-7.

Wood, D. B. 1988a. Costs of prescribed burning in southwestern ponderosa pine. *Western Journal of Applied Forestry* 3:115–119.

Wood, G. W. 1988b. Effects of prescribed fire on deer forage and nutrients. *Wildlife Society Bulletin* 16:180–186.

Index